126 Advances in Polymer Science

Springer
*Berlin
Heidelberg
New York
Barcelona
Budapest
Hong Kong
London
Milan
Paris
Santa Clara
Singapore
Tokyo*

Biopolymers
Liquid Crystalline Polymers
Phase Emulsion

With contributions by
B. A. Armitage, D. E. Bennett, N. R. Cameron,
H. G. Lamparski, D. F. O'Brien, T. Sato, D. C. Sherrington,
A. Teramoto, T. Tsuruta

With 85 Figures and 23 Tables

 Springer

ISBN 3-540-60484-7 Springer-Verlag Berlin Heidelberg NewYork
ISBN 0-387-60484-7 Springer-Verlag NewYork Berlin Heidelberg

This work is subject to copyright. All rights are reserved, whether the whole or part of the material is concerned, specifically the rights of translation, reprinting, re-use of illustrations, recitation, broadcasting, reproduction on microfilms or in other ways, and storage in data banks. Duplication of this publication or parts thereof is only permitted under the provisions of the German Copyright Law of September 9, 1965, in its current version, and a copyright fee must always be paid.

© Springer-Verlag Berlin Heidelberg 1996
Library of Congress Catalog Card Number 61-642
Printed in Germany

The use of registered names, trademarks, etc. in this publication does not imply, even in the absence of a specific statement, that such names are exempt from the relevant protective laws and regulations and therefore free for general use.

Typesetting: Macmillan India Ltd., Bangalore-25
SPIN: 10508856   02/3020 - 5 4 3 2 1 0 - Printed on acid-free paper

Editors

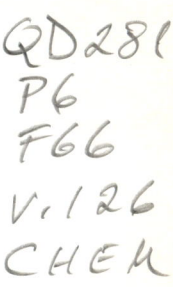

Prof. Akihiro Abe, Department of Industrial Chemistry, Tokyo Institute of Polytechnics, 1583 Iiyama, Atsugi 243-02, Japan

Prof. Henri Benoit, CNRS, Centre de Recherches sur les Macromolécules, 6, Rue Boussingault, 67083 Strasbourg Cedex, France

Prof. Hans-Joachim Cantow, Freiburger Materialforschungszentrum, Stefan Meier-Str. 31a, D-79104 Freiburg i. Br., FRG

Prof. Paolo Corradini, Università di Napoli, Dipartimento di Chimica, Via Mezzocannone 4, 80134 Napoli, Italy

Prof. Karel Dušek, Institute of Macromolecular Chemistry, Czech Academy of Sciences, 16206 Prague 616, Czech Republic

Prof. Sam Edwards, University of Cambridge, Department of Physics, Cavendish Laboratory, Madingley Road, Cambridge CB3 OHE, UK

Prof. Hiroshi Fujita, 35 Shimotakedono-cho, Shichiku, Kita-ku, Kyoto 603 Japan

Prof. Gottfried Glöckner, Technische Universität Dresden, Sektion Chemie, Mommsenstr. 13, D-01069 Dresden, FRG

Prof. Dr. Hartwig Höcker, Lehrstuhl für Textilchemie und Makromolekulare Chemie, RWTH Aachen, Veltmanplatz 8, D-52062 Aachen, FRG

Prof. Hans-Heinrich Hörhold, Friedrich-Schiller-Universität Jena, Institut für Organische und Makromolekulare Chemie, Lehrstuhl Organische Polymerchemie, Humboldtstr. 10, D-07743 Jena, FRG

Prof. Hans-Henning Kausch, Laboratoire de Polymères, Ecole Polytechnique Fédérale de Lausanne, MX-D, CH-1015 Lausanne, Switzerland

Prof. Joseph P. Kennedy, Institute of Polymer Science, The University of Akron, Akron, Ohio 44 325, USA

Prof. Jack L. Koenig, Department of Macromolecular Science, Case Western Reserve University, School of Engineering, Cleveland, OH 44106, USA

Prof. Anthony Ledwith, Pilkington Brothers plc. R & D Laboratories, Lathom Ormskirk, Lancashire L40 SUF, UK

Prof. J. E. McGrath, Polymer Materials and Interfaces Laboratory, Virginia Polytechnic and State University Blacksburg, Virginia 24061, USA

Prof. Lucien Monnerie, Ecole Superieure de Physique et de Chimie Industrielles, Laboratoire de Physico-Chimie, Structurale et Macromoléculaire 10, rue Vauquelin, 75231 Paris Cedex 05, France

Prof. Seizo Okamura, No. 24, Minamigoshi-Machi Okazaki, Sakyo-Ku, Kyoto 606, Japan

Prof. Charles G. Overberger, Department of Chemistry, The University of Michigan, Ann Arbor, Michigan 48109, USA

Prof. Helmut Ringsdorf, Institut für Organische Chemie, Johannes-Gutenberg-Universität, J.-J.-Becher Weg 18-20, D-55128 Mainz, FRG

Prof. Takeo Saegusa, KRI International, Inc. Kyoto Research Park 17, Chudoji Minamima-chi, Shimogyo-ku Kyoto 600 Japan

Prof. J. C. Salamone, University of Lowell, Department of Chemistry, College of Pure and Applied Science, One University Avenue, Lowell, MA 01854, USA

Prof. John L. Schrag, University of Wisconsin, Department of Chemistry, 1101 University Avenue. Madison, Wisconsin 53706, USA

Prof. G. Wegner, Max-Planck-Institut für Polymerforschung, Ackermannweg 10, Postfach 3148, D-55128 Mainz, FRG

Table of Contents

Contemporary Topics in Polymeric Materials for Biomedical Applications
T. Tsuruta 1

Polymerization and Domain Formation in Lipid Assemblies
D. F. O'Brien, B. A. Armitage, D. E. Bennett, H. G. Lamparski 53

Concentrated Solutions of Liquid-Crystalline Polymers
T. Sato, A. Teramoto 85

High Internal Phase Emulsions (HIPEs) - Structure, Properties and Use in Polymer Preparation
N. R. Cameron, D. C. Sherrington 163

Author Index Volumes 101 - 126 215

Subject Index 223

Contemporary Topics in Polymeric Materials for Biomedical Applications

Teiji Tsuruta
Department of Industrial Chemistry, Faculty of Engineering, Science University of Tokyo, 1-3 Kagurazaka, Shinjuku-ku, Tokyo, 162 Japan

Contemporary topics in polymeric materials for biomedical applications are reviewed, with the emphasis on the modes of interaction of water soluble polymers as well as their conjugates, and those of microdomain-structured polymers with proteins and cells. The nature of biocompatibility of these polymer surfaces is discussed in terms of the random-network concept of water molecules on the polymer surface. Several recent studies on biohybridized and biomimicking materials are also reviewed.

1 Introduction – Multifaceted Aspects of Research on Biomedical Polymers 2
2 Hydrophilicity or Hydrophobicity of Polymeric Materials and Their Behavior toward Protein Adsorption 6
  2.1 Two Models of the Protein-Adsorption Processes 6
  2.2 Comparison of the Adsorption of Four Proteins to Five Biomaterials with Different Hydrophilicity 13
  2.3 Polymeric Materials with Ionic Functional Groups and Their Protein Adsorptive Behavior 14
3 Water-soluble Polymers, Their Conjugates and Hydrophilic Polymers 15
  3.1 Interaction of Poly(ethylene oxide) and its Conjugates with Cell and Other Biological Elements 16
  3.2 Poly(N-isopropylacrylamide) and Its Copolymers as Thermoresponsive Hydrogels 18
4 Microdomain-Structured Materials 21
  4.1 Segmented Polyurethanes 21
  4.2 HEMA-Styrene Triblock Copolymers and Polyether-Segmented Polyamides 25
  4.3 Polyamine-graft Copolymers 28
  4.4 Random Network Concept of Water Molecules on the Surface of Materials. 33
5 Biohybridized and Biomimicking Materials 35
  5.1 Cell-adhesive Peptide Conjugate Materials 35
  5.2 Other Bio-conjugate (or Bio-mimicking) Materials 41
6 Summary 45
7 References 47

1 Introduction – Multifaceted Aspects of Research on Biomedical Polymers

Biomaterials are defined as materials which are used in contact with biological tissue: blood, cells, protein, and any other living substances [1]. Polymeric biomaterials – along with ceramic and metallic ones – have long been recognized to be versatile materials, which have wide-ranging applications in the medical, biomedical, and biological fields: e.g. non-thrombogenic materials for artifical organs, devices and systems for diagnosis and therapy, cell cultivation substrata, cell-sorters, membranes, adsorbents, biosensors, and controlled delivery systems for bioactive agents [1].

Multifaceted aspects of research on biomedical polymers are shown in Table 1. End-use devices are manufactured starting from their original concept. To approach the target, materials design is carried out so that the materials can exhibit the desirable property when they are brought into contact with any particular biological element. Fundamental studies are carried out in order to elucidate structure-property relationships in the interaction of materials with biological elements.

In Fig. 1, the author schematically shows how device manufacturing and fundamentals are positioned relative to concept, property and materials design. Some researchers may be interested mostly in device manufacturing for end-use, while others are more concerned with fundamentals. It is evident, however, these five elements play a complementary role with respect to one another.

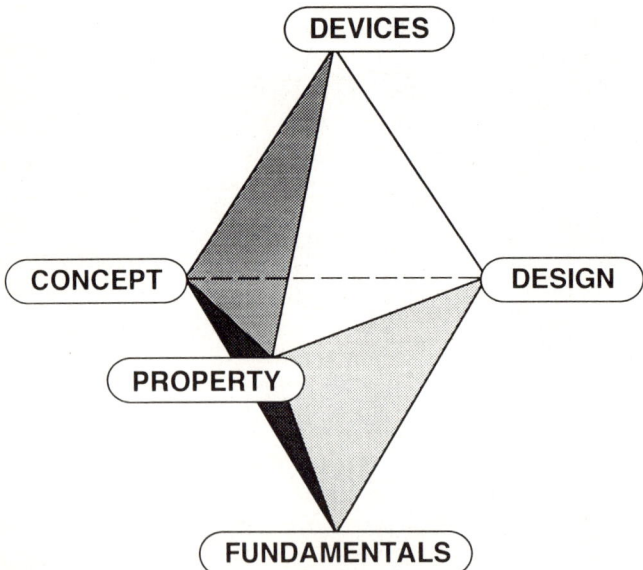

Fig. 1. Five elements of researches on biomedical polymers

Biocompatibility (See Table 1), which is a phenomenological concept, is the essential property of biomaterials. For instance, the inner surface of an implanted vascular graft or blood pump (artificial heart) must be blood-compatible, while its outer surface must be tissue-compatible. In other words, the material surfaces must not exert any adverse effects upon blood or tissue, or upon other biological elements at the interfaces.

Physical or physico-chemical capability (Table 1), including mechanical strength, permeation, or sieving characteristics, is another important requirement of biomaterials. Cuprammonium rayon, for instance, maintains its dominant position as the most popular material for hemodialysis (artificial kidney). Thanks to its good mechanical strength, cuprarayon can be fabricated into much thinner membranes than synthetic polymer membranes; as a consequence, much better clearance of low-molecular-weight solutes is achieved.

The hydrophilicity and hydrophobicity of materials are the most fundamental properties to be controlled whenever they are utilized in biomedical devices. In Sect. 2, the author will review the role of hydrophilicity or hydrophobicity of polymeric materials in protein adsorption processes on their surfaces. It is well-known that protein adsorption is the first event when any of the body fluids encounters artificial materials.

Water soluble polymers have long been believed to be inert at the interface with any biological element. For example, an aqueous solution of poly(ethylene glycol) in an appropriate concentration is known to form an immiscible two-phase liquid system with an aqueous solution of dextran. The two-phase system is used for partitioning two or more types of living cell, based on subtle differences in their properties for attachment to the two types of substrate. In the partitioning process, poly(ethylene glycol) normally had no adverse effect onto the living cells [2–4].

A large number of researches have been carried out in order to obtain biocompatible hydrophilic surfaces by introducing water-soluble polymer grafts, including poly(ethylene glycol), polyacrylamide and poly(vinyl alcohol). It was, however, occasionally observed that there is an optimum in the number of grafted chains for the surface to exhibit the best biocompatibility. The discussions in Sect. 3 will present an approach to a possible elucidation of this problem. A very unique property of poly(N-isopropylacrylamide) as thermoresponsive hydrogel will also be discussed in this chapter.

Biomer is a commercially available, medical grade segmented poly(ether urethane urea) (PEUU), which is prepared by treating an isocyanate-capped polyetherurethane with diamine (e.g. ethylenediamine) as chain extender:

$$\text{MDI} - [(CH_2CH_2CH_2CH_2O)_m - \underset{\underset{O}{\|}}{C} - \underset{H}{N} - \langle\text{Ar}\rangle - CH_2 - \langle\text{Ar}\rangle - \underset{H}{N} - \underset{\underset{O}{\|}}{C}O]_n - (\text{polyether}) - \text{MDI}$$

(polyether)   (urethane)

Here, MDI is p-diphenylmethanediisocyanate.

Table 1. Multifaceted aspects of research on biomedical polymers

	Factors to be Controlled	Biologically Derived Materials	Synthetic Materials		Biohybridized and Biomimicking Materials
Materials Design	• hydrophilic • hydrophobic • microdomain • cationic • anionic	• cellulosics • chitin, chitosan • dextran • agarose • collagen	• polysiloxane • polyester • polyurethane • PEO • poly(methyl methacrylate) • polyacrylamide • PTEF • polyolefin, polystyrene • newly designed materials		• heparin conj. materials • urokinase conj. materials • receptor mimicking materials • biomimicking materials

	Implants	Prosthesis	Blood Purification	Cell Separation	Biohybridization	Cell Cultivation	Drug Delivery	Bioanalysis	Biosensing	Enzyme Immobilization

Concept or Methodology / **Devices**	• bone cement
	• dental cement
	• dressing
	• IOL
	• contact lense
	• ligament
	• pacemaker
		• vascular graft (I)
		• AO heart
		• LVAD
		• IABP
			• AP long
			• AO kidney
			  (dialyzer
			  filter
			• heart valve
			• LDL apheresis
				• cell sorter	• AO liver
					• artificial skin
					• vascular graft(2)
					• AO pancrea
										• bioreactor

	Biocompatibility	Biospecificity	Physical & Physicochemical Capability
Properties Demanded	• blood compatibility • tissue compatibility • bioinertness	• specific adsorption • specific adhesion • specific recognition • cytokine production • cell cycle determination	• mechanical strength • selective permeation • non-specific adsorption (adhesion) • degradability

Fundamentals: Elucidation of structure-property relationship in interaction with biological elements
    at molecular level
    at cellular level
    at molecular assembly level

(AO: Artificial Organ)

There are several other commercial products containing segmented polyurethane (SPU) such as pellethane or cardiothane. These SPUs are widely recognized to possess notable biomedical properties as materials for artificial heart and intra-aortic balloon pumping (IABP), and also as coating materials for pacemaker-lead insulators (See Sect. 4.1).

The most important structural features of these SPUs are the microphase-separated (or microdomain) structures which are formed in their molecular assembly systems. In 1972, Lyman [5] suggested that the antithrombogenic property of his PEUU may be brought about by its "microarchitectural effect." In the same year (1972), Imai and Masuhara [6, 7] reported that "microheterogeneous surface" structure was responsible for its antithrombogenicity.

Since the 1970s, a number of reports on biomaterials other than SPU have also been presented, providing us with evidence which shows the important role played by microdomain structures in realizing excellent biomedical properties. For instance, an A–B–A type block copolymer (HEMA–St—HEMA) (See Sect. 4.2) was shown to form microdomain structure and to exhibit excellent blood compatibility in both in vitro and in vivo tests.

A series of polyamine-graft copolymers (See Sect. 4.3) were found to form microdomain structure and to exhibit unique biomedical behavior at the interface with living cells, e.g. blood platelets or lymphocytes. Although a number of postulates were proposed to explain the unique behavior of microdomain-structured surface, mechanisms for the mode of interaction of living cells with any of the domain-structured materials have not been adequately explained. In Sect. 4, the author will review results on the biomedical behavior of SPUs, HEMA–STY, and polyamine-graft copolymers, and discuss their interfacial properties in terms of the random network concept of water molecules on the material's surface.

Biohybridized surfaces which are conjugated with biological elements (e.g. heparin, prostaglandin, urokinase etc.) have been considered to be biocompatible for many years. Endothelial cell lining (or sodding) on the luminal surface of vascular grafts is one of the leading concepts for realizing blood-compatible surfaces. Ishihara et al. reported that a copolymer having phospholipid polar groups exhibited excellent blood compatibility (See Sect. 5.2). Wildevuur [8–10] developed a biodegradable vascular graft, that was intended to function as a temporary scaffold on which autologous cells would complete the regeneration of a new biological system for their own blood vessel.

Cell-adhesive oligopeptides (e.g. Arg–Gly–Asp–Ser (RGDS), Arg–Glu–Asp–Val (REDV), etc.) are often immobilized on the surface of artificial materials, in order to bring about effective adhesion and/or proliferation of a particular cell species (See Sect. 5.1). Oligosaccharides also are known to function as specific receptors toward particular cell species. For instance, the lactose-carrying polystyrene derivative, poly(N-p-vinylbenzyl-β-D-lactonamide), PVLA, was reported to exhibit a highly specific affinity for hepatocytes. β-Galactose residues of PVLA interact with the asialo-glycoprotein receptors on the surface of hepatocytes (See Sect. 5.2).

Oligopeptide- or oligosaccharide-conjugate materials are the typical examples for receptor-mimicking materials, the essential feature of which, of course, is biospecificity. In addition to these examples, a number of other processes and methodologies in ongoing biomaterials research are also based upon the concept of biospecificity; some explicitly and others implicitly (See Table 1). Several of these examples will be reviewed in Sect. 5 under the title of biohybridized and biomimicking materials.

2 Hydrophilicity or Hydrophobicity of Polymeric Materials and Their Behavior toward Protein Adsorption

Protein adsorption is the first event that takes place on material surfaces when blood or other body fluids are brought into contact with any material. Therefore, cell – material interactions must be discussed by taking into consideration the species and the nature of the protein adsorbed on the material surfaces. For instance, a series of cell-attachment and spreading experiments [11] of fibroblasts on the surface of modified polystyrene (TCP and Primaria) carried out in the presence of fetal calf serum (FCS) showed that FCS contains components which tend to decrease the attachment and spreading of fibroblast cells. The effect of these nonadhesive components was only evident when the FCS was depleted of vitronectin, showing that vitronectin overcomes the effect of these nonadhesive components and promotes cell-attachment and spreading on the polystyrene surface. Fibronectin, on the other hand, does not play a principal role in this fibroroblast adhesive process (Fig. 2).

In Sect. 2, the author discusses protein adsorptive behavior of various polymeric materials in terms of their hydrophilicity or hydrophobicity.

2.1 Two Models of the Protein-Adsorption Process

More than 10 years ago, Ikada et al. evaluated the free energy (γ_{1w}) of the interface between polymer and water, and the work of adhesion ($W_{12,w}$) in water of bovine serum albumin (BSA) to the polymer surface [12]. According to a well-known concept of surface chemistry, Ikada derived the following equations.

$$W_{12,w} = \gamma_{1w} + \gamma_{2w} - [\gamma_{12}]_w \tag{2.1}$$

where γ_{2w} is the free energy of the interface between BSA and water, and $[\gamma_{12}]_w$ is the interfacial energy of polymer/BSA in water.

Ikada evaluated γ_{1w} by using Eq. 2.2.

$$\gamma_{1w} = \gamma_{1w}^d + \gamma_{1w}^p \tag{2.2}$$

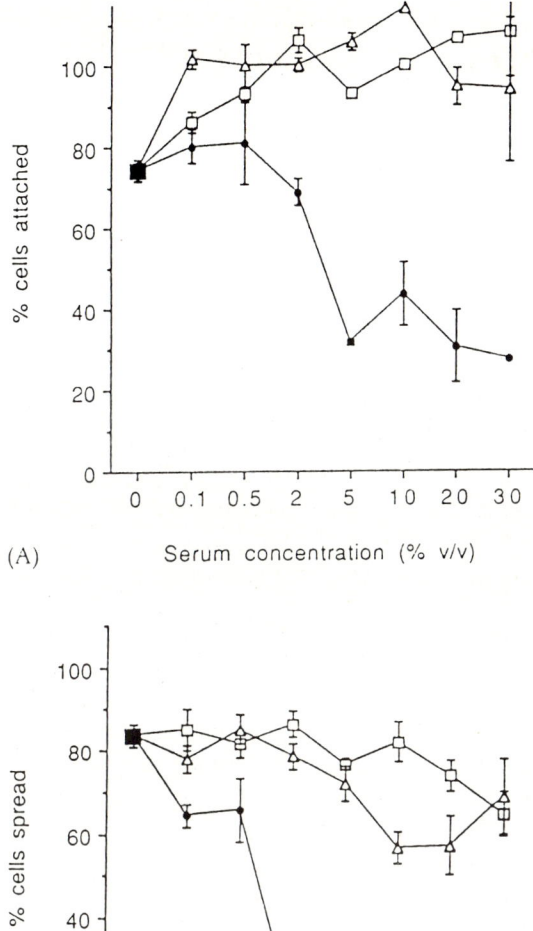

Fig. 2. Effect of serum concentration on the attachment and spreading of BHK-21* cells onto TCP[2]* surface. BHK-21 cells were seeded in media containing the indicated concentrations of intact serum (*open squares*), Fn-depleted serum (*triangles*), Vn-depleted serum (*circles*), or serum-free medium alone (*the single closed square*) and the attachment panel (**A**) and spreading panel (**B**) of the cells were determined after 90 min culture on TCP (panel (**A, B**)) Mean ± SEM. (Reproduced from J. Biomed. Mater. Res. [Ref. 11: Role of serum vitronection and fibronectin in adhesion of fibroblasts following seeding onto tissue culture polystyrene] through the courtesy of John Wiley & Sons, Inc.)
*BHK-21: Fibroblast cell lines from Baby Hamster Kidney
2*Similar results on Primaria are also presented in [Ref 11]

where γ_{1w}^d and γ_{1w}^p are, respectively, dispersive and polar components of γ_{1w}. For evaluating γ_{1w}^d, data reported in the literature for γ_1^d and γ_w^d were inserted in Eq. 2.3.

$$\gamma_{1w}^d = \gamma_1^d + \gamma_w^d - 2(\gamma_1^d \gamma_w^d)^{1/2} \tag{2.3}$$

As for γ_{1w}^p, Ikada used Eq. 2.4.

$$\gamma_{1w}^p = \gamma_1^p + \gamma_w^p - 2(\gamma_1^p \gamma_w^p)^{1/2} \tag{2.4}$$

where γ_1^p (or γ_w^p) was evaluated as the difference between observed value (γ_1) and γ_1^d (or γ_w and γ_w^d). (It is to be noted that objections have often been raised against the concept of Eq. 2.4 [13]). Evaluation of γ_{2w} was carried out similarly by inserting reported data (31.4 and 33.7 erg·cm^{-2}, respectively) for γ_2^d and γ_2^p of BSA.

The values of $W_{12,w}$ were calculated by inserting values obtained for γ_{1w}^d, γ_{1w}^p, γ_{2w}^d and γ_{2w}^p in Eq. (2.5).

$$W_{12,w} = [2(\gamma_{1w}^d \gamma_{2w}^d)^{1/2} + 2(\gamma_{1w}^p \gamma_{2w}^p)^{1/2}]_w \tag{2.5}$$

where the subscript w of the bracket means that the values within it refer to an aqueous medium. Ikada et al. carried out their calculation using a drastic approximation.

Partial results of the calculation are cited in Table 2, which shows that the work of adhesion is expressed as a bell-shape curve. In other words, the maximal adsorption of protein takes place on polymer surface having intermediate hydrophilicity. Ikada et al. confirmed this by the results of BSA adsorption on 8 polymer surfaces [12], and also by those of albumin and fibronectin adsorption on 13 polymer surfaces [14]. L-Cell attachment also showed a bell-shape profile [15].

The bell-shape-profile concept for protein adsorption may be a useful guideline for researchers when they consider the adsorptive behavior of proteins on polymers with different degrees of hydrophilicity. As will be discussed in later Sect. (2.3), when material surfaces carry ionic groups, the contribution from

Table 2. Surface free energies and work of adhesion with BSA in water for different polymers (erg·cm^{-2})

Polymer	γ_{1w}	$W_{12,w}$
Polyethylene (HD)	52.6	1.40
PTFE	51.2	0.25
Plystyrene	43.7	3.21
Poly(ethylene terephthalate)	29.8	4.59
PMMA	27.4	3.88
Polyurethane	6.20	3.04
PVA	3.1	2.5
Canine artery	0.02	0.18

(from Ref. 12)

electrostatic interactions becomes significant, so that deviations from the bell-shape profile are marked.

Peppas et al. [16] presented a new method for calculating protein adsorption on polymeric surfaces. In their model, protein adsorption is regarded as an equilibrium reaction, which takes place on the polymer surface in competition with the adsorption of water.

$$[\text{water, water}] \xrightarrow{g_{13}} [\text{water, surface}] \quad (2.6)$$

$$[\text{protein, water}] \xrightarrow{g_{23}} [\text{protein, surface}] \quad (2.7)$$

where g_{13} and g_{23} are free energy change in the processes of (2.6) and (2.7), subscripts 1, 2 and 3 being those for water, protein and polymer surface, respectively.

The global interaction parameter, g_s, can be expressed as Eq. (2.8):

$$g_s = g_{23} - g_{13} \quad (2.8)$$

Protein adsorption can take place only when g_s is negative; the larger the absolute value of g_s, the more adsorption on the relevant polymer surface is anticipated. In addition, process (2.9) should be taken into consideration along with process (2.7):

$$[\text{protein, water}] \xrightarrow{-g_{12}} [\text{water, water}] \quad (2.9)$$

The parameter g_{12} is the change of the free energy of water molecules when they are driven back to the bulk water system. The exergonic character of process (2.9) will favor the concurrent process (2.7) of protein adsorption onto the material.

Peppas et al. divided the free energy change, g_{23}, into three components originating from: (i) dispersive forces (g_{23}^d), (ii) the acid-base interactions (g_{23}^{ab}) and (iii) the hydrophobic effect (g_{23}^h).

For the evaluation of g_{23}^{ab} values, Peppas et al. first calculated the acid-base interaction parameter of every constituent amino acid residue, $\bar{g}_{23}^{ab}$, according to the method of Lee and Richards [17].

$$\bar{g}_{23}^{ab} = K\Delta H_a^3 \quad (2.10)$$

where ΔH_a^3 means the enthalpy change originating from the acid-base interaction when a constituent amino acid residue, a, is adsorbed to the polymer surface. K is a coefficient representing the probability of exposure (or relative accessibility) of an amino acid residue, a. Each amino acid residue has a different value of K. Summation over all the amino acid residues of the protein gives Eq. (2.11):

$$g_{23}^{ab} = \sum_a KK'(\Delta H_a^3 - \Delta H_a^1) \quad (2.11)$$

where K' is the fraction of the protein surface in contact with the polymer surface, and ΔH_a^1 is the enthalpy change ($\bar{g}_{12}^{ab}$) originating from the acid-base interaction between an amino acid residue and its surrounding water molecules located in the contact area.

Peppas et al. calculated the values for ΔH_a^3 and ΔH_a^1 by using Drago's acid-base coefficients of amino acid polar groups and of polymer surfaces [13, 18–20].

The dispersive force contribution (g_{23}^d) is expressed as Eq. (2.12):

$$g_{23}^d = -2[(\gamma_2^d \gamma_3^d)^{1/2} - (\gamma_1^d \gamma_2^d)^{1/2}]K'S_2 \tag{2.12}$$

where S_2 represents the outer surface area of the protein, and γ_1^d, γ_2^d and γ_3^d are, respectively, the dispersive components of the surface tension of water, protein and the polymer.

The hydrophobic component (g_{23}^h) contains a large entropy contribution, which becomes more significant when the adsorptive substance is a macromolecule such as protein. Peppas et al. regarded the quantity, $N_2^c S_2^c$, as the hydrophobic exposure area (S_2^h) of the protein. Here, N_2^c is the total number of carbon atoms contained in the protein, and S_2^c is their surface accessibility. Thus, the fraction of the hydrophobic exposure area of protein, which is in contact with polymer surface, is expressed as $K'S_2^h$.

For the polymer surface, on the other hand, it was assumed that only a fraction (γ_3^d/γ_3), is effective for the hydrophobic interaction. Peppas et al. also assumed that 24 cal/(Å)² is released when an exposed hydrophobic group of the protein comes into contact with a hydrophobic surface. Thus, they obtained, Eq. 2.13 for the evaluation of g_{23}^h.

$$g_{23}^h = 24K'S_2^h(\gamma_3^d/\gamma_3) \tag{2.13}$$

To evaluate the global interaction parameter (g_s), the value of g_{13} was also needed. It was calculated by Eq. (2.14):

$$g_{13} = -(W_A^{13} - W_A^{11})K'S_2 \tag{2.14}$$

where W_A^{13} is the work of adhesion at the water-polymer interface and W_A^{11} is the (artificial) work of water-water adhesion (i.e. in bulk water). Peppas et al. evaluated W_A^{13} with Eq. (2.15).

$$W_A^{13} = 2(\gamma_1^d \gamma_3^d)^{1/2} + I_{13}^p \tag{2.15}$$

where I_{13}^p is the contribution of the work of adhesion from polar interactions.

On the basis of the aforementioned considerations, Peppas et al. carried out an evaluation of the interaction parameters which are operative in ternary systems consisting of water, protein (BSA, γ-Ig and FGN), and polymers (PMMA, PE, PVC, PS, PVA, PVDF, PDMS and PEO). Their results are cited, with some simplification, in Tables 3, 4 and 5.

It is seen from Tables 3, 4, and 5 that almost all of the values of the global interaction parameters (g_s) are negative ($g_s < 0$). This is compatible with our knowledge concerning the protein-adsorption behavior of polymeric materials.

Table 3. Calculated values of the interaction parameter g_s, and its components g_{23} and g_{13} (all in kcal/mol) between BSA and different polymeric surfaces

	PMMA	PE	PVC	PS	PVA	PVDF	PDMS	PEO
g_{23}^d	−16.8	−10.9	−17.9	−20.7	−8.4	−1.5	2.0	−12.5
g_{23}^h	−12.4	−12.9	−12.8	−15.2	−6.8	−10.3	−12.4	−8.8
g_{23}^{ab}	52.0	58.3	31.8	57.4	13.0	27.5	49.5	51.6
Total: g_{23}	22.8	34.5	1.1	21.5	−2.2	15.7	39.1	30.3
g_{13}^d	−14.1	−9.2	−15.0	−17.2	−7.1	−1.3	1.5	−10.5
g_{13}^p	82.1	92.3	89.2	93.0	28.1	67.3	88.5	45.0
Total: g_{13}	68.0	83.1	74.2	75.8	21.0	66.0	90.0	34.5
$g_s = g_{23} - g_{13}$	−45.2	−48.6	−73.1	−54.3	−23.2	−50.3	−50.9	−4.2

Reprinted with some simplification from Ref. [16] through the courtesy of VSP BV.

Table 4. Calculated values of the interaction parameter g_s and its components g_{23} and g_{13} (all in kcal/mol) between γ-Ig and different polymeric surfaces

	PMMA	PE	PVC	PS	PVA	PVDF	PDMS	PEO
g_{23}^d	−28.1	−18.3	−29.9	−34.5	−14.0	−1.6	3.3	−20.9
g_{23}^h	−28.6	−29.9	−29.7	−35.9	−15.9	−23.7	−28.7	−20.4
g_{23}^{ab}	89.0	102.5	51.1	100.5	14.8	43.1	83.3	87.9
Total g_{23}	32.3	54.3	−8.5	30.1	−15.1	17.8	57.9	46.6
g_{13}^d	−24.5	−15.8	−25.8	−28.6	−12.1	−2.2	2.6	−18.1
g_{13}^p	141.3	158.9	153.8	154.0	48.3	115.9	152.5	77.5
Total: g_{13}	117.0	143.1	128.0	125.4	36.2	113.7	149.9	59.4
$g_s = g_{23} - g_{13}$	−84.7	−88.7	−136.5	−95.0	−51.3	−95.9	−92.0	−12.8

Reprinted with some simplification from Ref. [16] through the courtesy of VSP BV.

Table 5. Calculated values of the interaction parameter g_s and its components g_{23} and g_{13} (all in kcal/mol) between Fibrinogen and different polymeric surfaces

	PMMA	PE	PVC	PS	PVA	PVDF	PDMS	PEO
g_{23}^d	−101.4	−65.7	−119.1	−124.8	−50.6	−9.2	11.5	−75.5
g_{23}^h	−57.8	−60.3	−60.0	−61.5	−32.1	−47.4	−57.4	−40.9
g_{23}^{ab}	451.9	502.0	265.8	494.9	99.4	228.4	431.1	448.6
Total g_{23}	292.7	376.0	86.7	308.6	16.7	171.8	385.2	332.2
g_{13}^d	−95.7	−62.0	−101.6	−116.8	−47.8	−8.6	10.2	−66.2
g_{13}^p	556.8	624.8	604.5	629.9	190.4	456.6	600.6	310.5
Total: g_{13}	461.1	562.8	502.9	513.1	142.6	448.0	610.8	244.3
$g_s = g_{23} - g_{13}$	−168.4	−186.8	−416.2	−204.5	−125.9	−276.2	−225.6	87.9

Reprinted with some simplification from Ref. [16] through the courtesy of VSP BV.

On the other hand, the values for g_{13} or g_{23} are positive, with a few exceptions. In other words, the adsorption processes of water and of protein onto polymer surfaces are for the most part thermodynamically unfavorable. The negative values of g_s are the result of a remarkable decrease in free energy when adsorptive water molecules are driven back to the bulk water system, as seen

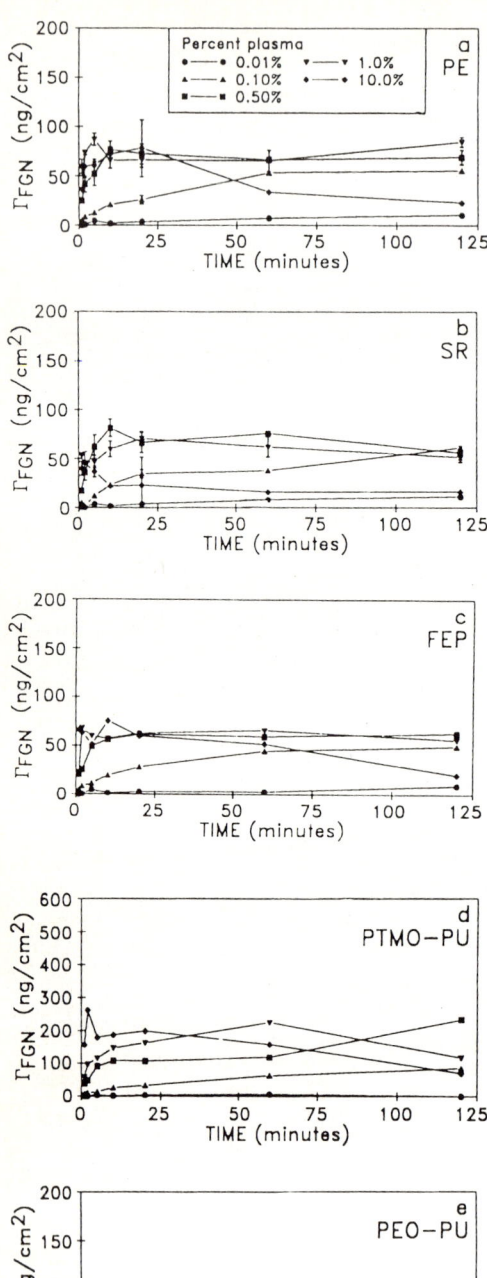

Fig. 3. Adsorption of FGN from various dilutions of plasma onto (**a**) PE, (**b**) SR, (**c**) FEP, (**d**) PTMO–PU and (**e**) PEO–PU. Γ represents the surface concentration of protein adsorbed (Reproduced from J Biomater Sci Polymer Edn [Ref 21] through the courtesy of VSP-BV)

from Eq. 2.14. The free energy change, g_{13}, is larger on hydrophobic surfaces than on hydrophilic ones. Furthermore, on the hydrophobic surfaces, g_{23}^h contributes to a greater extent to reducing the absolute value of g_{23}. Thus, the global interaction parameter, $g_s(=g_{23}-g_{13})$, becomes still smaller; in other words, protein adsorption on such a hydrophobic surface is more facilitated. This conclusion is supported by a variety of experimental results, but seems to be in conflict with the bell-shape correlation discussed above.

As Peppas pointed out, however, the values of the interaction parameter g_{23} reveal that protein 'dislike' hydrophobic materials more than they 'dislike' hydrophilic ones. It is anticipated that a similar result to the bell-shape profile may be obtained from the ternary systems where influence of parameter g_{13} becomes less significant, as will be discussed in connection with the results illustrated in Fig. 3.

On hydrophilic surfaces, such as PVA or poly(HEMA), OH-groups of the materials are incorporated in the network structure of adsorbed water molecules (see Sect. 4.4). In consequence, the absolute value of $W_A^{13} - W_A^{11}$ is considered to become still smaller, where – owing to the stabilization of water molecules on the hydrophilic surface – the water-removing-process (reverse reaction of Eq. (2.6)) proceeds slowly. Many experiments were carried out with water-adsorbed hydrophilic surfaces, the behavior of which was time-dependent. In a similar way, the water removal from the proteins [Eq. (2.9)] is also considered to proceed slowly. Thus, we must be careful in considering experimental results in comparison with the data in Tables 3, 4 and 5.

Another point to be discussed is the anormally small g_{23} values of PVA in the three Tables, which may suggest PVA to be a favorable surface for protein adsorption, in contrast with the reported data (Table 2). The small values of g_{23} arise from the small values of g_{23}^{ab}, which was calculated using the Drago equation. According to the Drago concept, PVA is an "acidic" component which can only interact with "basic" polar groups of constituent amino acid residues of the protein. This may be an unfortunate limitation of the Drago concept. If we also consider a contribution from interactions of PVA (as "base") with "acidic" polar groups of protein, we should obtain larger values for g_{23}^{ab} and g_{23}.

2.2 Comparison of the Adsorption of Four Proteins to Five Biomaterials with Different Hydrophilicity

Cooper et al. [21, 22] reported in detail the results of their laborious work on the adsorption of four proteins; human serum albumin (HSA), fibrinogen (FGN), fibronectin (FN), and vitronectin (VN), on five biomaterials: polyethylene (PE), silicone rubber (SR), Teflon-FEP (FEP), poly(tetramethylene oxide)-polyurethane (PTMO-PU), and poly(ethylene oxide)-polyurethane(PEO-PU). Hard segments of these polyurethanes are composed of a methylene-bis(p-phenylisocyanate) (MDI) chain extended wih 1,4-butanediol.

The adsorption experiments were carried out by quantifying each of proteins adsorbed on the material from mono-component protein solutions, from four-component protein solutions, and from plasma and diluted plasma. Adsorption profiles of protein were largely different, depending on the aforementioned experimental conditions. For instance, the behavior of any particular protein from diluted plasma varied in response to the extent of plasma dilution. Cooper's results are illustrated in Fig. 3, on fibrinogen adsorption onto five polymer surfaces. It is seen that the adsorption profiles are different one another, being influenced by the different nature of the polymer surfaces. The surface concentrations of adsorbed protein are mostly time-dependent, and maxima in the adsorption profiles were observed. This is interpreted in terms of replacement of adsorbed fibrinogen molecules by other proteins later in time (Vroman effect). Corresponding profiles were also presented for FN and VN.

From an analysis of the Cooper results the present author found that the surface concentrations of FGN, FN and VN at 120 min. correlate with the contact angles of the five polymer surfaces, to give a bell-shape profile with a maximum at the PTMO-PU surface. Under the experimental conditions of Fig. 3, the final quantity of adsorption (Γ) is probably determined by the g_{23} term, but not by the g_{13} term (see Sect. 2.1).

2.3. Polymeric Materials with Ionic Functional Groups and Their Protein Adsorptive Behavior

The hydrophilicity of polymeric materials carrying ionic functional groups varies in response to their electric charge under the pH of biological surroundings, because the electric charge is closely correlated with the extent of hydration around the key functional groups. Whitesides et al. [23] examined the correlation between the nature of functional groups introduced in a PE film and its wettability in terms of advancing contact angle θ_a. They reported results of their "contact angle titration", in which advancing contact angles θ_a of buffered water on PE-CO$_2$H, PE-NRR′ and several other derivatives including carbonyl or hydroxyl groups were plotted as a function of pH. According to their results, only surfaces containing ionizable functionality, $RCO_2H \rightleftharpoons RCO_2^{(-)}$, $RNH(CH_2)_4(+) \rightleftharpoons RN(CH_2)_4$, showed a pH dependence of the contact angle.

Using frontal chromatography, we examined the adsorptive behavior of the polyamine-HEMA graft copolymer HAx (x means polyamine content in wt.%) (See Sect. 4.3) against BSA [24]. Glass beads (average ϕ, 87.4 µm) coated with HA copolymer were packed into a column (10 mm I.D.), and BSA in phosphoric buffer solution (PBS) was loaded at constant flow rate to obtain a frontal chromatogram. Part of the results obtained from experiments carried out using PBS of several pH levels are shown in Fig. 4. In the pH range of 6.4–7.4, almost no adsorption was observed on the surface of poly(HEMA). This was anticipated on the basis of the Ikada concept (See 2.1), if we recall that γ_{1w} for poly(HEMA) is less than 1.0 [25]. On the surface of HA7 and HA13, BSA adsorption became more significant as the pH value of PBS was lowered. It was

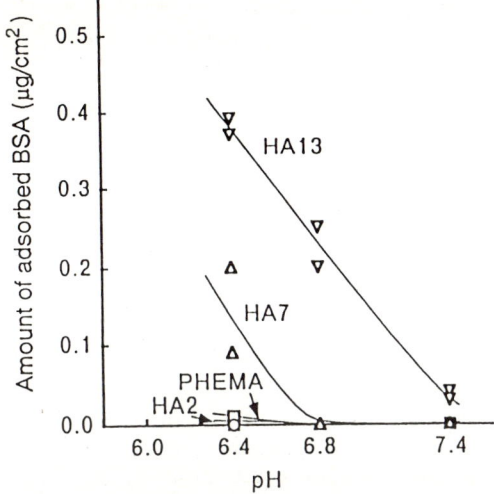

Fig. 4. The pH-dependence of the amount of adsorbed BSA on the surface of PHEMA and HA copolymers

also noticeable that the amount of adsorbed BSA increased with increasing polyamine content of the HA copolymers. More detailed study revealed that the amount of BSA adsorption decreased drastically when the pH value of the medium became high enough to reduce the degree of protonation of the polyamine portion of HA to 0.5. This result indicates that the contribution of the electrostatic force is influences the BSA adsorption, because the globular surface of BSA (isoelectric point: 4.7) carries negative electric charge in the range of pH 6.4–7.4.

We found also that the amino-content of the polyamine-styrene graft copolymer, SA_x (x is the polyamine content in wt.%) correlates closely with the adsorption behavior of FN or VN as it also does with the adhesivity of bovine aortic endo-thelial cells to the SA copolymer surface, as will be discussed in Sect. 4.4.

As for the effect of anionic group, there are a number of reports dealing with the antithrombogenic behavior of sulfonate-modified surfaces of segmented polyurethane (SPU). An interesting feature of the adsorptive behavior of fibrinogen on these material surfaces will be discussed in Sect. 4.1.

3 Water-soluble Polymers, Their Conjugates and Hydrophilic Polymers

As stated in the introductory chapter, water-soluble polymers, such as poly(ethylene oxide), poly(N-vinylpyrrolidone), polyacrylamide, poly(vinyl alcohol), dextrans etc., have been believed to be inert to any of the biological elements. In fact, a number of trials have been carried out to improve the biocompatibility of polymeric materials by conjugating water soluble polymers, especially poly(ethylene oxide) (PEO), to the base-material surfaces. Although improved biocompatibility of PEO-modified surfaces were reported in many cases, difficulties associated with the PEO-modification were observed in several other cases. Recently, Sefton et al. [26] published a review entitled "Does polyethylene oxide possess a low thrombogenicity?". Amiji and Park [27] also reviewed this problem with extensive references.

3.1 Poly(ethylene oxide) and its Conjugates

The bioinert property of PEO chains is often ascribed to their flexibility and mobility in aqueous media. It is also known that the biomedical behavior of PEO-modified surfaces is remarkably changeable, depending on molecular weight and the density of the PEO chains conjugated to the material surface. For instance, Merrill et al. [28] synthesized interpenetrating networks (IPN) by end linking PEO to poly(glycidoxy-propyl methyl/dimethylsiloxane) (PGPM/DMS) and carried out, by using ex vivo baboon-shunt assay, several biomedical examinations in terms of platelet deposition, fibrinogen deposition and complement C_3 activation. They found that the platelet and fibrinogen depositions as well as complement activation were accelerated on the surface of IPN conjugated with lower molecular weight PEO (MW = 2000), whereas these biological responses were suppressed on IPN with higher molecular weight PEO (MW = 20 000). A number of researchers reported molecular weight dependency of PEO in terms of its biomedical behavior [29–33]. Many of these researches, however, present seemingly controversial results. Apparent inconsistencies in their results may arise from different conditions in the biological assay and in the physicochemical properties of base-materials to which the PEO chains were conjugated.

Ikada [34] introduced methoxy-poly(ethylene glycol) methacrylate, $CH_2 = C(CH_3)-CO-(OCH_2CH_2)_n-OCH_3$, into a polyurethane, PTMO-PEU, by a glow discharge technique (see Sect. 4.1). IgG adsorbed on the modified surface was found to decrease drastically compared to the unmodified surface. However, higher density of grafting of PEG chains did occasionally enhance the IgG adsorption when the degree of polymerization (n) of PEG chain was 23. No such an enhancement of IgG adsorption was observed with shorter PEG chains (n = 4 or 9).

Imanishi et al. [33] pointed out that there was an optimum point at which the tethering density of PEO chains makes a polybutadiene urethane surface biocompatible. They also reported that bovine plasma FGN and bovine γ-globulin adsorbed on the PEO-modified polyurethane were completely denatured, in comparison with 4–38% denaturation of BSA adsorbed.

S.W. Kim et al. [35] examined blood compatibility of heparin-conjugated SPUU-PEO samples, which were prepared by introducing PEO (MW = 1000, 3350 and 7500, respectively) into Biomer, followed by conjugation of heparin. The surfaces of B-PEO 3.4 K and B-PEO 7.5 K were found to suppress markedly the adsorption of BSA, bovine serum IgG and bovine FGN from diluted plasma (1% of normal plasma concentration), even when heparin was not conjugated. Platelet adhesion from rabbit platelet-rich-plasma (PRP) was also suppressed *more or less* on the B-PEO surfaces. However, an ex vivo study, in which intra-arterial A-A shunts (1.5 mm ID tubing) were implanted in the arteries of male rabbits, showed that both of the non-heparinized surfaces, B-PEO 3.4 K and B-PEO 7.5 K, failed to prolong the occlusion time, while the heparin-conjugated surfaces, B-PEO 3.4 K-Hep and B-PEO 7.5 K-Hep, did prolong the occulusion time by 400 percent and 350 percent, respectively.

Sefton et al. concluded their review [26] with the following sentence: "It is clear that more evidence needs to be gathered before definitive conclusions about the long-term non-thrombogenic capability of PEO can be reached".

Tanzawa et al. [36–39] synthesized several hydrogels consisting of a polymethacrylate (or PVC) backbone with one of a variety of structurally different poly(ethylene glycol)s (PEGs) as side chain (Table 6). They carried out in vitro examinations (protein adsorption and platelet adhesion) of these hydrogels and found polymers C′ and D′ in particular to exhibit excellent blood compatibility. The ex vivo experiments with polymer D′ were then carried out, using rabbits to estimate changes in the biological functionality of blood coagulation system. Their results showed that the circulating blood was scarcely injured by contact with the surface of polymer D′.

Studies to elucidate the correlation between the structure of the polymer gels and their blood compatibility were carried out by means of ^{13}C-NMR (for mobility of the PEG chains) and ^{1}H-NMR and DSC (for the effect of water on their properties). Results are shown in Table 7. By comparing these results with one another, Tanzawa et al. concluded that material surfaces with the highest fraction of water molecules of intermediate mobility exhibit the best blood compatibility. This was supposed to come from a similar mobility of the intermediate water compared to that of oligosaccharides on the outermost surface of the cell.

The present author is inclined to consider the formation of entanglements to be a key event which determines the property of water molecules involved in water-soluble polymer matrices. If PEO chains are tethered densely enough on the material surface to form entanglements, the mobility of the surrounding

Table 6. Synthetic hydrogels containing PEG [36–39]

A	PMMA
B	PHEMA
C	Poly(MMA-*co*-9EGMA)
C′	Poly(MMA-*co*-100EGMA)
D	Poly(VC-*graft*-9EGMA)
D′	Poly(VC-*graft*-100EGMA)

*9EG: $-(O-CH_2CH_2)_9-OCH_3$
100EG: $-(O-CH_2CH_2)_{100}-OCH_3$

Table 7. NMR and DSC data for hydrogels containing PEG [36–39]

DSC freezing point (°C)		< −100	−60	−10	0	0
NMR correlation time τ_c(s)		> 10^{-6}	10^{-9}	10^{-10}	10^{-11}	10^{-12}

Hydrogel polymer	Total water content (%)	Fractions of water structure (%)				
		Unfrozen	Intermediate		Free	
					Capillary	Bulk
A PMMA	40	22	0	0	78	0
B PHEMA	39	54	22	0	11	23
C Poly(MMA-co-9EGMA)	44	39	3	16	10	22
C' Poly(MMA-co-100EGMA)	43	47	3	22	5	23
D Poly(VC-g-9EGMA)	38	70	1	17	8	4
D' Poly(VC-g-100EGMA)	44	62	0	25	8	5

water molecules decreases; possibly causing some enhancement of the interaction with proteins as well as with cells. In the blood circulating system, the blood flow causes a shear stress between the materials and the glycocalix of the cell; the shear stress increases with decreasing mobility of water molecules surrounding the tethered PEO chains, and this may result in the destruction of the glycocalix of disjoined cells.

The mobility concept for surrounding water molecules is closely related to the random-network concept, as will be discussed in Sect. 4.4.

3.2 Poly(N-isopropylacrylamide) and Its Copolymers as Thermoresponsive Hydrogels

It is widely known that poly(N-isopropylacrylamide), poly(IPAAm), in water has a lower critical solution temperature (LCST) at 32 °C. LCST was originally observed in PEG solutions a long time ago. Rowlinson et al. [40] (1957) explained the "lower consolute temperature" for PEG in water in terms of negative entropies. The first paper on the LCST of poly(IPAAm) at about 31 °C was presented by Heskins and Guillet in 1968 [41]. They reported that aqueous solution of poly(IPAAm) showed phase separation above this temperature, and ascribed it primarily to an entropy effect on the basis of thermodynamical considerations.

Tanaka et al. studied the volume transition of IPAAm-(sodium acrylate) copolymers gel as a function of temperature for various copolymer compositions. Gels that had been swollen at lower temperatures underwent a sharp collapse at different transition temperatures, depending on the ionic composition of each gel; the minimum transition temperature was approximately 34 °C. Tanaka et al. carried out a theoretical analysis of the data obtained and derived the equation of state for the phase and the equation for the phase transition of the poly(IPAAm) gel systems [42, 43].

A great number of researches have so far been carried on the incorporation of poly(IPAAm) and its copolymers in various biomedical devices, utilizing soluble/insoluble or swelling/deswelling processes in the temperature range of LCST. As overviewed by Okano et al. [44] these include: drug delivery system (DDS); solute separation; concentration of dilute solutions; immobilization of enzymes; detachment of cultured cells; coupling to biomolecules, and other aspects.

Hoffman et al. [45] synthesized thermo-reversible, soluble-insoluble polymer-enzyme conjugates in order to apply them to reactions with macromolecular or solid substrates. In contrast to usual solid phase enzymes, polymer-enzyme conjugates in soluble form are expected to react smoothly with solid substrates. Poly(IPAAm) with a carboxyl end group was prepared by radical polymerization of IPAAm in the presence of β-mercaptopropionic acid as the chain transfer agent. This polymer was conjugated to β-D-glucosidase with 1-ethyl-3-(3-dimethylaminopropyl)carbodiimide hydrochloride. After the reaction, the conjugated enzyme was easily isolated as a precipitate from the reaction medium by heating to about 40 °C. It was found in hydrolysis reaction of p-nitrophenyl-β-D-glucopyranoside that the conjugate enzyme retained a high percentage (> 90%) of its activity and – in addition – improved thermal stability over the native enzyme.

Hoffman et al. [46, 47] found that an LCST polymer remains strongly bound to a substrate, especially to cellulose acetate (CA), at a temperature above its LCST, whereas most of the adsorbed polymer molecules are easily rinsed off below the LCST. For instance, they synthesized a room-temperature-precipitable terpolymer (LCST = 7–13 °C), consisting of IPAAm, *N*-butylacrylamide (BAAm) and *N*-acryloxy succinimide (NASI), which was conjugated to a murine monoclonal antibody. They developed the membrane-affinity concentration immunoassay [48].

Okano et al. [49, 50] have been carrying out extensive studies on controlled drug delivery in response to temperature changes with poly(IPAAm-*co*-alkyl methacrylate) hydrogels. Their objective is to achieve complete 'on-off' regulation of drug release in response to stepwise temperature changes. They observed that the surface skin layer that is formed with increasing temperature stops drug release from the polymer matrix (Fig. 5).

The length of the alkyl side-chain of alkyl methacrylate (RMA) was influential in controlling the thickness and density of the surface skin layer. Poly(IPAA-*co*-HMA) or poly(IPAAm-*co*-LMA), were H is hexyl and L is lauryl, was able to form a thin and dense surface skin layer which was favorable for rapid 'on-off' control of drug release and for maintaining a constant release rate.

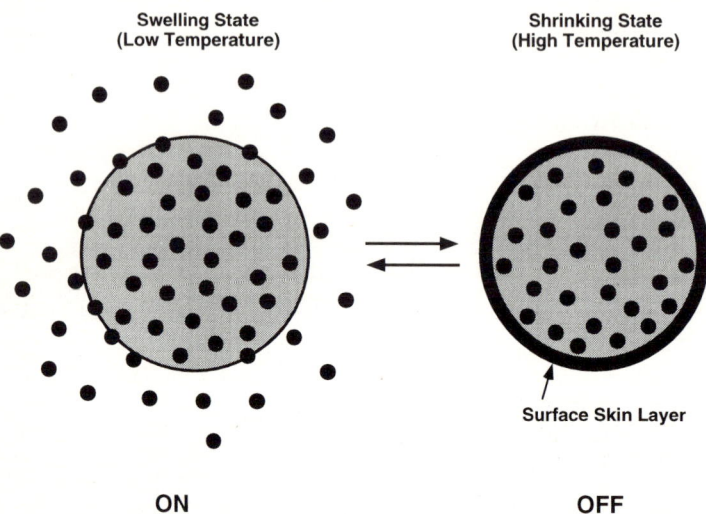

Fig. 5. On-off switching mechanism of drug release (Reproduced from J. Biomater. Sci. Polymer. Edn. [Ref. 49] through the courtesy of VSP BV)

Okano et al. [51] applied poly(IPAAm) to modifying the surface of commercial polystyrene[1] culture dishes for bovine aortic endothelial cells as well as for rat hepatocytes. Graft polymerization of IPAAm onto the polystyrene dishes was carried out by using an electron beam. The thickness of the poly(IPAAm)-grafted layer in aqueous systems was 0.5 µm at 37 °C and 0.6 µm at 15 °C. After endothelial cells were cultured at 37 °C for 2 days, the temperature was decreased to 10 °C and the number of cells that were detached from the surface was counted. Okano et al. observed that 100% of cells were detached from the poly(IPAAm)grafted surface.

Endothelial cells recovered from both the enzyme (trypsin) recovery system (ERS) and the poly(IPAAm) temperature recovery system (TRS) were subcultured and examined comparatively for cell adhesivity, cell morphology and cell growth activity. The most important finding in the subcultured system was that the TRS exhibited much a higher activity of prostacyclin generation than ERS. It is known that prostacyclin generation is an important function of endothelial cells.

Okano et al. [52] also reported the significant feature of cell assemblage of hepatocytes. Hepatocytes recovered by TRS were found to keep their original membrane sheet structure (Fig. 6).

---

[1] Falcon 3001 (polystyrene which is treated by glow discharge to introduce carbonyl and carboxyl groups, giving the surface a certain hydrophilicity).

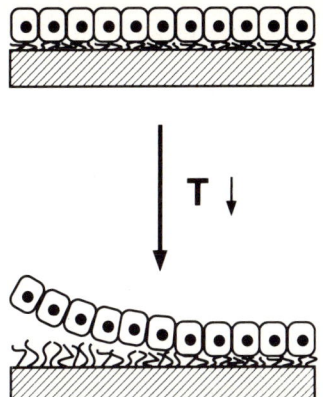

Fig. 6. The lowered temperature treatment enables recovery of hepatocytes keeping their membrane sheet structure. [Reproduced from Jinko-zoki (Jpn. J. Artif. Organs) [Ref. 52] through the courtesy of Jpn. Soc. for Artif. Organs]

The hepatocytes was subcultured for 2 days and their albumin secretion activity was measured. The TRS-harvested hepatocytes showed nearly the same albumin secretion activity as the primary culture, whereas the ERS-harvested ones showed only 20% activity. This finding is important because no subculture of hepatocyte has ever been successful, owing to a possible proteolytic disruption of cell-membrane structure (especially cell-cell adhesion) in the course of trypsin treatment.

Since the poly(IPAAm) layer in aqueous medium contains 25–30% water even at 37 °C, protein adsorption from the serum of the culture medium probably takes place at the network structure of water molecules in the interface of the poly(IPAAm) layer. The present author conjectures that there may be some similarity between this case and that discussed in Sect. 4.4.

4 Microdomain-Structured Materials

As stated in Chapter 1, microdomain-structured surfaces are believed to play an important role in their interactions with cells, proteins, and other biological elements. In this Chapter, the author will discuss biomedical behavior of three types of microdomain-structured materials: segmented polyurethanes, A-B-A block copolymers, and polyamine-graft copolymers.

4.1 Segmented Polyurethanes

A number of studies on the morphology and mechanical properties of segmented polyurethanes (SPU) are accessible in the literature [53–55].

In 1982, Takahara et al. [56] reported their estimates of the domain size (hard segment domain 9.2 nm; soft segment domain 0.9–2.1 nm) of their SPU samples based on low-angle X-ray scattering measurements. More recently, they [57, 58] carried out the surface analysis by using the freeze-etch XPS method to clarify the details of the domain structure of SPU surface under the biological environment. They also measured the dyamic contact angle at the surfaces. On the basis of their results from dynamic freeze-etch XPS and dynamic contact angle measurements, they presented surface structure models of their PEU samples under water as well as air; these are shown in Fig. 7.

It may be of interest to compare these results with those of Cooper, which were discussed in Section 2.2; PEO-PU and PTMO-PU correspond, respectively, to the two types of SPU samples in Fig. 7.

Surface Structure Model
Based on XPS and DCA

PEO based SPU

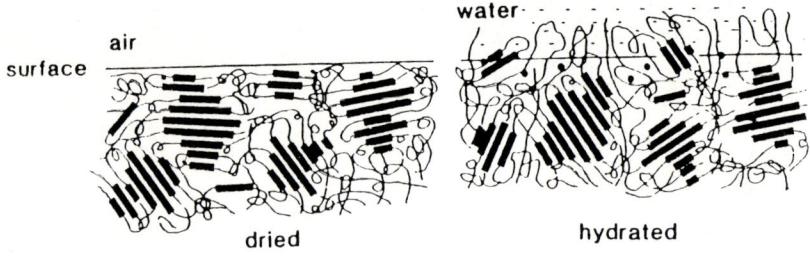

PTMO, PDMS based SPU

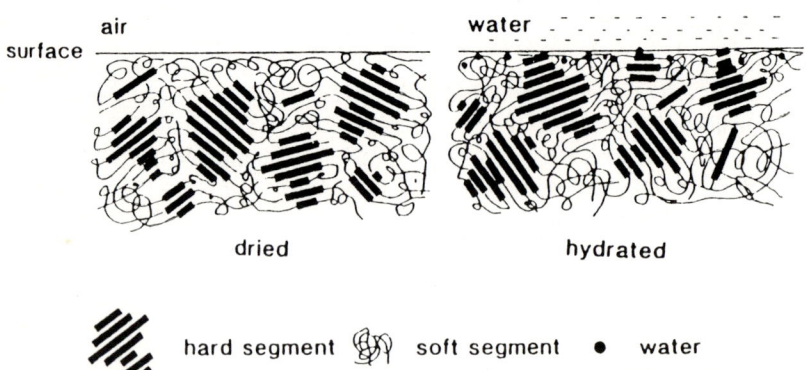

Fig. 7. Schematic representation of the surface reorganization process upon immersion of SPUs in water. (Reproduced from Kobunshi Ronbunshu [Ref. 57] through the courtesy of Soc. of Polymer Science, Japan)

As stated in Sect. 1, segmented PEU and PEUU are incorporated in various types of biomedical device. The most distinguished advantage of SPUs is their excellent mechanical strength as elastomers, but none of them posseses satisfactory in vivo durability. For instance, oxidative degradation took place on SPU in the biological environment, as confirmed by the analysis with explanted devices. Hydrogen peroxide generated from phagocyte cells activation in inflammatory sites is considered to be a major player in the oxidation process. Usually, Biomer and other SPUs have to have additive antioxydants, such as Santowhite powder (4,4'-butylidene-bis(6-tert-butyl-m-cresol)) [59].

Takahara et al. [60, 61] examined the effect of polyol chemistry on the in vitro biostability of SPU. Hydrolytic degradation studies [62] were carried out using a papain solution at 310 K for 14 days. Lipid sorption studies were carried out using a liposome solution (composed of phosphatidylcholine and cholesterol) at 310 K for 28 days. They found that the SPU with the hydrophilic PEO soft segment showed a large decrease in weight, which suggested its high susceptibility to hydrolytic decomposition via enzymatic reactions. On the other hand, PDMS-SPU showed the largest decrease in its durability against the lipid solution owing to its strong lipophilic property. Oxidation of PEU, as well as of PEUU, was confirmed to be initiated by a hydrogen abstraction from a polyether chain. Thus, hydrogenated polybutadiene (HPBD) (2100)SPU and "Biostable PUR" (a commercial product which contains no ether-linkage) were less susceptible to oxidation than Biomer or other SPUs having a polyether soft segment. Takahara reported that HPBD (2100)SPU and "Biostable PUR" also showed excellent properties against hydrolysis and lipid sorption.

Some of recent papers by Ratner et al. [63, 64] revealed that there are significant differences in the surface chemistry of Biomer lots. The surface of some lots was dominated by poly(diisopropylaminoethyl methacrylate) (DPAEMA or DIPAM), a high molecular weight UV stabilizer, which was absent from some older lots [65]. Ratner et al. carried out comparative studies on in vitro enzymatic and oxidative degradation of two lots of Biomer, BSU 001 and BSP 067. Lot BSU 001 contains both DPAEMA and an antioxidant, Santowhite powder, while BSP 067 contains only the antioxidant. It was found that DPAEMA retarded the enzymatic degradation process, but accelerated oxidative degradation.

From the above discussion, it is understood that available data in the literature on the blood-compatibility of SPUs are in many cases contradictory, because the extent of structural destruction of any particular SPU sample may vary, depending on the particular biological environment (contact time, temperature, pH, in vitro, ex vivo, in vivo etc.) in which the experiments were carried out.

Though we must be careful in discussing the structure-property relationship in terms of the blood-compatibility of SPUs, PEUs – such as PTMO–PU or PEO–PU – normally exhibited worse compatibility than PEUU (e.g. Biomer). For instance, Cooper et al. [21] reported a serious thrombogenicity of PTMO–PU as well as PEO–PU when these materials were brought into contact with ex vivo canine blood for 5 minutes. On the other hand, Takahara et al. [66] confirmed that his PEUU (PTMO–PU–ethylene diamine) exhibited a relatively non-thromobogenic response to ex vivo canine blood.

To improve the biomedical properties of SPUs, a number of attempts have so far been proposed. In particular, surface modifications by mobile, hydrophilic poly(ethylene glycol) (PEG) chains were extensively studied, and some of them proved to give fairly good results in terms of the antithrombogenicity. Nevertheless, as discussed in Sect. 3.1, the effect of PEG chains tethered onto the SPU surface should be carefully evaluated in detail with regard to clinical application as well as to fundamental considerations.

A number of researches have been carried out on surface modification by sulfonate groups in order to obtain antithrombogenic surfaces of PEU or PEUU. This strategy is based on the heparin-mimicking concept, but the methods by which sulfonate is introduced to the materials are different from one another. In some cases, 4,4'-diamino-2,2'-biphenyl disulfonic acid [67] and N,N'-bis(2-hydroxyethyl)-2-aminoethane sulfonic acid [68] were used as chain extenders for surface-modified PEUU. In other cases [69, 70], the sulfonic group was introduced by reacting γ-1,3-propane-sultone with urethane hydrogen after preliminarily treatment after the latter with sodium hydride. The propane-sultone [71, 72] was also used to prepare sulonated PEO, which was introduced covalently into PEU (Pellethane) to form a sulfonated PEO-grafted polyurethane surfaces. Imanishi et al. [73] carried out graft-polymerization of sodium vinylsulfonate onto a PTMO–PEUU sample.

Surface chemistry of these modified PEUs and PEUUs were extensively studied, and their enhanced hydrophilicity was confirmed. Biological examinations (in vitro and ex vivo) showed that most of these modified SPU surfaces possess more or less improved antithrombogenicity. Interestingly, plasma fibrinogen seems to exhibit a remarkable affinity with sulfonate groups, regardless of differences in the surrounding structure of the sulfonyl group. In most cases – with one exception [73] – fibrinogen deposition increases with increasing density of sulfonic groups on the polyurethane surface. We find it hard to understand how the blood compatibility of the fibrinogen-adsorbed surface can be improved. Y.H. Kim suggested [71] that the interaction of fibrinogen with the sulfonate group causes a conformational change of the protein such that the binding fibrinogen for the platelet receptor (i.e. GPIIb/IIIa, see Sect. 5.1) is buried. In consequence, platelets cannot recognize fibrinogen; in other words, the surface can maintain its blood compatibility.

Stanislawski, Jozefowicz et al. [74] reported another example which showed the role played by sulfonate groups in the conformational change of fibronectin adsorbed to polystyrene substituted with sulfonate group ($PSSO_3$). The surface of $PSSO_3$ coated with FN supported the growth of human umbilical vein endothelial cells. The surface of $PSSO_2$-Asp (substituted with aspartic acid sulfamide group) exhibited a similar affinity for FN, but the FN-precoated $PSSO_2$–Asp surface did not support endothelial cell growth.

Anderson et al. [59, 75, 76] have been pursuing their extensive researches on the biomedical behavior of PEUUs having various formulations modified with hydrophobic acrylate (or methacrylate) polymer or copolymer additives. The most distinguished additive was Methacrol 2138F, which is a copolymer between diisopropylaminoethyl methacrylate and decyl methacrylate [co(DIPAM/DM)] (in a 3-to-1 ratio). The protein adsorption assay showed that PEUU (Biomer-type) films loaded or coated with Methacrol or poly(DIPAM) adsorbed significantly lower amounts of human blood proteins (Fb, IgG, factor VIII, Hageman factor and Alb) than the base PEUU or PEUUs modified by other additives. It was revealed from their experiments that poly(DIPAM) as well as Methacrol exhibited a prominent suppressing effect on the protein adsorption process.

In this connection, it is to be noted that polyamine-modified poly(HEMA) surfaces exhibit surprisingly reduced interaction with blood proteins and cells (e.g. erythrocyte, platelet, lymphocyte etc.), as will be discussed in Sects. 4.3 and 4.4. The present author considers that there are probably closely related mechanisms between the suppressing effect of the poly(DIPAM) or (Methacrol)-modified SPU and that of the polyamine-modified poly(HEMA) surfaces, with regard to their mode of interaction with the biological elements.

4.2 HEMA-Styrene Triblock Copolymers and Polyether-Segmented Polyamide

In 1976, Okano, Shinohara et al. [77] synthesized, by radical polymerization processes, a triblock (A-B-A) copolymer, consisting of HEMA (A) and styrene (B):

$$H + \left[\begin{array}{c} CH_3 \\ | \\ C-CH_2 \\ | \\ C=O \\ | \\ HOCH_2CH_2O \end{array} \right]_m \boxed{J} \left[CH_2-CH \atop \phi \right]_n \boxed{J'} \left[\begin{array}{c} CH_3 \\ | \\ CH_2-C \\ | \\ C=O \\ | \\ OCH_2CH_2OH \end{array} \right]_m H$$

where J is $-SCH_2CH_2NHCONH-\phi-S-$

and J' is $-S-\phi-NHCONHCH_2CHS-$

Okano et al. [78–80] found that HEMA–STY triblock copolymer formed a typical domain structure when it was cast into a film. The domain structure changed with the HEMA mole fraction, and the lamella structured surface (mole fraction of HEMA 0.61; lamella width 30–50 nm) exhibited the best blood-compatibility.

To control the sequence length of HEMA–STY triblock copolymer, Okano et al. [81] later adopted the anionic living-polymerization method. Junction groups J and J' can be eliminated when HEMA–STY block copolymer is prepared anionically. A transmission electron micrograph of HEMA–STY block copolymer (HEMA molar fraction: 0.5), stained with osmium tetroxide, showed alternating lamellar microstructures with a repeating width of 14 nm.

Nojiri, Okano, S.W. Kim et al. [82, 83] carried out long-term in vivo examination of the surfaces of HEMA–STY block copolymer (anionically prepared; HEMA mole fraction 0.5) and poly(ethylene oxide) of M.W. 4000 (PEO 4 K) grafted Biomer (B-PEO 4 K) in comparison with that of Biomer. The luminal surfaces of Biomer vascular grafts (6 mm I.D., 7 cm in length) were modified by PEO-grafting or coated with HEMA-STY copolymer – or Biomer itself – as control. These vascular grafts were implanted in the abdominal aortas of dogs for evaluation of graft patency and protein adsorption.

The luminal surface of the HEMA–STY graft was bare, without detectable thrombi, after 3 months' implantation. In contrast, the entire surface of the B-PEO 4 K graft was covered with red thrombi, loosely attached to the surface, PEO chains exerting a rather adverse effect on the blood compatibility (see Sect. 3.1). Surface protein-layer thickness was measured by cross-sectional transmission electron micrography of the modified Biomer graft lumens. As can be seen in Table 8, B-PEO 4 K and Biomer showed thick multilayers of adsorbed proteins (with high concentrations of fibrinogen and IgG), whereas HEMA–STY showed a monolayer protein thickness (with high concentrations of albumin and IgG) even after 3 months. Nojiri et al. suggested that the microdomain structures of the surface enabled the formation of a stable, thin adsorbed protein layer on HEMA–STY surfaces. The thin, stable adsorbed proteins are considered to keep their native molecular conformation and to play an important role in long-term in vivo blood compatibility, as will be discussed in Sect. 4.4.

Okano et al. [84] measured changes in cytoplasmic Ca^{2+} concentrations in platelets adhering to HEMA–STY block copolymer (HSB) surfaces by means of fluorescence microscopy combined with a high performance image processor. Comparative studies were also carried out with the HEMA–STY random (HSR) copolymer of poly HEMA and polystyrene. Their results showed that cytoplasmic free calcium levels in platelets that were in contact with the HEMA–STY

Table 8. Surface protein thickness

Polymer	Implantation time	Protein layer thickness (Å)*
Biomer	21 days	980 ± 56
B-PEO 4K	1 month	1860 ± 203
HEMA-STY	3 months	190 ± 54

*mean ± S.D.
n = 2 samples × 10 views.
(Reprined from J. Biomed. Mater. Research [Ref. 82: Blood compatibility of PEO grafted polyurethane and HEMA/styrene block copolymer surfaces] through the courtesy of John Wiley & Sons. Inc.)

block compolymer remained relatively constant, in contrast to the significant increase observed for other four polymer surfaces, as shown in Fig. 8.

Okano et al. observed from fluorescence images that the elevation of Ca^{2+} levels was followed by the spreading of platelets. They regarded the increases in cytoplasmic Ca^{2+} levels to be an initial signal of platelet activation, which promotes morphological changes to enhance platelet adhesivity to the polymer surface. Adhering platelets on HSB did not change their morphology.

Recent results [85] from carotid replacement by HEMA–STY-coated dacron graft (3 mm ID) were noteworthy. Luminal surface was very clean after 372-day implantation.

Yui et al. [86–89] have previously reported another type of microdomain-structured polymer, poly(propylene oxide) (PPO)-segmented nylon 610, which has a crystalline-amorphous microdomain structure:

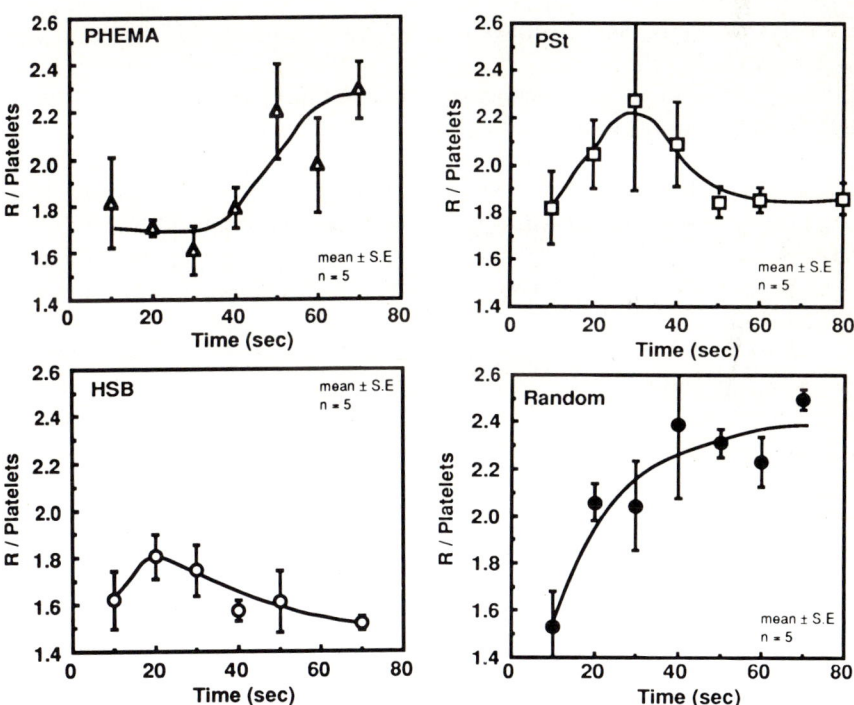

Fig. 8. R/Platelet* in individual platelets adhering to polymer surfaces. HSB data were statistically confirmed to be different from PSt ($P < 0.5$), HSR ($P < 0.5$) and PHEMA ($P < 0.5$) after 40 s
*R/Platelet (an index of cytoplasmic free calcium concentration) is the ratio of fluorescence emission intensitie of a Ca^{2+} indicator dye (Fura 2) loaded in platelets when they are excited at 340 nm and 380 nm. (Reproduced from J Biomed Mater Res [Ref 84: Prevention of changes in platelet cytoplasmic free calcium levels by interaction with 2-hydroxyethyl methacrylate/styrene block copolymer surfaces] through the courtesy of John Wiley & Sons, Inc.)

The structural parameters of PPO-segmented nylon were determined by small-angle X-ray scattering, wide-angle X-ray diffraction and density measurement. Yui et al. found, in a series of PPO-segmented nylon 610, that the copolymer which minimized platelet adhesion in vitro and exhibited nonthrombogenicity in vivo was 61P3-25 (nylon 610 copolymer containing 25 wt% of PPO of mol. wt. 3000).

To examine the cell-material interaction, Yui et al. [90, 91] used a column method, in which the Fura 2-loaded platelet suspension was passed through a column packed with glass beads precoated with PPO-segmented nylon 610, and the column effluent was subjected to fluorescence measurement. It was noticeable in their results that there was the least increase in cytoplasmic free calcium on the 61P3-25 surface, even when it had been pretreated with rabbit plasma fibrinogen, though fibrinogen molecules adsorbed on polymer surfaces normally undergo conformational changes that are responsible for the subsequent processes of cell adhesion (see Sect. 4.1).

4.3 Polyamine-graft copolymers

A series of polymine-graft copolymers of styrene [92–95] and hydroxyethyl methacrylate [96–98] were found to form a microdomain structure and exhibit unique biomedical behavior at the interface with living cells, such as blood platelets and lymphocytes. The most intensive studies were made with poly(hydroxyethyl methacrylate)-*graft*-polyamine copolymers (HA):

$$-(CH_2-CH)_m------(CH_2-\underset{\underset{O=COCH_2CH_2OH}{|}}{\overset{\overset{CH_3}{|}}{C}})_n-$$

$$\underset{C_2H_5}{\overset{|}{CH_2}}\ \underset{C_2H_5}{\overset{|}{CH_2\ (NCH_2CH_2NCH_2CH_2}}-\!\!\!\!\bigcirc\!\!\!\!-CH_2CH_2)_p\ \underset{C_2H_5}{\overset{|}{NCH_2CH_2NH}}\ \underset{C_2H_5}{\overset{|}{}}$$

HA copolymers were prepared by radical copolymerization of HEMA with polyamine macromonomer which had been synthesized by a self-polyaddition reaction of N,N'-diethyl-N-(vinylphenethyl)ethylenediamine (EDAS) in the presence of lithium diisopropylamide [98–100]

$$m\ CH_2=CH-\!\!\!\!\bigcirc\!\!\!\!-CH_2CH_2\overset{\overset{C_2H_5}{|}}{N}CH_2CH_2\overset{\overset{C_2H_5}{|}}{N}H \quad \overset{>NLi}{\Longrightarrow}$$

EDAS

$$CH_2{=}CH{-}\langle\bigcirc\rangle{-}CH_2CH_2{-}\left[\underset{}{N(C_2H_5)CH_2CH_2N(C_2H_5)CH_2CH_2}{-}\langle\bigcirc\rangle{-}CH_2CH_2\right]_{m-1}{-}N(C_2H_5){-}CH_2CH_2N(C_2H_5){-}H$$

Polyamine Macromonomer

Beautiful transparent films were easily prepared from HA copolymers by solution-casting technique onto a carbon-coated collodion membrane on a copper grid. TEM photographs of these films, stained with OsO_4 vapor, clearly showed the formation of island-shaped microdomains of polyamine dispersed in the continuous phase of poly(HEMA) [97]. In the course of our biomedical studies on the interaction between cells (blood platelets or lymphocytes) and synthetic polymers, we became interested in difference in the adhesion properties of cells on polymer surfaces having different structures. Thus, our research was carried out with emphasis on cell separation, paying special attention to the different adhesion behavior of the cells on HA copolymers [101, 102]. Biomedical examinations were carried out by using a column method [103–105] and its modification [93–95, 97, 98]: a sample of the cell suspension (concentration: 1×10^7, in some cases 2×10^7 cells/ml) was passed through a column in which polymer-coated glass beads (average diameter: 275 μm) were densely packed. A lymphocyte suspension is loaded into the column with a syringe pump at a constant flow rate. The lymphocytes were obtained from the mesenteric lymph nodes of Wister male rats aged 5 weeks, and suspended in Ca^{2+}- and Mg^{2+}-free Hanks' balanced salt solution (HBSS). A spectrophotometric detector was installed at the column outflow to monitor the concentration of lymphocytes effusion.

It is known that lympocytes are composed of two major subpopulations: T-cell and B-cell, each of which plays an essential role in all of the immunological reactions in living systems. There are strong medical demands for the separation of T- and B-cells, which is essential for the therapy and diagnosis of autoimmune diseases and cancer, as well as for HLA typing in transplantation.

It had been found previously [106] that cell adhesion on the surface of HA copolymers decreased with increasing ambient pH. With polyHEMA, on the other hand, no pH-dependency was observed. We considered the adhesion to be caused primarily by ionic interaction between the lymphocytes and the HA surfaces. The degree of protonation (α) of amino groups in HA copolymer was estimated by acid-base titration of an HCl solution of polyamine macromonomer with NaOH solution. In physiological conditions (pH 7.2–7.4) about 50% of the amino groups of the macromonomer are protonated. In this pH range, the polyamine macromonomer was found to be insoluble.

The length of the polyamine grafts is an influential factor in determining the mode of pH dependence of the cell on the surface of HA copolymers [107, 108]. The extent of retention of B and T cells on HA13 of varying polyamine graft chain length (11000, 6600, 3750 and 875 in M_n, respectively) was measured at different values of the surrounding pH, adjusting the cell concentration to

2×10^7 cells/ml. The pH dependence of B- and T-cell retention on HA13 copolymers is very variable, depending on the length of the polyamine chain. At physiological condition, about 50% of the amino groups are deprotonated and some of the polyamine chains of HA are forced to be deposited on the copolymer surface. In other words, the polyamine chain changes its conformation in this pH range from a protonated extended state to a random coil, forming a compact aggregate. This morphological change is assumed to vary in response to the length of polyamine chain of HA copolymer.

The isoelectric points of B cells and T cells are reported to be 3.8 and 4.6, respectively. This may explain the higher adhesivity of the former to the HA surface. For this reason, adhesivity of T cells are assumed to be more susceptible than B cells to the decrease in ionic character of the HA surface.

Our target was to design a copolymer surface which would be suitable for B cells to exhibit their maximal adhesivity but would be minimally adhesive for T cells. After further detailed studies, the most effective method for B/T separation was established using a HA13 (3750) column, where the recovery of T cell at pH 7.3 was almost qunatitative. The viability of the cells in effluent was larger than 95%. It was also possible to obtain an enriched B cell population with low contamination by T cells. To facilitate the detachment of the adhered cells from the beads, the latter were taken out of the column and put into HBSS containing 1% bovine serum albumin (BSA) at an adjusted pH. By gentle pipetting applied to the HBSS system, we obtained B cells in 92.4% recovery and 97.3% purity.

Scanning electron micrographs of adhering lymphocytes on the polymer surfaces showed that most of the lymphocytes adhering to the surface of the HA copolymers of varying polyamine content and chain length of the polyamine grafts keep their native shape, in contrast to the extensive shape change observed for lymphocytes bound to the surface of poly(HEMA). This is the most distinct biomedical behavior of microdomain surfaces, and a prerequisite for the materials to be applied to cell separation, because selective separation is impossible on surfaces which induce an irreversible shape change to the adhering cells. A possible rationale for the unique biomedical behavior of microdomain surfaces will be discussed in Sect. 4.4.

The efficacy of B/T separation was very low on HA13 (11000) because of the extremely high retention of T cells on this surface, even at $\alpha < 0.5$. To obtain more information about the HA surfaces having a varying chain length of polyamine graft, surface analyses were carried out by ESCA and streaming potential measurements. Results showed that – with the exception of HA13 (246) (a copolymer of EDAS with HEMA) – the polyamine grafts in HA copolymers were concentrated in the surface layer (air side) of the HA films. It was noticeable too that the angular dependence of N/O ratio becomes significant in the order: HA13 (890) < HA13 (3750) < HA13 (6600) < HA13(11000), which means that the increase in the chain length of the polyamine grafts resulted in an enhanced concentration of amino groups in the surface layer. The surface concentrations of amino groups were estimated in terms of the weight percent of polyamine portions of HA copolymers. Results are shown in Table 9.

Table 9. The surface concentration of amino groups of HA copolymers

HA copolymer	Weight% of polyamine on the surface of HA
HA13(11000)	36.2
HA13(6600)	31.1
HA13(3750)	26.6
HA13(890)	14.3
HA13, bulk	13.0

(Reprinted from Makromol Chem Macromol Symp [102] through the courtesy of Hüthig & Wepf Verlag)

Table 10. ζ-Potential for HA13 copolymer coated glass beads

Sample	Potential (mV) pH 6.0	pH 7.4
HA13 (11000)	$+ 3.50 \pm 0.41$	$- 9.83 \pm 0.68$
HA13 (6600)	$- 10.10 \pm 0.27$	$- 20.34 \pm 0.5$
HA13 (3750)	$- 10.21 \pm 0.15$	$- 20.96 \pm 0.74$
HA13 (890)	$- 11.47 \pm 0.63$	$- 21.35 \pm 0.69$
HA13 (246)	$- 15.32 \pm 0.33$	$- 23.98 \pm 0.90$
poly-HEMA	$- 29.63 \pm 1.36$	$- 31.93 \pm 1.82$

(Reprinted from Makromol Chem Macromol Symp [102] through the courtesy of Hüthig & Wep Verlag)

Although HA13 (11000) showed the highest weight % of polyamine portion on the surface (36.2), this value may not be enough to account for the anomalous behavior of HA13 (11000), which exhibited much higher adhesivity to T lymphocyte, in comparison with the other HA13 copolymers. It is seen from Table 9 that there is no abrupt increase in the weight % of the polyamine portions on the copolymer surface.

To be certain that ionic character of the HA surface is operative in its contact with the cell in aqueous media, streaming potential measurements were carried out at pH 6.0 and 7.4. The ζ potential for the polymer-coated glass beads (48–60 mesh) was calculated from the measured values of the streaming potential. The results obtained are shown in Table 10.

As seen in Table 10, the ζ potential at pH 7.4 takes on more negative values than those at pH 6.0, which can probably be explained by the change in the degree of protonation of the polyamine grafts in HA copolymers. Changes in the conformation of polyamine chains from the extended state to the aggregated one may also be responsible for the different values of ζ potential. The ζ potential of polyHEMA shows only a slight negative shift in response to the pH increase. It is to be noted that the potential of HA copolymers is extremely sensitive to the change in chain length of polyamine grafts. Especially, the ζ potential of HA13 (11000) is significantly more positive than that of the other HA13 copolymers.

This means that the electric charge distribution in the slipping plane on the HA13(11000) surface was shifted to the more positive side, owing to the long polyamine graft chains. It is probably reasonable to assume that the slipping plane has enough positive charge to exhibit adhesivity both to B cells and T cells. Actually, a less densely grafted copolymer HA6 (11000) proved to show fairly good selective adhesivity to B cells.

The previous discussion was based on the results for lymphocytes obtained from rat mesenteric lymph node. Resolution of lymphocyte subpopulations derived from rat spleen were also possible by using HA copolymer column [109].

In the course of our study with HA copolymers, we frequently observed that some of HA copolymer surfaces which have a small number of polyamine graft chains became sufficiently inert against blood platelets – as well as against lymphocytes – to reject their attachment [97, 108–113].

Another study [114] on the role of protonated amine of HA copolymers in platelet retention showed that minimum retention was again observed on the surface of the copolymer containing N^+ in 0.1–0.3 wt%. In other words, cell attachment scarcely takes place on the surface of HA2, in contrast to polyHEMA, on the surface of which more than 90 percent cell attachment was observed.

A similar cell-retention profile was also observed on the surface of poly[N-methyl-N-(4-vinylphenethyl)ethylenediamine(AVEMA)-co-HEMA] (HAV) [115].

$$—(CH_2—CH)_m ————(CH_2—\underset{\underset{\underset{O-CH_2CH_2OH}{|}}{O=C}}{\overset{\overset{CH_3}{|}}{C}})_n—$$

with phenyl ring bearing $CH_2CH_2NCH_2CH_2NH_2$ substituted with CH_3 on the nitrogen.

In Fig. 9, the percent retention of lymphocytes, platelets and erythrocytes are plotted againsts AVEMA content in the HAV copolymers.
A range of minimal retention is observed with every cell species.

Interesting swelling behavior was also found with HAV copolymers containing one and two mol% of AVEMA. The total water content of the swollen copolymers HAV1 and HAV2, were increased by 150% and 175% respectively, compared with that of polyHEMA itself. The free water content of the HAV copolymers was found to be 4 to 5 times larger than that of polyHEMA. This result can probably be explained by the destabilization of the random-network structure of water molecules (4.4) as a result of introducing a small number of amino groups into the polyHEMA matrix.

The inertness of polyHEMA surfaces, which contain a small number of amino groups, was also confirmed by analysing frontal chromatograms of

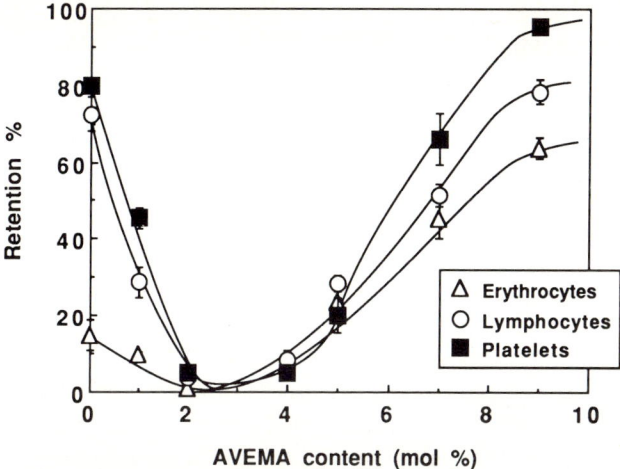

Fig. 9. Retention behavior of blood cells* on poly(HEMA–co[N-methyl-N-(4-vinylphenethyl) ethylenediamine] (HAV) (pH 7.2, 23 °C; flow rate, 0.4 ml/min; flow time, 3,5 min). *Open circles*, lymphocytes; *closed squares*, platelets; *open triangles*, erythrocytes; AVEMA, N-methyl-N-(4-vinylphenethyl)ethylenediamine [Ref. 115]
*Cell suspensions in HBSS in the absence of plasma proteins

lymphocytes [97]. Another quantitative evaluation of the affinity of lymphocyte subpopulations toward the HAV copolymer surfaces was carried out by using a novel technique: hybrid field-flow fractionation/adhesion chromatography (FFF/AC) [116–18].

4.4 The Random Network Concept of Water Molecules on Materials Surface

Beside the above-mentioned examples, there are available several others that show the unique biomedical effect of a small number of amino groups as surface modifier.

Asahi Medical Co. (Japan) developed Sepacell-PL filter as a device for removing white blood cells from whole blood [119, 120]. Sepacell-PL is a nonwoven fabric (PE or PET), coated with a copolymer of HEMA with diethylaminoethyl methacrylate (e.g. in 5 mole%). It is noticeable that platelets in the blood can pass through the filter without suffering any damage during the filtration processes, owing to the presence of the amino group.

Another commercial product containing a small number of amino groups is Hemophan (ENKA, Germany) which was developed to ameliorate disadvantageous properties (e.g. causing leucopenia and activation of complements) of the cuprarayon hemodialyzers. It is known that Hemophan contains one diethylamino group for every five glucose units.

The third example to be mentioned is the protein-adsorption behavior of PEUU samples which are coated or loaded with DIPAM or Methacrol, as discussed in Section 4.1.

The unique behavior of amino groups introduced in a small proportion may be evident from the examples discussed above, but the role of amino groups at the biointerfaces has not been thoroughly elucidated. Though much must be done before we can reach a final conclusion, the author believes that it is very reasonable to assign an important role to the random network structures of water molecules which are constructed through hydrogen bonding to the surface of the polymeric materials [121–126]. Protein adsorption processes are considered to start with protein-trapping by the random networks of water molecules on the material's surface. The longer the residence time of a protein molecule on the network, the greater the chance for the protein molecule to interact with the material's surface and thus undergo a conformational change, and – as the consequence – to be irreversibly adsorbed on the material's surface. On the surface of polyHEMA, as well as that of cellulose, OH-groups of the material are incorporated in the network structure of water molecules, spreading over the interfaces with the biological liquid.

In contrast, on the surface of the amino-containing polymeric materials, protonated amino groups introduced in a small proportion under physiological conditions, destroy their surrounding hydrogen bonds to produce, here and there, gaps in the network [127, 128]. Thus, the network structures are considered to become more or less unstable. As a consequence, the residence time of protein molecules trapped by these defective networks will be shorter than in the case of polyHEMA or cellulose. On the surface of these amino-containing materials, reversible protein adsorption and desorption, and also replacement (Vroman effect) – or even protein rejection – will become possible.

If we consider that cell adhesion under biological circumstances is mainly brought about with the aid of preadsorbed protein on the material's surface, we may explain the unique behavior of amino-containing materials against the cell-adhesion process in terms of the reduced residence-time of protein molecules at the interface. Actually, a recent study [129] revealed that the surface of polyamine-*graft*-polystyrene copolymer (SA) containing 6 wt.% polyamine portion exhibited a minimal adsorptive property against bovine plasma fibronectin (FN) and vitronectin (VN), both of which are known to mediate cell-adhesion processes.

It must be mentioned here that some of the results discussed in Sect. 4.3 were obtained with cell suspensions in Hanks'-balanced salt solution (HBSS) in the absence of plasma proteins. The present author believes that the random-network concept can be applied to the events which happen to lymphocytes, platelets or erythrocytes when they come into direct contact with hydrated material surfaces in the absence of interverning protein.

The random-network concept discussed above may be applicable, in principle, to the surfaces of some microdomain-structured materials. As stated in Sect. 4.2, the HEMA-STY copolymer showed the best blood compatibility after

Table 11. Results of adsorption and desorption behavior of IgG on PHEMA and HA copolymers [130]

Run	Polymer	Adsorbed protein[1] µg/cm^2	Recovery[2] %
1	PHEMA	0.55	70
2	HA2	0.22	~100
3	HA7	0.18	~100
4	HA13	0.14	~100

[1] Protein solution was prepared by 0.02M PBS at pH 6.4
[2] Absorbed protein was eluated by 0.1M PBS at pH 6.4.

a long-term in vivo examination. The present author draws attention to the mode of protein adsorption of HEMA-STY surface, as shown in Table 8. The random-network structure of water molecules on the HEMA-STY surface must be very labile under the influence of its hydrophilic-hydrophobic lamellar structure, where the protein residence time will be too short to cause the trapped protein molecules to undergo conformational change. The Vroman effect will operate easily on the surface. This may be the reason why albumin, the most abundant blood protein, was found in the form of monomolecular layer, thus keeping its native conformation. Since albumin is not a cell-adhensive protein, there was presumably no chance for blood cells to adhere on the HEMA-STY surface.

Our previous study [130] on the adsorption and desorption behavior of bovine IgG has shown that the protein adsorbed to HA surface would be eluted quantitatively by 0.1 M PBS (phosphoric buffer solution) as shown in Table 11. Presumably, IgG molecules had been trapped in the destabilized network of water molecules on the surface of the HA copolymer.

5 Biohybridized and Biomimicking Materials

As stated in Sect. 1, a variety of studies on biohydridized and biomimicking materials have been extensively reported, though the approach of the various investigators are quite diverse, depending on their objectives (see Table 1).

5.1 Cell-adhesive Peptide Conjugate Materials

Pierschbacher and Ruoslahti [130] in 1984 reported that the ability of fibronectin to bind cells can be accounted for by the tetrapeptide L-arginyl–glycyl–L-aspartyl–L-serine (RGDS), a sequence which is part of the cell attachment domain of fibronectin. They examined the cell-attachment activity of several synthetic peptides consisting of 5–, 6–, 9–, 14– and 30–amino acid residues, and found all the peptides that promoted the attachment of rat kidney fibroblast to contain the sequence RGDS (or RGDC). Further study revealed that the RGD tripeptide sequence was essential to preserve activity, while conservative substitution of serine residues (e.g. serine → threonine) did not abrogate the cell-attachment-promoting activity [130, 131].

The presence of the tripeptide RGD sequence, and its crucial role in the cell-attachment process, were found subsequently in vitronectin [32, 133], fibrinogen [134], von Willebrand factor [134], type I collagen [135], and many other proteins present in body fluids and extracellular matrices [136].

Beside the RGD tripeptide sequence, a number of peptide sequences have been found to promote cell-adhesive activity, as shown in Table 12 [137].

Imanishi, Ito et al. [138] immobilized GRDS tetrapeptide on a modified silicone rubber film and observed enhanced cell attachment to this film when mouse fibroblast cells STO were incubated in a modified Eagle's medium in contact with the silicone film. They [139] also immobilized RGDS to poly(acrylic acid) which had been grafted to a polystyrene film with the aid of a glow discharge technique. Comparative studies of the cell-adhesive activity (toward mouse-fibroblast-cells STO) of the RGDS-immobilized film as against

Table 12. Oligopeptide cell-adhesive signals

Signals	Origin	Cell types
a. RGD-related sequences		
RGDS	FN, vWF, OP, FG	FB, many cells
RGDV	VN	FB, END
RGDF	FG	PL
RGDA	COL, TSP?	FS, PL
RGDN	LN	END, NB
b. Heparin-binding sequences		
YEKPGSPPREVVPRPRPGV	FN	MEL
RYVVLPRPVCFEKGMNYTVR	LN	MEL, FS, PC
c. Others		
EILDVPST	FN	MEL, LY
REDV	FN	MEL
KQAGDV	FG	PL
LGTIPG	LN	FB
IKVAV	LN	PC, MEL, FS
YIGSR	LN	MEL, PC, FS
PDSGR	LN	MEL, FS
HAV	Cadherin	EPI
YKLNVNDS	gp80	Discoidium

FN, fibronectin; vWF, von Willebrand Factor, OP, osteopontin;
FG, fibrinogen; VN, vitronectin; COL, collagen;
TSP, thrombospondin; LN, laminin; FB, fibroblast;
END, endothelial cell; PL, platelet; FS, fibrosarcoma;
NB, neuroblastoma; MEL, melanoma cell; PC, phenochromocytoma;
LY, lymphocyte; EPI, epithelial cell.
(Reprinted from [137] through the courtesy of Kagaku-Dojin, Kyoto)

fibronectin (FN)-immobilized film showed that the former was more stable against heat treatment and pH variation than the latter. It was also reported that cell growth was enhanced on the RGDS-immobilized film in comparison with the FN-immobilized film.

Brandley and Schnaar [140] immobilized a synthetic nonapeptide, Tyr–Ala–Val–Thr–Gly–Arg–Gly–Asp–Ser, on a polyacrylamide gel, which had been prepared by a ternary copolymerization of acrylamide. bisacrylamide and the acrylic ester of N-hydroxysuccinimide. They reported that Balb/c 3T3 mouse fibroblast cells (in Hepes-buffered Dulbecco's modified Eagle medium) adhered readily to the peptide-derivatized surfaces, even in the absence of serum, although long-term cell growth required the presence of serum. It was noticed that reference nonapeptide. Tyr–Arg–Leu–Glu–Asp–Pro–Ala–Met–Trp, which has no RGD sequence, failed to promote cell-attachment.

Matsuda, Akutsu, et al. [141] carried out chemical fixation of GRGDSP onto a poly(vinyl alcohol) film which had been activated by carbonyl diimidazol in dimethylformamide. For comparison, fibronectin was also immobilized on a poly(vinyl alcohol) film by the same technique. Matsuda et al. found that bovine endothelial cells (ECs) adhered well to the peptide- and fibronectin-immobilized film; furthermore, after a three-day incubation, aggregated ECs floated on nontreated PVA film, while the peptide-immobilized surface greatly enhanced cell spreading and growth, which surpassed that on fibronectin-bound PVA.

Massia and Hubbell [142] coupled two synthetic peptides, Gly–Arg–Gly–Asp–Tyr and Gly–Tyr–Ile–Gly–Ser–Arg–Tyr, covalently onto glycophase glass [143], which has anchoring groups, $-(CH_2)_3-O-CH_2-CH(OH)CH_2OH$, on the surface. Their peptides contain RGD and YIGSR sequences, respectively. Hubbell et al. examined the adhesive and spreading behavior of cultured human foreskin fibroblast (HFFs) on the peptide immobilized glass. They found that cells formed focal contacts, in the presence or absence of serum, on the glycophase glass which had been derivatized with the peptide containing an RGD sequence. On the other hand, the surface coupled with the peptide containing YIGSR was found to require the presence of serum for the formation of focal contacts region.

Massia and Hubbell [144] also examined the cell-adhesive behavior of GRGDY- and GYIGSRY-immobilized glycophase glass, poly(ethylene terephthalate) (PET), and polytetrafluoroethylene (PTFE). They observed the attachment, spreading, spreading rate, focal contact formation, and cytoskeletal organization of human umbilical-vein endothelial cells (HUVECs) on peptide-immobilized surfaces. Before the peptide-immobilization, PET- and PTFE films had been hydroxylated according to known procedures to yield PET-OH [145] and PTFE-OH [146], respectively. After converting these hydroxyl groups to sulfonyl ester with 2,2,2-trifluoroethanesulfonyl chloride (tresyl chloride) [147], the N-terminal primary amine of the glycine residue of GRGDY or GYIGSRY was allowed to displace the sulfonyl ester and form linkages with the polymeric substrates. Massia and Hubbell observed that the peptide-immobilized surface supported cell-adhesion and spreading even when only albumin was present in the medium.

More recently, Hubbell et al. [148] found the highly specific adhesivity of the REDV sequence toward the endothelial cells, HUVEC. The cell-spreading assays were carried out in a serum-containing medium. The results obtained are cited in Table 13.

It is seen from Table 13 that the GREDVY-immobilized surfaces exhibited highly specific cell-spreading activity toward the endothelial cell, HUVEC, in contrast with rather non-specific activity of the GRGDY immobilized surfaces, on which not only HUVEC, but also HFF and HVSMC, showed high percentage of spreading. Another study by Hubbell et al. [149] revealed that REDV-, RGD- and YIGSR-supported endothelial monolayers all appeared to maintain their non-thrombogenic properties, while PDSGR-supported monolayers were very thrombogenic.

Hirano et al. [150, 151] immobilized several peptides, RGDS, on ethylene-acrylic acid copolymer (EAA, acrylic acid content: 20 wt%) film by reacting the amino-terminal of the peptide with the carboxylic acid of the copolymer with the aid of a water-soluble carbodiimide, to form EAA-co–NH–RGDX. Their objective was to examine effect of the fourth residue, X, on the cell-attachment activity of the tetrapeptide, RGDX, where X is S, V and T. They also examined the activity of RGD, YIGSR and YIGSR–NH$_2$ for comparison. The cell lines used were ovary CHO–K1 cell (chinese hamster), kidney NRK cell

Table 13.

A) Base material–Glycophase glass (% spread cells)

Grafted peptide	HFF	HVSMC	PRP untreated	PRP treated with 5 µM ADP	HUVEC
GRGDY	85 ± 15	93 ± 3	0	0	90 ± 9
GREDVY	9 ± 4	7 ± 2	0	0	89 ± 6
None	9 ± 2	10 ± 4	0	0	8 ± 1

B) Base material–PET/18.5 kD PEO SPIN (% spread cells)

Grafted peptide	HFF	HVSMC	PRP untreated	PRP treated with 5 µM ADP	HUVEC
GRGDY	93 ± 5	0	0	0	91 ± 2
GREDVY	0	0	0	0	88 ± 4
None	0	0	0	0	0

HFF: human foreskin fibroblast
HVSMC: human vein smooth muscle cells
PRP: platelet in platelet-rich plasma
SPIN: surface physical interpenetrating network.
(Reprinted from 17th Annual Meeting Transactions [148] through the courtesy of Society of Biomaterials U.S.A)

(rat), epidermoid carcinoma A431 cell (human), cervix HeLa S3 cell (human), liver RLC-16 cell (rat), and fibroblast L-929 cell (rat). The cell-attachment examinations were carried out in the absence of serum. The largest cell-attachmet activities were always found in the RGDS-, RGDV- and RGDT-immobilized surfaces, regardless of the nature of the cell-line. In contrast, the promoting effect of the RGD-immobilized surface was very slight. Hirano et al. considered that the fourth residue X may contribute to holding the RGDX conformation so as to fit the cell attachment receptor. The YIGSR-immobilized surface exhibited less attachment activity than the RGDX surfaces against every cell-line, though a slightly enhanced activity was observed on the surface of YIGSR–NH_2-immobilized copolymer.

Kugo et al. [152] studied the behavior of human fibroblast attachment to RGDS- and GRGDS-peptide-immobilized poly(γ-methyl L-glutamate) (PMLG) film via three types of spacer group.

Cooper et al. [153] synthesized a series of PTMO-based polyetherurethane (PEU) containing covalently grafted GRGDSY-, GRGDVY- and GRGESY-peptides. Their synthetic route is as follows:

$$\text{PEU} \xrightarrow[\beta\text{-propiolactone}]{NaH} \text{PEU–COOH} \xrightarrow[\text{carbodiimide}]{H_2N\text{-GRGDY*}} \text{protected peptide grafted PEU} \xrightarrow{\text{deprotection}} \text{peptide grafted PEU}$$

Here GRDSY* means GRDSY-peptide in which the functional side groups of R, D, S, and Y, respectively, are chemically protected.

Cell-attachment and spreading on polymer-coated materials were measured in vitro, using HUVECs, because Cooper et al. were interested in the development of novel biomaterials that may improve long-term endothelial cell attachment and growth. They found the GRGDSY- and GRGDVY-grated PEU greatly enhanced cell adhesion and spreading in the absence of adhesive plasma proteins in the culture medium. In contrast, the GRGESY-grafted PEU did not support endothelial cell adhesion and spreading. It is to be noted that these peptide-grafted polyetherurethanes can be applied to solution casting techniques.

In another paper [154] dealing with platelet adhesion to synthetic surfaces (PEUU-block copolymers), Cooper et al. revealed that pretreatment of platelets with RGD-containing peptides did not affect the number of adherent platelets, but produced a substantial decrease in the extent of spreading, in terms of the spread areas on the polymer surface as well as the formation of organized microfilament structures in the adhering platelets. Cooper et al. suggested that platelet integrins, possibly GPIIb–IIIa, are involved in spreading on the material surface but not in the initial adhesion.

Beer et al. [155] immobilized a series of RGD peptides containing variable numbers of glycine residues, $(G)_n$-RGDF, on polyacrylonitrile beads (1–3 μm diameter) with N-hydroxysuccinimide residues, and evaluated the ability of the beads to interact with platelets. The degree of interaction was estimated in terms of the extent of agglutination of the $(G)_n$-RGDF bead-suspension by platelet-rich plasma (PRP). Beer et al. also reported the effect of adenosine diphosphate (ADP), which is known to stimulate blood platelets. Remarkably, G_3-RGDF beads showed the greatest increase in the extent of agglutination with PRP that had been preactivated by ADP stimulation. On the basis of their studies, including deactivation of PRP with several monoclonal antibodies, Beer et al. concluded that GPIIb–IIIa is the dominant RGD-binding receptor on the platelet surface, and that most of the RGD binding sites can be reached by peptides that extend 32 Å outwards from the surface of the beads. Further information on the nature of integrin receptors containing RGD-binding site can be obtained from several publications [156–158].

According to Urry et al. [159], the polypentapeptide cross-linked by 20 Mrad γ-irradiation, χ^{20}-poly (GVGVP), has elastic properties that are appropriate for medical prostheses. They synthesized χ^{20}-poly [n(GVGVP), (GRGDSP)] and tested for its ability to support the adhesion and growth of bovine aortic endothelial cells and of bovine ligamentum nuchae fibroblasts. A cell-adhesion experiment carried out in albumin-containing media showed that matrices containing 60:1, 40:1 and 20:1 ratios of (GVGVP):(GRGDSP) supported maximal cell attachment, while poly(GVGVP) itself was a very poor support for cell attachment. Endothelial cell spreading was also observed on these matrices. For cell spreading of the fibroblast the RGD-content of the 60:1 matrix was insufficient.

In a more recent paper [160], Urry et al. reported that χ^{20}-poly (GGAP) was nonadhesive to both bovine ligamentum nuchae fibroblast and human umbilical vein endothelial cells (HUVECs), even in the presence of serum. Urry et al. interpreted this behavior in terms of their hydrophobicity scale. They suggested that χ^{20}-poly[50(GGAP), (GRGDSP)], for example, could provide the inner lamina for a vascular prosthesis.

Kawaguchi et al. [161] immobilized protected RGDS–OH and protected RGES–OH on amino-containing microspheres (294 nm av. diameter) which had been prepared by soap-free emulsion-copolymerization of styrene, acrylamide, and divinylbenzene, followed by partial conversion of the surface amide groups to amino groups. They found that extensive and tightly packed platelet aggregation, accompanied by drastic ATP release, was induced by the RGDS-carrying microspheres when the microsphere dispersion was added to a platelet suspension. This behavior of the RGDS-microspheres contrasted with that of the RGES-microspheres.

By using the same RGDS-carrying microspheres, Kawaguchi et al. [162] examined whether these microspheres could activate polymorphonuclear leukocytes (PMNs) to cause their phagocytic response. Kawaguchi found that there was little or no difference in phagocytosis among RGDS-carrying, RGES-carrying, and the parent microsphere system, in contrast to findings reported previously [163, 164]. In the latter studies, enhancement of phagocytosis by RGD-containing molecules was observed in the system of opsonized sheep erythrocytes versus monocytes [163] or PMNs [164].

Kawaguchi et al. [162] found that PMNs exhibited unique oxygen consumption and enhanced liberation of reactive oxygen, when RGDS (especially RGDSGG) carrying microspheres were phagocytosed. On the basis of this finding, together with supplemental experiments, Kawaguchi proposed a mechanism in which the biospecific activation of PMNs is assumed to be induced by signal transduction via RGDS-integrin binding, without alteration of the degree of phagocytosis.

5.2 Other Bio-conjugate (or Bio-mimicking) Materials

A variety of researches on bio-conjugate (or bio-mimicking) materials have been carried out during the last few years. As seen in Table 1, the aims and scope of many of the researchers are directly connected with clinical applications. For instance, endothelial-cell seeding (or sodding) on the luminal surface of vascular grafts is a widely-known technique [165–169] for improving the blood compatibility of polymetric materials. On the other hand, not a few of researchers are oriented to the exploitation of future possibilities of biomaterials.

Imanishi et al. [170] immobilized insulin, transferrin, and collagen onto polyurethane membranes and tubes containing primary amino groups, which had been introduced into the polyurethane surface by preliminary treatment with a glow discharge in the presence of ammonia gas. Bovine endothelial cells were cultivated in Dulbecco's modified Eagle medium supplemented with fetal bovine serum. Imanishi et al. found that the rate of cell growth was higher on the immobilized insulin or transferrin than on free insulin or transferrin, respectively. It was also found that the coimmobilization of insulin or transferrin with collagen brought about greater acceleration of cell growth, and that cells on the collagen immobilized tubes remained stable after 270 days of cultivation. They also observed more prostacyclin secretion on the collagen-immobilized tubes.

Ishihara et al. [171, 172] have been studying hemocompatible polymers with phospholipid polar groups. Their idea was to synthesize a polymer possessing a strong affinity for phospholipids from blood, which could be organized to form a biomembrane-like assemblage on the polymer surface. Phospholipid polymers were prepared by copolymerization of 2-methacryloyloxyethyl phosphorylcholine (MPC) with an alkyl methacrylate. Among the copolymers examined, poly(MPC–co–BMA) (MPC mole fraction, 0.26; BMA, butyl methacrylate) exhibited the best result in terms of antithrombogenicity. The adsorption of phospholipids and protein from human plasma onto the surface of poly(MPC–co–BMA) was examined, in order to clarify the mechanism of the excellent hemocompatibility of the copolymer surface. Ishihara et al. found that the amount of adsorbed phospholipid was much larger on the surface of poly(MPC-co-BMA), where the mole fraction of BMA was 0.26, than on the surfaces of poly(HEMA) and poly(BMA). On the other hand, the extent of protein adsorption was smallest on the phospholipid copolymer.

$$-(CH_2-\underset{\underset{\underset{O^-}{|}}{\overset{\overset{CH_3}{|}}{C}}}{\overset{\overset{CH_3}{|}}{C}})_a\text{---}(CH_2-\underset{OCH_2CH_2CH_2CH_3}{\overset{\overset{CH_3}{|}}{C}})_b$$

Poly(MPC-co-BMA)

On the basis of these results, Ishihara et al. postulated that the mechanism of the observed non-thrombogenicity of poly(MPC–co–BMA) involves formation of a biomembrane-like surface as shown in Fig. 10.

Ishihara [173] also synthesized poly(MPC–graft–BMA). This graft copolymer could be dissolved in ethanol and was useful as a non-thrombogenic coating material.

Akaike et al. [174–182] synthesized a new type of polystyrene derivative, poly(N-p-vinylbenzyl-D-lactonamide) (PVLA), having lactose residues as side

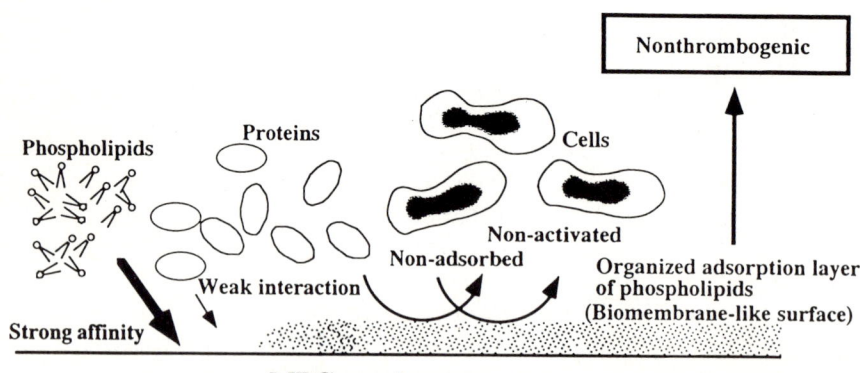

Fig. 10. Schematic representation of mechanism of nonthrombogenicity observed on poly(MPC-co-MBA). (Reproduced from J Biomed Mater Res [Ref 171: Hemocompatibility of human whole blood on polymers with a phospholipid polar group and its mechanism] through the courtesy of John Wiley & Sons. Inc.)

chains which exhibit a highly specific affinity for hepatocytes. β-Galactose residues of PVLA interact with the asialo-glycoprotein receptors on the surface of hepatocytes. In aqueous physiological media, the hydrophobic polystyrene main-chain of PVLA forms a hydrophobic core that extends its hydrophilic carbohydrate side chains into the aqueous medium. Owing to the folded conformation of the hydrophobic core, the density of the carbohydrate residues in the hydrophilic outer shell becomes high enough to induce a carbohydrate clustering effect, which is known to enhance the recognition ability of receptors. Akaike et al. found that rat hepatocytes on dishes coated with PVLA at high concentration retained their round shape even after 48 hours, whereas hepatocytes on dishes coated with PVLA at low concentration were flattened,

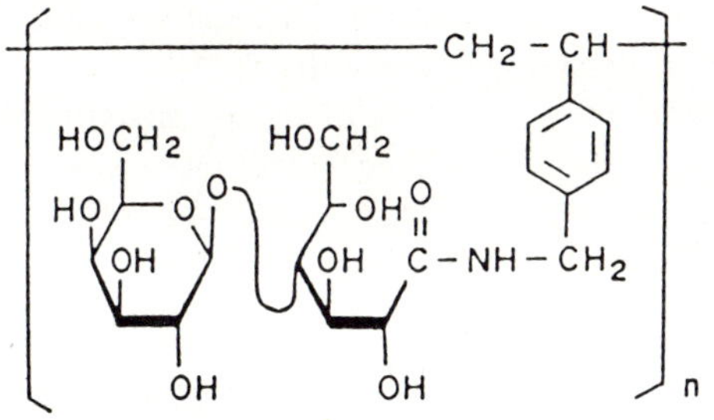

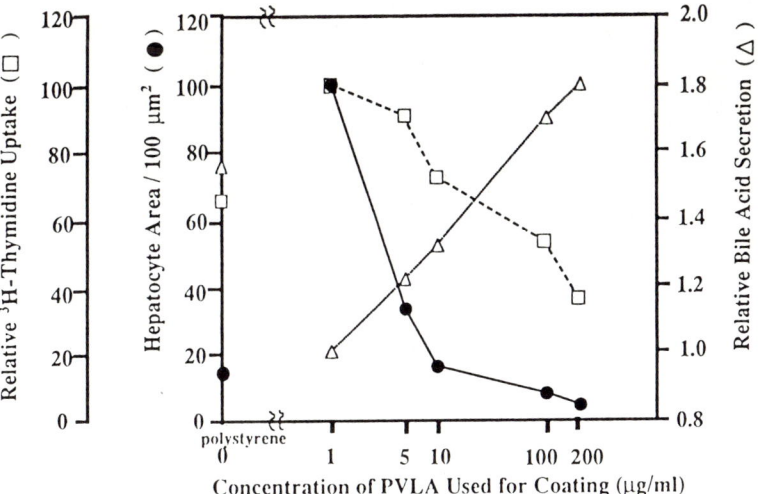

Fig. 11. Latose-substituted polystyrene (PVLA)

Fig. 12. The relation between hepatocyte shape (●), ^{3}H-thymidine uptake (□) expressed as ratio to hepatocytes cultured on polystyrene dishes coated with 1 μg/ml PVLA solution and hepatic function, bile acid secretion (△), expressed as ratio to bile acid secretion of hepatocytes cultured on dishes coated with 1 μg/ml of PVLA solution (Reproduced from New Functionality Materials, Vol. B [Ref. 181] through the courtesy of Elsevier Science Publisher B.V.)

similarly to those cultured on collagen. It was found also that the shape of hepatocytes correlates closely with the biological functions (differentiation and proliferation) of hepatocytes adhered on the PVLA surface, as shown in Fig. 12, where the respective activities of differentiation and proliferation were estimated, by measuring bile-acid secretion and thymidine uptake.

Hepatocytes attached to PVLA were found to form anchored multilayer aggregates when epidermal growth factor (EGF) and insulin were added to the culture medium. The aggregated hepatocytes exhibited a higher level of albumin secretion and longer-term preservation of this function, in comparison with those in monolayer culture on collagen. The culture system using PVLA is expected to be applied to the development of hybrid artificial liver and also to the studies of the liver regeneration processes.

Liposome is widely recognized to be a potent carrier for drug delivery systems, with biomedical application of liposome has scarcely been successful, owing to its instability or fragility under physiological conditions. Sunamoto et al. [183] have been developing a new concept and methodology to substantiate both the stability and active-targetability of the liposome in vitro and in vivo. They utilized a polysaccharide-bearing cholesterol (CH) moiety for this purpose. The CH-moieties penetrate into the liposomal bilayer and stabilize it as an anchor when the outer-most surface of liposome was covered with the polysaccharide. Sunamoto et al. [184] reported that appreciable amount of egg-phosphatidyl choline (PC) liposome coated with CH- and 1-aminolactose-bearing pullulan were accumulated at tumors. Their experiments were carried out with rats which had been implanted with 9L-gliosarcoma. They also developed a liposomal vaccine against adult T-cell leukemia by using a macrophage-specific mannan-coated liposome [185]. All of the approaches mentioned above assume the liposome to be internalized by the cell by means of endocytosis, which may cause undesirable decomposition of fragile substances (e.g. proteins, plasmids) encapsulated in the liposome. Sunamoto et al. [186] designed PEG-lipid-modified egg-PC liposome, which could undergo direct fusion with cytoplasm membrane, to deliver such a fragile substance into the cytosol. Here, PEG-lipid is 2-[ω-hydroxy(oxyethylene)-α-yl]-1,3-bis-(dodecyloxy)propane (molecular weight: 1000–1500). A series of experiments on cell-fusion activity of the PEG-lipid-containing liposomes with carrot protoplasts showed that these liposomes seemed to be fusogenic enough to fulfill their objectives.

Kataoka, Yokoyama et al. [187, 188] succeeded in developing a stable multimolecular micelle system with a drug-binding inner core, and verified its desirable behavior as the vehicle for targeting therapy. The key mateial is poly(ethylene oxide)–poly(aspartic acid) block copolymer (PEO/PASP), and a hydrophobic anti-cancer drug, adriamycin (ADR), was conjugated to the pendant carboxylic groups of the PASP block of the copolymer to form PEO/PASP (ADR). Although ADR is widely recognized to be very effective in cancer chemotherapy, it is also known that ADR exerts serious toxicity espeically on heart and bone. Kataoka's strategy was to prevent the toxic side-effects of ADR by conjugating it to the polymeric carrier systems as shown in Fig. 13.

$$CH_3-(OCH_2CH_2)_n-NH-(COCHNH)_x-(COCH_2CHNH)_y-H$$
$$\overset{|}{CH_2COR}\overset{|}{COR}$$

R=OH or [anthracycline/doxorubicin structure]

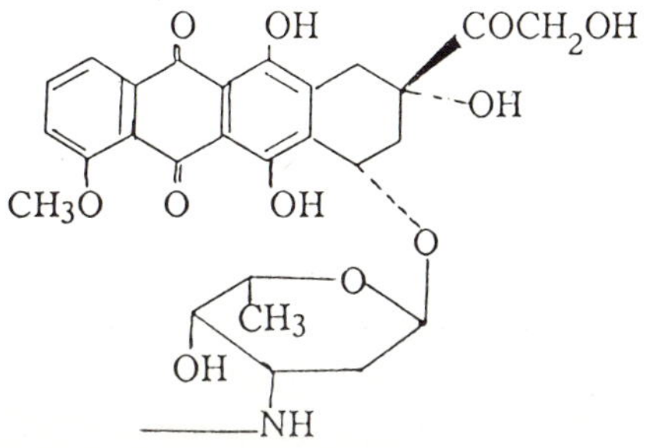

Fig. 13. Chemical structure of PEO-PASP (ADR) conjugates

Here, the β-amide linkage of PASP was formed by ca. 80% conversion of the α-amide linkage in the deprotection process of poly(β-benzyl L-aspartate) to PASP. Detailed studies [189] were carried out to find a correlation between physicochemical features of micelle formation and the primary structures of PEO/PASP (ADR), in which the PEO block length was adjusted to range from 1000 to 12000 in molecular weight and a PASP block length from 10 to 80 in the repeating unit. Conjugation of ADR can be performed without precipitation even in the case of an installing capacity as high as 77%. Micelle formation was confirmed by gel-filtration chromatography and micelle size was measured by dynamic laser light scattering. These micelles were revealed to have a core-shell structure, the core of which was formed by non-covalent hydrophobic interactions among ADR residues. The outer shell, composed of PEO segments, made the micelles behave like stealth vehicles. It is to be noted that the diameter of micelles ranged from 10 to 100 nm, and, mostly around 50 nm. This size range corresponds approximately to that of viruses and lipoproteins. Kataoka suggested that this may be the reason why the PEO/PASP (ADR) micelles can escape recognition by reticuloendothelial systems. An appropriately designed PEO/PASP (ADR) micelle (molecular weight of the segments: PEO 4,300, PASP 1,900) was actually found to achieve much longer circulation in blood and to exhibit the highest anticancer activity [190, 191].

6. Summary

1. Multifaceted aspects of biomaterials research was surveyed in Table 1. Enduse devices are manufactured starting from their original concept. To

approach the target, materials design should be carried out so that the materials can exhibit a desirable property when they are brought into contact with some particular biological element. Fundamentals are studied in order to elucidate structure-property relationship in interaction of materials with biological elements.
2. Hydrophilicity and hydrophobicity are the most fundamental properties to be controlled for materials whenever they are utilized in biomedical devices. Protein-adsorption behavior on several biomaterials of different hydrophilicity was discussed by comparing available data with two modellings (Ikada and Peppas) for the protein-adsorption process. The adsorptive behavior of poly(HEMA) carrying polyamine functional groups was also discussed. It is well-known that protein adsorption is the first event when any of the body fluids encounters an artificial material.
3. Water soluble polymers were believed to be inert at the interface with any of the biological elements. A large number of researches have been carried out to obtain biocompatible hydrophilic surfaces by introducing water-soluble polymer grafts including poly(ethylene glycol), polyacrylamide and poly(vinyl alcohol). It has however, occasionally been observed that there is an optimum in the number of grafted chains for the surface to exhibit the best biocompatibility. It should also be pointed out that excellent results of a PEG-modified surface in terms of low adhesivity of platelets and/or blood protein are not always maintained when the surfaces are examined by ex vivo (e.g. intra-arterial A–A shunt) assay.

    Blood compatibility of PEG-modified surfaces was discussed in terms of the mobility of water molecules at the interface of hydrogel materices. The property and application of poly(N-isopropylacrylamide) and its copolymers as thermoresponsive hydrogel were also reviewed.
4. Segmented polyurethanes (SPUs), such as Biomer, Pellethane or Cardiothane, are widely recognized to possess notable biomedical properties as materials for artificial heart, intra-aortic balloon pumping (IABP), and also as coating materials for pacemaker-lead insulators. The most important structural features of these SPUs are the microphase-separated (or microdomain) structures which are formed in their molecular assembly systems. None of them, however, possesses satisfactory properties in terms of in vivo durability. Several attempts to improve the durability of SPUs were reviewed. The surface of some Biomer lots is dominated by poly(diisopropylaminoethyl methacrylate). Significant suppression of protein adsorption onto several amino-group modified PEUU is to be noted.
5. An A–B–A-type block copolymer (HEMA–St–HEMA) was shown to form a microdomain structure and to exhibit excellent blood compatibility in both in vitro and ex vivo examinations. For instance, the luminal surface of the HEMA–STY coated vascular graft was bare without detectable thrombi after 372-day implantation in dog carotid aortas. The excellent blood compatibility was discussed by taking results of the unique mode of protein adsorption of HEMA–STY surface into account.

6. A series of polyamine-graft copolymers were found to form microdomain structures and to exhibit unique biomedical behavior at the interface with living cells, e.g. blood platelets or lymphocytes. Poly(HEMA)-*graft*-polyamine (HA) copolymers, for instance, exhibited different adhesivity against lymphocyte B cell and T cell subpopulations, which enabled us to separate B cells effectively from T cells. The morphological change of polyamine chains in the physiological pH-range was closely related to the cell recognition mechanism. Some of HA copolymer surfaces containing a small proportion of polyamine became inert enough against lymphocytes, as well as against blood platelets, to reject their attachment. This was discussed in terms of the random-network concept of water molecules on the HA surface.
7. Several biohybridized and biomimicking materials were reviewed. Since Piersbacher and Ruoslahti (1984) reported that RGDS (or RGDC)-containing peptides promoted attachment of rat kidney fibroblast, many researches have been carried out to prepare RGDX containing materials, which are expected to be useful as cell-attachment and/or cell-cultivating materials.

Other bio-conjugate (or bio-mimicking) materials that were reviewed in the text include (i) insulin- (or transferrin-) and collagen-coimmobilized polyurethane for promoting endotheliazation, (ii) methacrylic copolymers with phospholipid polar groups as excellent antithrombogenic materials, (iii) poly(*N*-*p*-vinylbenzyl-D-lactonamide) (PVLA) as the key material in the culture system for hepatocytes, (iv) liposomes which are reinforced by polysaccharide for DDS and vaccine, and (v) a multimolecular micelle system composed of a PEO–poly(Asp) block copolymer as a vehicle for anticancer drugs.

7 References

1. Bamford CH, Cooper SL, Tsuruta T (1989) J Biomater Sci Polymer Edn 1: 1
2. Albertsson PÅ (1986) Partition of Cell Particles and Macromolecules. 3rd Ed. Wiley, New York
3. Walter H (1977) in Catsimpoolas (ed) Method of Cell Separation Vol. 1, Plenum, New York, p 307
4. Water H (1985) Partitioning in Aqueous Two-phase Systems – Theory, Methods, Uses and Application to Biotechnology, Academic Press, Orlando
5. Lyman DJ, Hill DW (1972) Trans Am Soc Artif Intern Organs 18: 19
6. Imai Y (1972) Kobunshi (High Polymers, Jpn 21: 569 (in Japanese)
7. Imai Y, Watanabe A, Masuhara E (1973) Jinko-zoki (Jpn J Artif Organs 2: 95 (in Japanese)
8. Wildevuur ChRH, Lei B vander, Schakenraad JM (1987) Biomaterials 8: 418
9. Yue X, Lei B vander, Schakenraad JM, Oene GH van, Wildevuur ChRH (1988) Surgery 103: 206
10. Lei B vander, Wildevuur ChRH (1989) Thorac Cardiovasc Surg. 37: 337
11. Steele JG, Johnson G, Underwood PA (1992) J Biomed Mater Res 26: 861
12. Ikada Y, Suzuki Y, Tamada Y (1985) In: Hoffman AS, Ratner BD, Horbett TA (eds) Polymer as Biomaterials, Plenum, New York, p. 135
13. Berg JC (1993) in: Berg JC (ed) Wettability (Surfactant Series) Vol. 49, Marcel Dekker, New York, p. 75

14. Tamada Y, Ikada Y (1993) J Colloid Interface Sci 155: 334
15. Tamada Y, Ikada Y (1993) Polymer 34: 2208
16. Straaten Jvan, Peppas NA (1991) J Biomater Sci Polymer Edn 2: 91
17. Lee B, Richards RM (1970) J Mol Biol 55: 379
18. Drago RS, Vogel GC, Needham TE (1971) J Amer Chem Soc 93: 6014
19. Drago RS, Wong N, Bilgrien C, Vogel GC (1987) Inorg Chem 26: 9
20. Fowkes FM (1990) J Adhesion Sci Technol 4: 669
21. Fabrizius-Homan DJ, Cooper SL (1991) J Biomater Sci Polymer Edn 3: 27
22. Fabrizius-Homan DJ, Cooper SL (1991) J Biomed Mater Res 25: 953
23. Holmes-Farley SR, Bain CD, Whitesides GM (1988) Langmuir 4: 921
24. Ueda Y (1990) Master's Thesis (Supervised by Tsuruta T) Sci Univ Tokyo p 34
25. Andrade JD (1974) J Colloid Interf Sci 72: 488
26. Llanos GR, Sefton MV (1993) J Biomater Sci Polymer Edn 4: 381
27. Amiji M, Park K (1993) J Biomater Sci Polymer Edn 4: 217
28. Chaikof EL, Merril EW, Callow AD, Connolly RJ, Verdon SL, Ramberg K (1992) J Biomed Mater Res 26: 1163
29. Okkema AZ, Grasel TG, Zdrahala RJ, Solomon DD, Cooper SL (1989) J Biomater Sci Polymer Edn 1: 43
30. Grainger DW, Knutson K, Kim SW, Feijen J (1990) J Biomed Mater Res 24: 403
31. Desai NP, Hubbell JA (1991) J Biomed Mater Res 25: 829
32. Bergström K, Holmberg K, Safranj A, Hoffman AS, Edgell MJ, Kozlowski A, Hovanes BA, Harris JM (1992) J Biomed Mater Res 26: 779
33. Liu SQ, Ito Y, Imanishi Y (1989) J Biomater Sci Polymer Edn 1: 111
34. Fujimoto K, Inoue H, Ikada Y (1993) J Biomed Mater Res 27: 1559
35. Lin SC, Jacobs HA, Kim SW, J Biomed Mater Res 25: 791
36. Tanzawa H, First Japan-US Workshop on Biomedical Polymer Science (1985) Kyoto, Japan
37. Tanzawa H (1986) Jino-zoki (Jpn J Artif Organs) 15: 16 (in Japanese)
38. Yamada-Nosaka A, Ishikiriyama K, Todoki M, Tanzawa H (1990) J Appl Polymer Sci 39: 2443
39. Yamada-Nosaka A, Tanzawa H (1991) J Appl Polymer Sci 43: 1165
40. Malcolm GN, Rowlinson JS (1957) Trans Faraday Soc 53: 921
41. Heskins M, Guillet JE (1968) J Macromol Sci-Chem A2: 1441
42. Tanaka T (1987) in: Mark H, Bikales NM, Overberger CG, Menges G, Kroschuwitz JI (eds) Encycl Polym Sci Eng (Wiley) 7: 514
43. Sato-Matsuo E, Tanaka T (1988) J Chem Phys 89: 1695
44. Yoshida R, Sakai K, Okano T, Sakurai Y (1994) J Biomater Sci Polymer Edn 6: 585
45. Chen GH, Hoffman AS (1994) J Biomater Sci Polymer Edn 5: 371
46. Monji N, Cole CA, Tam MR, Goldstein L, Hoffman AS, Nowinski RC (1988) Trans 3rd World Biomater Congress 11: 298
47. Cole CA, Monji N, Tam MR, Goldstein L, Hoffman AS, Nowinski RC (1988) Trans 3rd World Biomater Congress 11: 299
48. Monji N, Cole CA, Hoffman AS (1994) J Biomater Sci Polymer Edn 5: 407
49. Yoshida R, Sakai K, Okano T, Sakurai Y, Bae YH, Kim SW (1991) J Biomater Sci Polymer Edn 3: 155
50. Okano T, Yui N, Yokoyama M, Yoshida R (1994) Japanese Technology Reviews (Gordon and Breach Science Publisher) 4: 67
51 Okano T, Yamada N, Sakai H, Sakurai Y (1993) J Biomed Mater Res 27: 1243
52. Yamada N, Sakai H, Okano T, Sakurai Y (1992) Jinko-zoki (Jpn J Artif Organs) 21: 206 (in Japanese)
53. Sung CPS, Hu CB (1981) Macromolecules 14: 212
54. Wilkes GL, Abouzar S (1981) Macromolecules 14: 456
55. Wang CB, Cooper SL (1983) Macromolecules 16: 775
56. Takahara A, Tashita J, Kajiyama T, Takayanagi M (1982) Kobunshi Ronbunshu 39: 203 (in Japanese)
57. Takahara A, Korehisa K, Takahashi K, Kajiyama T (1992) Kobunshi Ronbunshu 49: 275 (in Japanese)
58. Takahara A, Takahashi K, Kajiyama T (1993) J Biomater Sci Polymer Edn 5: 183
59. Renier M, Anderson JM, Hiltner A, Lodoen GA, Payet CR (1993) J Biomater Sci Polymer Edn 5: 231
60. Takahara A, Hergenrother RW, Coury AJ, Cooper SL (1991) in: Akutsu T, Koyanagi H (eds) Artificial Heart 3, Springer-Tokyo, p 77

61. Takahara A, Hergenrother RW, Coury AJ, Cooper SL (1991) J Biomed Mater Res 25: 341
62. Takahara A, Hergenrother RW, Coury AJ, Cooper SL (1992) J Biomed Mater Res 26: 801
63. Tyler BJ, Ratner BD, Castner DG, Briggs D (1992) J Biomed Mater Res 26: 273
64. Tyler BJ, Ratner BD (1993) J Biomed Mater Res 27: 327
65. Ratner BD, Gladhill KW, Horbett TA (1988) J Biomed Mater Res 22: 509
66. Takahara A, Okkema AZ, Wabers H, Cooper SL (1991) J Biomed Mater Res 25: 1095
67. Santerre JP, tenHove P, VanderKamp NH, Brash JL (1992) J Biomed Mater Res 26: 39
68. Okkema AZ, Visser SA, Cooper SL (1991) J Biomed Mater Res 25: 1371
69. Silver JH, Lewis KB, Ratner BD, Cooper SL (1993) J Biomed Mater Res 27: 735
70. Silver JH, Marchant JW, Cooper SL (1993) J Biomed Mater Res 27: 1443
71. Han DK, Ryu GH, Park KD, Jeong SY, Kim YH, Min BG (1993) J Biomater Sci Polymer Edn 4: 401
72. Han DK, Jeong SY, Ahn KD, Kim YH, Min BG (1993) J Biomater Sci Polymer Edn 4: 579
73. Ito Y, Iguchi Y, Kashiwagi T, Imanishi Y (1991) J Biomed Mater Res 25: 1347
74. Stanislawski L, Serne H, Stanislawski M, Jozefowicz M (1993) J Biomed Mater Res 27: 619
75. Brunstedt MR, Ziats NP, Schubert M, Hiltner PA, Anderson JM, Lodoen GA, Payet CR (1993) J Biomed Mater Res 27: 255
76. Brunstedt MR, Ziats NP, Robertson, SP, Hiltner A, Anderson JM, Lodoen GA, Payet CR (1993) J Biomed Mater Res 27: 367
77. Okano T. Katayama M, Mogi S, Shinohara I (1977) Nippon Kagaku Kaishi 88 (in Japanese)
78. Okano T, Katayama M, Shinohara I (1978) J Appl Polymer Sci 22: 369
79. Okano T, Nishiyama S, Shinohara I, Akaike T, Sakurai Y, Kataoka K, Tsuruta T (1979) Jinko-zoki (Jpn J Artif Organs) 8: 292 (in Japanese)
80. Okano T, Nishiyama S, Shinohara I, Akaike T, Sakurai Y, Kataoka K, Tsuruta T (1981) J Biomed Mater Res 15: 393
81. Hirao A, Kato H, Yamaguchi K, Nakahama S (1986) Macromolecules 19: 1294
82. Nojiri C, Okano T, Jacobs HA, Park KD, Mohammad SF, Olsen DB, Kim SW (1990) J Biomed Mater Res 24: 1151
83. Nojiri C, Okano T, Koyanagi H, Nakahama S, Park KD, Kim SW (1992) J Biomater Sci Polymer Edn 4: 75
84. Okano T, Suzuki K, Yui N, Sakurai Y, Nakahama S (1993) J Biomed Mater Res 27: 1519
85. Nojiri C, Okano T, Takemura N, Senshu K, Kido T, Koyanagi H, Kim SW (1993) in: Akutsu T, Koyanagi H (eds) Artificial Heart 4, Springer-Tokyo, p. 53
86. Yui N, Tanaka J, Sanui K, Ogata N, Kataoka K, Okano T, Sakurai Y (1984) Polymer J 16: 111
87. Yui N, Sanui K, Ogata N, Kataoka K, Okano T, Sakurai Y (1986) J Biomed Mater Res 29: 929
88. Yui N, Kataoka K, Sakurai Y, Sanui K, Ogata N, Takahara A, Kajiyama T (1986) Makromol Chem 187: 943
89. Yui N, Kataoka K, Sakurai Y, Aoki T, Sanui K, Ogata N (1988) Biomaterials 9: 225
90. Yui N, Kataoka K, Okano T, Sakurai Y (1991) in: Akutsu T, Koyanagi H (eds) Artificial Heart 3, Springer-Tokyo p 23
91. Takei YG, Yui N, Okano T, Maruyama A, Sanui K, Sakurai Y, Ogata N (1994) J Biomater Sci Polymer Edn 6: 149
92. Nishimura T, Maeda M, Nitadori Y, Tsuruta T (1980) Makromol Chem Rapid Commun 1: 573
93. Kataoka K, Okano T, Sakurai Y, Nishimura T, Maeda M, Inoue S, Tsuruta T (1982) Biomaterials 3: 237
94. Kataoka K, Okano T, Sakurai Y, Nishimura T, Inoue S, Watanabe T, Tsuruta T (1982) Makromol Chem Rapid Commun 3: 275
95. Kataoka K, Okano T, Sakurai Y, Nishimura T, Inoue S, Watanabe T, Maruyama A, Tsuruta T (1983) Eur Polymer J 19: 979
96. Maruyama A, Senda E, Tsuruta T, Kataoka K (1986) Makromol Chem 187: 1895
97. Maruyama A, Tsuruta T, Kataoka K, Sakurai Y (1988) Biomaterials 9: 471
98. Maruyama A, Tsuruta T, Kataoka K, Sakurai Y (1988) J Biomed Mater Res 22: 555
99. Nitadori Y, Tsuruta T (1979) Makromol Chem 180: 1877
100. Nishimura T, Maeda M, Nitadori Y, Tsuruta T (1982) Makromol Chem 183: 29
101. Tsuruta T (1987) Makromol Chem Macromol Symp 12: 323
102. Tsuruta T (1990) Makromol Chem Macromol Symp 33: 243
103. Kataoka K, Akaike T, Sakurai Y, Tsuruta T (1978) Makromol Chem 179: 1121
104. Kataoka K, Maeda M, Nishimura T, Nitadori Y, Tsuruta T, Akaike T, Sakurai Y (1980) J Biomed Mater Res 14: 817
105. Kataoka K, Tsuruta T, Akaike T, Sakurai Y (1980) Makromol Chem 181: 1363

106. Maruyama A, Tsuruta T, Kataoka K, Sakurai Y (1987) Makromol Chem Rapid Commun 8: 27
107. Nabeshima Y, Maruyama A, Tsuruta T, Kataoka K (1987) Polymer J 19: 593
108. Nabeshima Y, Tsuruta T, Kataoka K, Sakurai Y (1989) J Biomater Sci Polymer Edn 1: 85
109. Kikuchi A, Mizutani S, Kataoka K, Tsuruta T (1991) Polymers for Advanced Technologies 2: 245
110. Kataoka K, Sakurai Y, Hanai T, Maruyama A, Tsuruta T (1988) Biomaterials 9: 218
111. Maruyama A, Tsuruta T, Kataoka K, Sakurai Y (1989) Biomaterials 10: 291
112. Kikuchi A, Maruyama A, Tsuruta T, Kataoka K, Yui N, Sakurai Y (1988) Jinko-zoki (Jpn J Artif Organs) 17: 487 (in Japanese)
113. Kataoka K, Sakurai Y, Nabeshima Y, Sasaki Y, Maruyama A, Tsuruta T (1991) Kobunshi Ronbunshu 48: 201 (in Japanese)
114. Kikuchi A, Kataoka K, Tsuruta T (1992) J Biomater Sci Polymer Edn 3: 355
115. Kikuchi A, Karasawa M, Kataoka K, Okuyama K, Tsuruta T (1993) in: Akutsu T, Koyanagi H (eds) Artificial Heart 4, Springer-Tokyo, p 29
116. Bigelow JC, Giddings JC, Nabeshima Y, Tsuruta T, Kataoka K, Okano T, Yui N, Sakurai Y (1989) J Immunol Methods 117: 289
117. Kikuchi A, Karasawa M, Kataoka K, Tsuruta T (1992) Jinko-zoki (Jpn J Artif Organs) 21: 1212 (in Japanese)
118. Kikuchi A, Karasawa M, Tsuruta T, Kataoka K (1993) J Colloid Interface Sci 158: 10
119. Nishimura T, Kuroda T, Umegae M, Miyamoto M, Ishikawa Y, Sasakawa S (1988) Trans 3rd World Biomater Congress 11: 481
120. Miyamoto M, Sasakawa S, Ishikawa Y, Ogawa A, Nishimura T, Kuroda T (1989) Vox Sang 57: 164
121. Stillinger FH (1980) Science 209: 451
122. Rice SA, Sceats MG (1981) J Phys Chem 85: 1108
123. Belch AC, Rice SA (1983) J Chem Phys 78: 4817
124. Green JL, Racey AR, Sceats MG (1986) J Phys Chem 90: 3958
125. Green JL, Racey AR, Sceats MG (1987) J Chem Phys 86: 1841
126. Hare DE, Sorensen CM (1992) J Chem Phys 96: 13
127. Maeda Y, Tsukida N, Kitano H, Terada T, Yamanaka J (1993) J Phys Chem 97: 13903
128. Terada T, Maeda Y, Kitano H (1993) J Phys Chem 97: 3619
129. Taira H, Tanaka A, Kataoka K, Tsuruta T, Hayashi M (1994) Jinko-zoki (Jpn J Artif Organs) 23: 695 (in Japanese)
130. Pierschbacher MD, Ruoslahti E (1984) Nature 309: 30
131. Pierschbacher MD, Ruoslahti E (1984) Proc Natl Acad Sci USA 81: 5985
132. Hayman EG, Pierschbacher MD, Ruoslahti E (1985) J Cell Biol 100: 1948
133. Pytela R, Pierschbacher MD, Ruoslahti E (1985) Proc Natl Acad Sci USA 82: 5766
134. Plow EP, Pierschbacher MD, Ruoslahti E, Marguerie GA, Ginsberg MH (1985) Proc Natl Acad Sci USA 82: 8057
135. Dedhar S, Ruoslahti E, Pierschbacher MD (1987) J Cell Biol 104: 585
136. Ruoslahti E, Pierschbacher MD (1987) Science 238: 491
137. Maeda T, Titani K, Sekiguchi K (1991) in: Okamura S, Tsuruta T, Imanishi Y, Sunamoto J (eds) Proceedings of 11th Taniguchi Conference on Polymer Research, Kagaku-Dojin, Kyoto, p 45
138. Imanishi Y, Ito Y, Liu LS, Kajihara M (1988) J Macromol Sci Chem A25: 555
139. Ito Y, Kajihara M, Imanishi Y (1991) J Biomed Mater Res 25: 1325
140. Brandley BK, Schnaar RL (1988) Analytical Biochemistry 172: 270
141. Matsuda T, Kondo A, Makino K, Akutsu T (1989) Trans Am Soc Artif Intern Organs 35: 677
142. Massia SP, Hubbell JA (1990) Analytical Biochemistry 187: 292
143. Ohlson S, Hanson L, Larsson P, Mosbach K (1978) FEBS Lett 93: 5
144. Massia SP, Hubbell JA (1991) J Biomed Mater Res 25: 223
145. Massia SP, Hubbell JA (1990) Ann NY Acad Sci 589: 261
146. Costello CA, McCarthy TJ (1987) Macromolecules 20: 2819
147. Nilsson K, Mosbach K (1981) Biochem Biophys Res Comm 102: 449
148. Massia SP, Hubbell JA (1991) Trans 17th Meeting Soc Biomater 14: 238
149. Drumheller PD, Desai NP, Hubbell JA (1991) Trans 17th Meeting Soc Biomater 14: 239
150. Hirano Y, Kando Y, Hayashi T, Goto K, Nakajima A (1991) J Biomed Mater Res 25: 1523
151. Hirano Y, Okuno M, Hayashi T, Goto K, Nakajima A (1993) J Biomater Sci Polymer Edn 4: 235
152. Kugo K, Okuno M, Masuda K, Nishino J, Masuda H, Iwatsuki M (1994) J Biomater Sci Polymer Edn 5: 325

153. Lin HB, Sun W, Mosher DF, García-Echeverría C, Schaufelberger K, Lelkes PI, Cooper SL (1994) J Biomed Mater Res 28: 329
154. Goodman SL, Cooper SL, Albrecht RM (1993) J Biomed Mater Res 27: 683
155. Beer JH, Springer KT, Coller BS (1992) Blood 79: 117
156. Dejana E, Colella S, Conforti G, Abbandini M, Gaboli M, Marchisio PC (1988) J Cell Biol 107: 1215
157. Hautanen A, Gailit J, Mann DM, Ruoslahti E (1989) J Biol Chem 264: 1437
158. Massia SP, Hubbell JA (1991) J Cell Biol 114: 1089
159. Nicol A, Gowda DC, Urry DW (1992) J Biomed Mater Res 26: 393
160. Nicol A, Gowda DC, Parker TM, Urry DW (1993) J Biomed Mater Res 27: 801
161. Kasuya Y, Fujimoto K, Miyamoto M, Juji T, Otaka A, Funakoshi S, Fujii N, Kawaguchi H (1993) J Biomater Sci Polymer Edn 4: 369
162. Kasuya Y, Fujimoto K, Miyamoto M, Kawaguchi H (1994) J Biomed Mater Res 28: 397
163. Brown EJ, Goodwin JL (1988) J Exp Med 167: 777
164. Gresham HD, Goodwin JL, Allen PM, Anderson DC, Brown EJ (1989) J Cell Biol 108: 1935
165. Herring MB, Gardner AL, Glover JA (1978) Surgery 84: 498
166. Stanley JC, Burkel WE, Ford JC, Vinter DW, Kahn RH, Whitehouse Jr WM, Graham LM (1982) Surgery 92: 994
167. Williams SK, Jarrell BE (1987) Ann NY Acad Sci 516: 145
168. Sharefkin JB, Latker C, Smith M, Gruss D, Clagett GP, Rich NM (1982) Surgery 92: 385
169. Williams SK, Rose DG, Jarrell BE (1994) J Biomed Mater Res 28: 203
170. Liu SQ, Ito Y, Imanishi Y (1993) J Biomed Mater Res 27: 909
171. Ishihara K, Oshida H, Endo Y, Ueda T, Watanabe A, Nakabayashi N (1992) J Biomed Mater Res 26: 1543
172. Ishihara K, Oshida H, Endo Y, Watanabe A, Ueda T, Nakabayashi N (1993) J Biomed Mater Res 27: 1309
173. Ishihara K, Tsuji T, Kurosaki T, Nakabayashi N (1994) J Biomed Mater Res 28: 225
174. Tobe S, Takei Y, Kobayashi K, Akaike T (1992) Biochem Biophys Res Commun 184: 225
175. Kobayashi A, Kobayashi K, Akaike T (1992) J Biomater Sci Polymer Edn 3: 499
176. Kugumiya T, Yagawa A, Maeda A, Nomoto H, Tobe S, Matsuda T, Onishi T, Kobayashi K, Akaike T (1992) J Bioactive Compatible Polym 7: 337
177. Akaike T, Tobe S, Kobayashi A, Goto M, Kobayashi K (1993) Gastroenterol Jpn Suppl 4: 45
178. Goto M, Yura H, Chang CW, Kobayashi A, Shinoda T, Maeda A, Kojima S, Kobayashi K, Akaike T (1994) J Controlled Release 28: 223
179. Kobayashi A, Goto M, Kobayashi K, Akaike T (1994) J Biomater Sci Polymer Edn 6: 325
180. Ohgawara H, Kobayashi A, Kawamura M, Karibe S, Fu Q, Omori Y, Akaike T (1994) Cell Transplantation 3: 83
181. Akaike T, Kobayashi A, Kobayashi K (1993) in: Tsuruta T, Doyama M, Seno M, Imanishi Y (eds) New Functionality Materials, Vol. B, Elsevier Amsterdam, p 303
182. Kobayashi K, Kobayashi A, Tobe S, Akaike T (1994) in: Lee YC,. Lee RT (eds) Neo-glyco-conjugates: Preparation and applications, Academic Press, San Diego, p 261
183. Sunamoto J, Sato T, Taguchi T, Hamazaki H (1992) Macromolecules 25: 5665
184. Shibata S, Ryu N, Ochi A, Jinnouchi T, Sato T, Sunamoto J, Hirai M, Matsuyama K (1990) Drug Delivery System 7: 109
185. Noguchi Y, Noguchi T, Sato T, Yokoo Y, Itoh S, Yoshida M, Yoshiki T, Akiyoshi K, Sunamoto J, Nakayama E, Shiku H (1991) J Immunol 146: 3599
186. Sunamoto J, Akiyoshi K, Sato T (1993) in: Tsuruta T, Doyama M, Seno M, Imanishi Y (eds) New Functionality Materials, Vol B, Elsevier Amsterdam, p 203
187. Yokoyama M, Miyauchi M, Yamada N, Okano T, Sakurai Y, Kataoka K, Inoue S (1990) J Controlled Release 11: 269
188. Kataoka K, Kwon GS, Yokoyama M, Okano T, Sakurai Y (1993) J Controlled Release 24: 119
189. Yokoyama M, Kwon GS, Okano T, Sakurai Y, Seto T, Kataoka K (1992) Bioconjugate Chem 3: 295
190. Yokoyama M, Miyauchi M, Yamada N, Okano T, Sakurai Y, Kataoka K, Inoue S (1990) Cancer Res 50: 1693
191. Yokoyama M, Okano T, Sakurai Y, Ekimoto H, Shibazaki C, Kataoka K (1991) Cancer Res 51: 3229

Editor: Prof. R. Reisfeld
Received: 1995

Polymerization and Domain Formation in Lipid Assemblies

Bruce A. Armitage, Doyle E. Bennett, Henry G. Lamparski, and David F. O'Brien*
C.S. Marvel Laboratories, Department of Chemistry, University of Arizona, Tucson, AZ 85721, USA

Lipid assemblies are arrays of noncovalently associated amphiphiles, i.e. supramolecular assemblies. They may be classified as supported or self-supported assmeblies. The advent of methods to polymerize these supramolecular assemblies has opened up opportunities for the creation of new materials. This review emphasizes the interaction of polymerization and lipid domain formation within supramolecular assemblies. The polymerization of amphiphilic assemblies can "lock in" preexisting lipid domains or create lipid domains from random mixtures depending on the nature of the polymerizable amphiphile. Lipid diacetylenes or fluorinated lipids provide a convenient means to form an unpolymerized immiscible mixture of reactive and nonreactive lipids in monolayers or bilayers. In contrast the polymerization of dienoyl-, sorbyl-, or acryloyl-substituted lipids can effectively induce the phase separation of unreactive lipids from the growing polymeric domains. Polymerization-induced lipid domains can endow bilayer vesicles with latent instability sites or can be used to concentrate membrane-associated electron or energy transfer cofactors. These polymeric materials suggest new approaches to the delivery of reagents as well as the transduction of light energy.

1	**Introduction**	54
	1.1 Domains in Lipid Assemblies	54
	1.2 Polymerization of Lipid Bilayers	56
2	**Polymerization of Phase-Separated Lipid Assemblies**	61
	2.1 Diacetylenic Lipids	61
	2.2 Other Polymerizable Lipids	64
3	**Polymerization-Induced Domain Formation**	69
	3.1 Polymerization of Bilayers of Neutral Lipids	69
	3.2 Polymerization of Bilayers of Neutral and Charged Lipids	79
4	**Summary**	82
5	**References**	82

1 Introduction

Lipid assemblies are arrays of noncovalently associated amphiphiles. They may be classified as supported assemblies, e.g. monolayers at the air water interface, self-assemblied monolayers; or self-supported assemblies, e.g. hydrated lipid bilayers, tubules, etc. During the past two decades or so, the understanding of the chemistry and biology of these supramolecular systems has advanced significantly. The fascination with the important phenomena of amphiphilic self-organization has fueled research at the boundaries of several disciplines, including materials and life sciences. The advent of methods to polymerize supramolecular assemblies; first monolayers in the 1970s [1], followed by bilayers in the early 1980 [1, 2], now nonlamellar phases [3], has led to the creation of new materials, experimental methods, perspectives, and even new journals. The potential uses of the materials created through this research include the controlled delivery of reagents and drugs, the separation and purification of biomolecules, the modification of surfaces, the efficient transduction of light energy, and the formation and stabilization of organic zeolites, among others.

This review emphasizes an intriguing and potentially useful aspect of the polymerization of lipid assemblies, i.e. polymerization and domain formation within an ensemble of molecules that is usually composed of more than one amphiphile. General aspects of domain formation in binary lipid mixtures and the polymerization of lipid bilayers are discussed in Sects. 1.1 and 1.2, respectively. More detailed reviews of these topics are available as noted. The mutual interactions of lipid domains and lipid polymerization are described in the subsequent sections. Given the proper circumstances the polymerization of lipid monolayers or bilayers can lock in the phase separation of lipids, i.e. pre-existing lipid domains within the ensemble as described in Sect. 2. Section 3 reviews the evidence for the polymerization-initiated phase separation of polymeric domains from the unpolymerized lipids.

1.1 Domains in Lipid Assemblies

The structure of biological and model membranes is frequently viewed in the context of the fluid mosaic model [4]. Since biological membranes are composed of a mixture of various lipids, proteins, and carbohydrates the suprastructure or lateral organization of the components is not necessarily random. In order to model biological membranes, lipid assemblies of increasing complexity were studied. Extensive investigation of multicomponent monolayers (at the air-water interface) as well as bilayers have been reported.

Epifluorescence microscopy has been fruitfully employed to characterize lipid domains in phospholipid monolayers. The sizes and shapes are dependent on the lipid composition, temperature, pressure, as well as electrostatic interactions. The smallest domains observed by fluorescence are at least a few microns in diameter. The domain shapes exhibit smooth boundary lines or fractal structures. Frequently an amorphous superlattice is observed. The dipolar lipid headgroups are sensitive to the ionic and pH conditions of the aqueous subphase. Reviews by Möhwald [5] of phospholipid monolayer domains at the air water interface, and by McConnell [6] of structures and transitions in these monolayers have appeared recently. The introduction of AFM and complementary microscopies has allowed the observation of submicron domains in some multilayer films. The observed geometry of stearic acid domains was sensitive to the addition of as little as 0.25 mole% of a fluorescent dye [7]. Thus investigators must consider the possibility that the shape of lipid domains can be altered by the added molecular probes.

The detection of domains in hydrated bilayers is dependent on their stability (reviewed by Klausner & Kleinfeld) [8]. Although non-equilibrium clusters of lipids probably exist for different periods of time, their detection depends on the experimental time scale. Most of the experimental literature is concerned with the equilibrium structure of lipid domains [9]. This is of most relevance here, since most of the research concerning lipid polymerization and domains involves bilayers at equilibrium. The detection of nonrandom lipid clusters by rapid chemical analysis has also been reported [10]. The best understood systems are those composed of two phospholipids. Both phosphatidylcholine(PC)-PC systems [8] and PC-phosphatidylethanolamine (PE) systems [11] have been reviewed. Binary lipid systems where the lipid chain lengths vary by less than four carbons in length exhibit nearly ideal mixing. Increased differences in chain lengths leads to immiscibility of the two lipids in the solid phase. This suggests the two lipids cannot co-crystallize. The size of lipid domains in bilayers can be estimated with the aid of electron spin resonance (ESR) spectrometry [12]. McConnell and coworkers determined the phase diagrams of several lipid pairs with the aid of ESR spin labels and freeze fractive electron microscopy. Mixtures of desaturated PCs of comparable chain length exhibit solid-solid miscibility, with some phase separation at the higher temperatures associated with the intermediate region, and complete miscibility in the fluidus region. When the lipids differ significantly in their T_m, the phase diagrams are more complicated due to the immiscibility of the lipids in the solidus region. The lipid chain substitution pattern, i.e. unsaturation, methyl branches, rigid entities such as diacetylenes, ester or ether groups will influence the lipid phase behavior. However, only the effects of unsaturation have been systematically studied to date. A binary system of a neutral and a charged lipid will tend to exhibit ideal behavior due to the electrostatic repulsion between charged lipids.

The most complete insights into the behavior of mixtures of nonpolymerizable lipids have come from the phase diagrams of these systems. These data have provided important reference points for the polymerizable lipid systems described next, even though few phase diagrams have been reported for polymerizable lipid mixtures. In spite of this deficiency the polymerization studies have revealed several important phenomena. A more thorough characterization of the lipid phase behavior of polymerizable lipids is expected to enhance the understanding of these phenomena.

1.2 Polymerization of Lipid Bilayers

Methods to polymerize lipid bilayer membranes were introduced in the early 1980s and have been most recently reviewed by O'Brien [2], Schnur and Singh [13], O'Brien and Ramaswami [14], and Ringsdorf et al. [1]. An important strategy utilizes the preparation and purification of polymerizable lipid monomer(s), the formation of lipid bilayers composed of this lipid either alone or in combination with other lipids, and the subsequent chain polymerization of the reactive lipids in the bilayer. It is also possible to polymerize appropriately designed amphiphilic monomers in solution and then form monolayers or LB multilayers from the polymers. This second strategy relies on the effective use of spacer groups to decouple the motions of the polymer chain from the associated amphiphiles. A great variety of reactive groups and lipid classes are described in the literature, consequently there is a wide latitude in the design of polymerized bilayers. Polymerizations may be initiated by light, heat, or redox chemistries.

Polymerizable amphiphiles are frequently classified according to the location of the reactive groups in the molecule. If the hydrophilic regions of lipids are represented by a circle, and the hydrophobic chains are represented by lines, then the approximate location of the polymerizable group can be indicated by an X (Fig. 1). The reactive group may be covalently attached at the hydrophobic chain terminus (type A), near the backbone of the amphiphile (type B), attached to the hydrophilic head group (usually via a spacer group) (type C), or electrostatically associated with an ionic lipid (type D). Each lipid in Fig. 1 is represented with a single polymerizable group, which yields linear polymer chains.

The reactive lipids are constrained to the bilayer, which may exist in a solid-like, L_β, phase or in a liquid-like, L_α, phase. The main phase transition between these phases is frequently designated the T_m. Several reactive groups have been usefully employed to make lipids polymerizable, including diacetylene, methacryloyl, acryloyl, dienoyl, sorbyl, lipoyl, and styryl. Most of these molecules can be polymerized at temperatures above or below the T_m. However the amphiphilic diacetylenes may only be polymerized in the L_β phase due to the topotactic nature of the reaction. Lipids in the L_α, phase exhibit a lateral diffusion coefficient (D) of ca. $1 \, \mu m^2 s^{-1}$ [15]. Rapid lipid diffusion facilitates the movement of reactive monomers to the growing chain end of the polymer although it is constrained in two-dimensions by the bilayer. Consequently high degrees of polymerization, x_n, are possible from radical chain polymerizations of lipids such as mono-acryloyl-substituted phosphatidylcholine (mono-AcrylPC) (**1**) or mono-sorbyl-substituted PC (mono-SorbPC) (**3**) in the L_α, phase. Sells and O'Brien found degrees of polymerization of 2×10^3 for the A1BN initiated

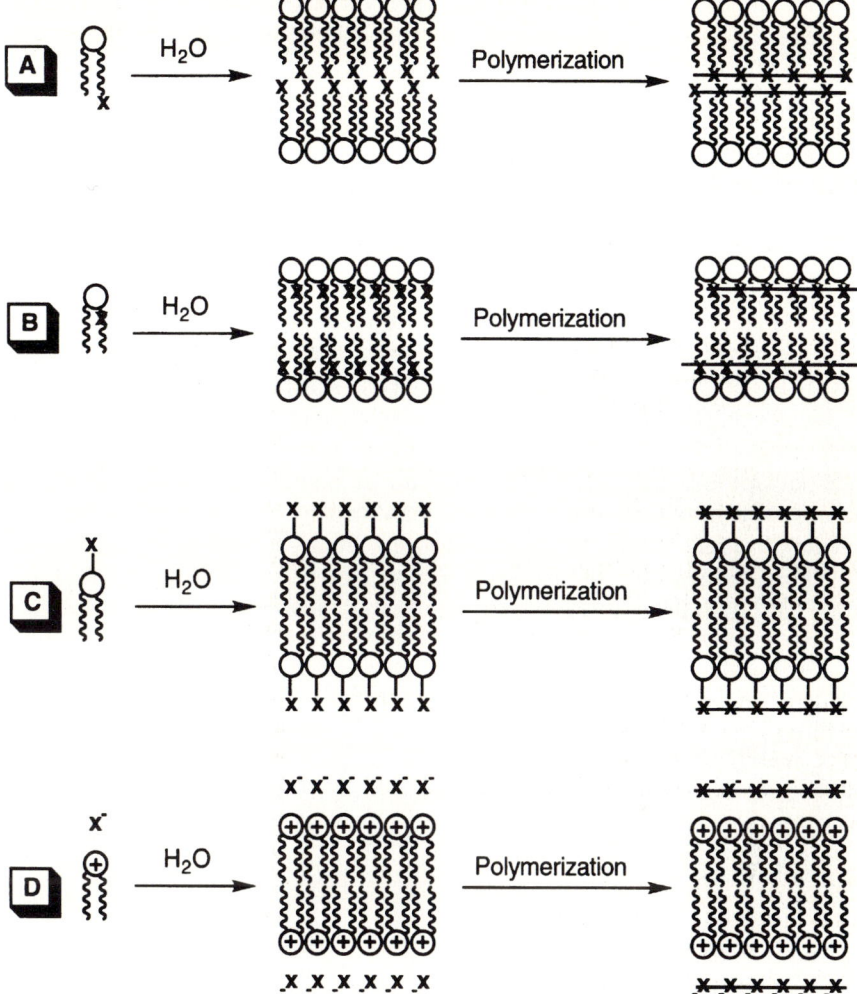

Fig. 1. Lipid substitution patterns for the polymerization of lipid bilayers, featuring polymerization of the lipid tails at (*A*) the chain terminus, (*B*) near the lipid backbone; or polymerization of reactive groups (*C*) covalently or (*D*) electrostatically associated with the hydrophilic headgroup.

polymerization of mono-AcrylPC [16, 17]. The polymer size was not limited by the lipids per vesicle, since the vesicles were composed of upwards of 10^5 lipids. The observed x_n was proportional to the square of the monomer concentration, $[M]^2$, and inversely proportional to $[I]$, the initiator concentration [17]. This dependence indicates the radical polymerization of mono-AcrylPC in bilayers does not terminate by bimolecular chain coupling or disproportionation, but more likely by primary termination. Analysis of the kinetic behavior of mono-AcrylPC polymerizations supports this hypothesis [18]. Similar behavior was found for the radical polymerization of mono-SorbPC in bilayers [19].

The conformational preference of phosholipids, i.e. phosphatidylethanolamines (PE) and PCs, in bilayer membranes has been determined from small angle X-ray diffraction [20] and NMR studies [21, 22]. The conformation shows the glycerol backbone is nearly perpendicular to the bilayer plane as shown in the structures in Fig. 3. Therefore the two equal length fatty acid chains, i.e. *sn*-1 and *sn*-2 positions, penetrate unequally into the lipid bilayer. Lopez et al. pointed out that this conformational preference may result in positional inequivalence of the reactive groups in the *sn*-1 and *sn*-2 chains [23]. To the extent that the reactive groups in the two chains are positionally inequivalent the polymerization of bis-substituted lipids will yield crosslinked polymer networks within the bilayer (Fig. 2) [24]. Presumably there is a separate crosslinked network in each monolayer half of the bilayer. However this hypothesis has not been rigorously proven to date. The crosslinking of lipid bilayers substantially changes their ease of dissolution with detergents or organic solvents. Unpolymerized lipid bilayer vesicles are converted into mixed micelles by treatment with a sufficient quantity of surfactant micelles [25, 26]. This solubilization of bilayer vesicles can readily be observed by changes in turbidity (quantified by light scattering), and the release of vesicle encapsulated molecules, e.g. fluorescent dyes.

The polymerization of bilayers is usually associated with the chemical and physical stabilization of the bilayer membranes. The polymerization of mono-

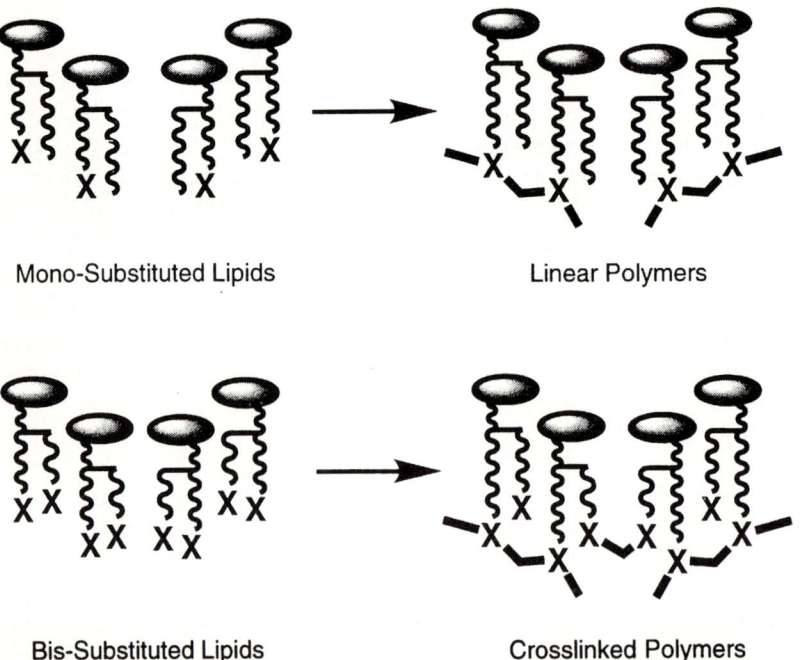

Fig. 2. Schematic representation of the polymers formed by reaction of either mono-substituted or bis-substituted lipids in a lipid bilayer.

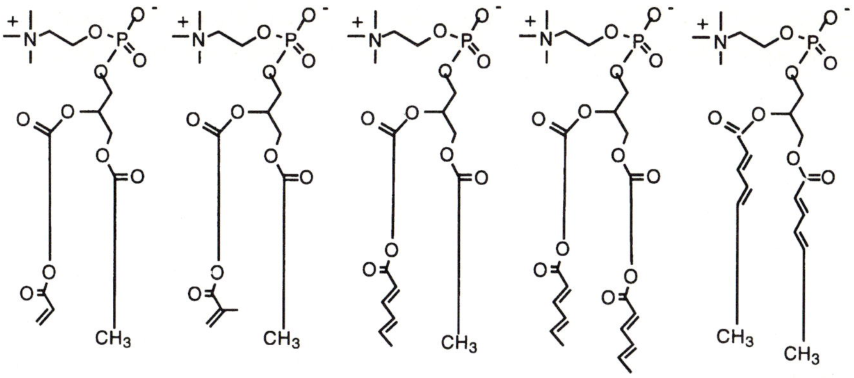

mono-AcrylPC 1 **mono-MethPC** 2 **mono-SorbPC** 3 **bis-SorbPC** 4 **bis-DenPC** 5

Fig. 3. Structures of selected polymerizable phosphatidylcholine (PC) lipids. The fatty acid chain lengths may vary from 12 to over 20 atoms.

substituted lipids in vesicles composed of tens of thousands lipid molecules into a structure that contains several tens of linear polymer chains produces moderate changes in stability. More significant changes are generally found upon the polymerization of bis-substituted lipids. For example Regen et al. [27] reported that the resistance of vesicles to ethanol solubilization increased from unpolymerized vesicles of polymerized vesicles of mono-methacryloylPC (mono-MethPC) (**2**) to polymerized bis-MethPC vesicles. Recently a significant enhancement of vesicle stability was described by Tsuchida et al., who reported that crosslinked vesicles prepared from a bis-DenPC (**5**) were not dissolved by Triton X-100 even at concentrations greater than 3 times the detergent critical micelle concentration (cmc) [28]. Similar results are found for the crosslinking polymerization of bis-SorbPC (**4**) vesicles [29].

The above data suggest that a crosslinked bilayer vesicle is essentially a single polymer molecule (really two, one in each half of the bilayer). In other words the polymerization of the lipid monomers exceeded a gel-point. This concept raises the question of what mole fraction of bis-substituted lipid is necessary to achieve a gel-point for a bilayer composed of a crosslinker lipid, i.e. bis-lipid, and a mono-substituted lipid. Approximately 30% of the lipids in a bilayer vesicle of SorbPCs must be bis-SorbPC (**4**) in order to produce a polymerized vesicle that could not be dissolved by detergent or organic solvent [29]. A complementary study of Kölchens et al. found that the lateral diffusion coefficient, D, of a small nonreactive lipid probe in a polymerized bilayer of mono- and bis-AcrylPC was dramatically reduced when the mole fraction of the bis-AcrylPC, was increased from 0.3 to 0.4 [24]. The decreased freedom of motion of the probe molecule indicates the onset of a crosslinked bilayer in a manner consistent with a 2-dimensional gel-point.

The permeability of solutes across lipid bilayers is a product of the partition coefficient and the transverse diffusion coefficient [30]. Bilayer polymerization can alter solute diffusion by modifying either or both of these processes. In order to examine the effect of polymerization on bilayer permeability a nonionic solute of moderate permeability, [^{3}H-glucose], was encapsulated in the vesicles prior to polymerization, removed from the exterior after polymerization, and its permeation across the bilayer was measured periodically [31]. Quantitative measurements of the ^{3}H-glucose leakage revealed that the formation of linear polymer chains from methacryloyl lipids reduced the permeability coefficient to 0.3 to 0.5 of that of the unpolymerized lipid vesicles. A larger reduction (two orders of magnitude) was only found when crosslinked polymer networks were formed [31].

The crosslinking of bilayers can stabilize vesicles to the point that they maintain their spherical shape even in the absence of water. Figure 4 shows a scanning electron micrograph of a poly(bis-DenPC) vesicle [32]. The polymerization was accomplished by 254 nm light in a manner that is now known to yield oligomers or short polymer chains for mono-substituted dienoyl lipids. The stabilization occurred as a consequence of the crosslinking nature of the polymerization. Thus although the polymerization was limited to the bilayer shell of the vesicle, i.e. a 2-D polymerization, a stable 3-D object was formed.

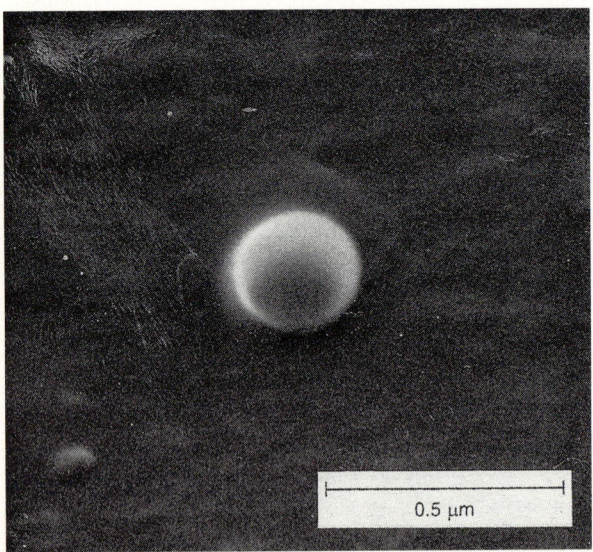

Fig. 4. Scanning electron micrograph of a crosslinked lipid vesicle prepared by photopolymerization of bis-DenPC. The vesicle is ca. 250 nm in diameter.

2 Polymerization of Phase-Separated Lipid Assemblies

2.1 Diacetylenic Lipids

Domain formation in binary mixtures of a polymerizable lipid and non-polymerizable lipid is well established for diacetylenic lipids. The rigid diacetylenic unit facilitates the formation of enriched domains in the condensed phase of monolayers or the solid-analogous phase of bilayers. Since diacetylenes polymerize most readily in solid-like states, most studies have focused on conditions that favor domain formation. Only in the case of a mixture of a charged diacetylenic lipid and a zwitterionic PC was phase separation not observed. Ringsdorf and coworkers first reported the polymerization of a phase-separated two-dimensional assembly in 1981 [33]. Monolayer films were prepared from mixtures consisting of a diacetylenicPC (**6**) (Fig. 5) and a nonpolymerizable distearoyl PE (DSPE).

Two experimental observations led to the conclusion that these lipids were phase-separated in the monolayer condensed state. First, the mean molecular area (A_m) varied linearly with the mole fraction of DSPE in the monolayer. Moreover, the data were fit well with a straight line which connected the A_m values for the individual lipids, indicative of ideal behavior [34]. These data show that the lipids were either completely phase separated or ideally mixed. The

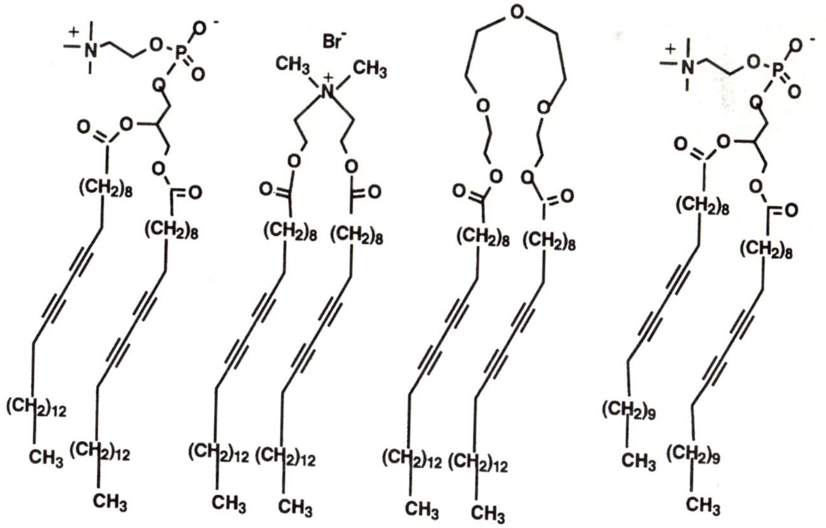

Fig. 5. Structures of polymerizable diacetylenic lipids.

observation of nonlinear deviation from linear behavior would occur only if the system exhibits partial miscibility.

Convincing evidence for phase separation was obtained from the photopolymerization behavior of **6** in the mixed **6**/DSPE monolayer films. Photopolymerization of diacetylenes is a topotactic process which requires the proper alignment of the 1,3-diyne moieties [35]. Thus diacetylenes typically polymerize rapidly in the solid state but not in solution. Polymerization is triggered by ultraviolet irradiation and proceeds via a 1,4-addition mechanism yielding a conjugated ene-yne backbone (Fig. 5). The reaction can be followed by the growth of the visible absorption band of the polymer.

Condensed monolayer films of pure **6** polymerized rapidly, as did mixed **6**/DSPE films of up to 75% DSPE, provided the monolayers were in the condensed state [33]. In the liquid-expanded state, polymerization did not occur. In the condensed state, lateral diffusion of individual lipids within the monolayer is severely restricted compared to the liquid-like state. This precludes initiation of polymerization by diffusive encounter between excited-state and ground-state diacetylene lipids. In order for polymerization to occur in the condensed state, the film must be separated into domains consisting of either "pure" **6** or "pure" DSPE. A demonstration that the rates of photopolymerization for pure **6** and mixed **6**/DSPE monolayers are equal would be a more stringent test for separate domains of the lipids, but no kinetic data have been reported for this system.

The polymerizable lipids used in the above research consisted of naturally occurring head groups and synthetic tails. A subsequent study by the same group utilized totally synthetic lipids as the polymerizable component [36]. Diacetylene lipids **7** and **8** were mixed with the naturally occurring lipids dioleoylPC (DOPC) and distearoylPC (DSPC) in monolayer films and vesicles. Mixtures of **7**/DSPC and **8**/DSPC showed drastically different behavior. In monolayers, plots of A_m versus composition suggested miscibility of **7** but not of **8** with DSPC. Importantly, mixed **7**/DSPC monolayers exhibited a single collapse point whereas **8**/DSPC monolayers featured two distinct collapse pressures similar to those of the pure components. Photopolymerization behavior also supported these conclusions. Films consisting of approximately equal amounts of **7** and DSPC were found to polymerize approximately two orders of magnitude more

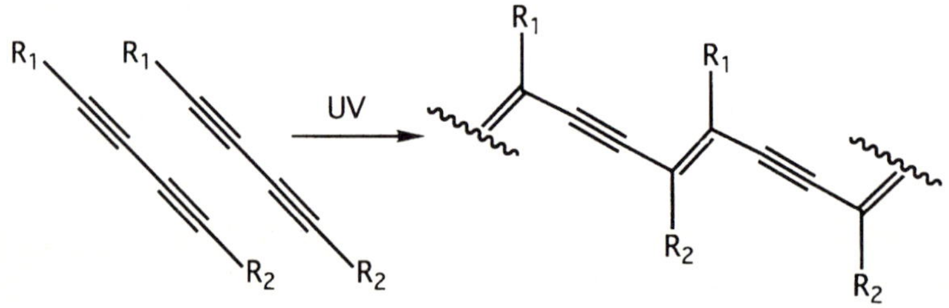

Fig. 6. Photopolymerization of oriented diacetylenes.

slowly than pure **7** films. Mixed **8**/DSPC films polymerized rapidly at all compositions. Similar results were obtained for films in which DOPC replaced DSPC. The **7**/DOPC films exhibited miscibility of the lipids whereas **8**/DOPC films were phase separated. Dependence of these effects on the physical state of the monolayer were not reported.

The different behavior of **7** and **8** is probably due to the charged head group in **7**. Phase separation to form enriched domains of this lipid in mixed monolayers would be inhibited by electrostatic repulsion. Interestingly, monolayer films of **7** mixed with the biologically important molecule cholesterol did exhibit phase separation at all compositions provided the temperature was maintained below the T_m of **7**. Presumably the significantly different shapes of the two molecules promotes the phase separation and overcomes the electrostatic barrier.

Phase separation was also investigated in bilayer systems [36]. Bilayers of **8**/DSPC (1:1) were studied by differential scanning calorimetry (DSC). This technique measures the excess heat capacity of the bilayer system as it passes through a thermotropic phase transition. For the mixed **8**/DSPC vesicles, two transitions were observed in the thermograms corresponding to those of the individual components. Polymerization of the bilayers caused a large decrease in the enthalpy of the transition for **8** but had very little effect on that of DSPC. Mixed **7**/DOPC bilayers were studied in the same manner. Multiple transitions were observed in the DSC thermograms, none of which could be assigned to the individual components, indicative of miscibility. Polymerization of a 1:1 mixture of these two lipids was possible at 0 °C suggesting that some phase separation had occurred, although the corresponding thermogram showed no transitions for the pure components. These results indicate that neither complete mixing nor complete phase separation occurs at 0 °C. Overall, the results from bilayer systems and monolayer films are consistent: in two-dimensions, **7** is miscible with the neutral lipids whereas **8** phase separates from them.

A similar study by O'Brien and coworkers utilized bilayers composed of a shorter chain diacetylenicPC (**9**) and DSPC or DOPC [37]. Phase separation was demonstrated in bilayers by calorimetry and photopolymerization behavior. DSC of the **9**/DSPC (1:1) bilayers exhibited transitions at 40 °C and 55 °C, which were attributed to domains of the individual lipids. Polymerization at 20 °C proceeded at similar rates in the mixed bilayers and pure **9** bilayers. A dramatic hysteresis effect was observed for this system, if the bilayers were first incubated at T > 55 °C then cooled back to 20 °C, the DSC peak for the diacetylenicPC at 40 °C disappeared and the bilayers could no longer be photopolymerized. The phase transition and polymerizability of the vesicles could be restored simply by cooling to ca. 10 °C. A similar hysteretic behavior was also observed for pure diacetylenicPC bilayers. Mixtures of **9** and DOPC exhibited phase transitions for both lipids (T = − 18 °C and 39 °C) plus a small peak at intermediate temperatures. Photopolymerization at 20 °C initially proceeded at a similar rate as observed for pure **9** but slowed after 10% conversion. These results were attributed to the presence of mixed lipid domains as well as enriched domains of each lipid. These mixed domains would give rise to the third DSC peak and also exhibit a much slower rate or photopolymerization than the pure domains. This explanation is reasonable since at 20 °C DOPC is in the liquid-like phase which facilitates mixing with other lipids. The 9/DSPC bilayers provided no evidence of partial miscibility since at 20 °C the DSPC is in the solid-like phase.

Collaborative studies between Sackmann and Ringsdorf employed electron microscopy to visualize the phase separation of polymerizable and nonpolymerizable lipids in giant (>0.5 µm diameter) vesicles [38]. Mixed vesicles composed of a diacetylene lipid and DPPC were used. At 20 °C and below 50 mole % of the diacetylenicPC the vesicles exhibited domain formation. In electron micrographs, the surface texture exhibited corrugated or wavy patches characteristic of the P'_b or "ripple" phase of DPPC and smooth patches assigned to domains of the diacetylenicPC. Above 50 mole % of the diacetylenicPC, phase separation was not observed in the micrographs. Enhanced phase separation in polymerized vesicles was attributed to the sample preparation rather than to polymerization-induced phase separation. In order to observe photopolymerization, the vesicles had to be cooled to 5 °C. After polymerization, the vesicles were warmed back up to 20 °C and then rapidly frozen for preparation of the sample for microscopy. Thus the polymerized vesicles reflect the distribution of lipids in the membrane at 5 °C (to the extent that polymerization "freezes" this distribution) whereas the unpolymerized vesicles reflect the distribution at 20 °C. These correspond to different points in the phase diagram for the lipid mixture, precluding direct comparison between the polymerized and unpolymerized samples.

2.2 Other Polymerizable Lipids

The extensive studies of the behavior of mixed monolayers or bilayers of diacetylenic lipids and other amphiphiles parallel to some degree the studies of dienoyl-substituted amphiphiles. Since the dienoyl lipids do not contain a rigid diacetylenic group in the middle of the hydrophobic chains, they tend to be miscible with other lipids over a wide range of temperatures and compositions. In order to decrease the lipid miscibility of certain dienoyl amphiphiles, Ringsdorf and coworkers utilized the well-known insolubility of hydrocarbons and fluorocarbons. Thus two amphiphiles were prepared, one with hydrocarbon chains and the other with fluorocarbon chains, in order to reduce their ability to mix with one another in the bilayer. Of course it is necessary to demonstrate that the lipids form a mixed lipid bilayer rather than independent structures. Elbert et al. used freeze fracture electron microscopy to demonstrate that a molar mixture of 95% DMPC and 5% of a fluorinated amphiphile formed phase-separated mixed bilayers [39]. Electron micrographs showed extensive regions of the ripple phase (P'_b phase) of the DMPC and occasional smooth patches that were attributed to the fluorinated lipid. In some instances it is possible to observe a mixture of morphologies in the micrographs. Thus an attempt to form vesicles by standard detergent dialysis procedure [40], from a 9:1 molar mixture of a bis-dienoylPC (**5**) and a fluorinated cystine (**10**) produced a population of vesicles and tubular structures observed by electron microscopy (Fig. 7) [32]. Therefore researchers should be aware that under some conditions two immiscible lipids can form independent structures rather than phase-separated bi-

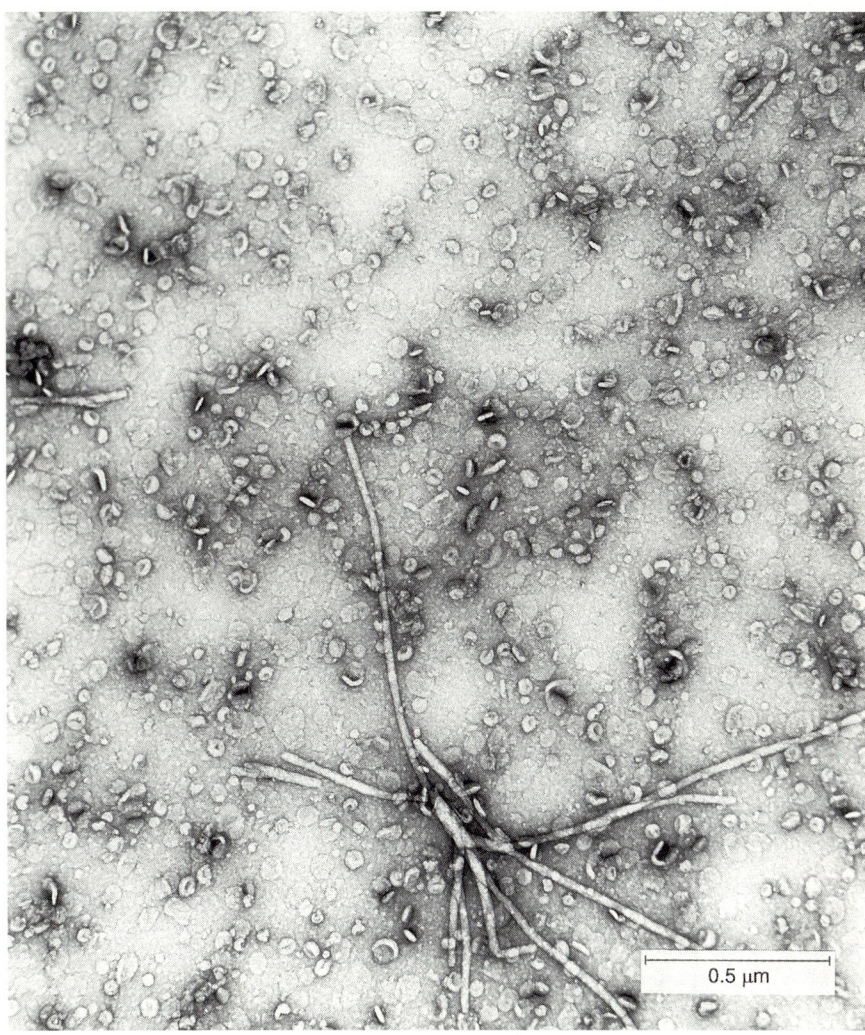

Fig. 7. Transmission electron micrograph of vesicles and tubules formed from a 9:1 molar mixture of a bis-dienoylPC (**5**) and **10**. The sample was stained with 1% uranyl acetate. The tubules are 25 nm in diameter.

layers. These data indicate that more extensive information on the phase diagrams of the desired lipid mixtures is critical to the design and utilization of the types of lipids and phenomena described in this review.

Ringsdorf and coworkers used electron microscopy and vesicle permeability studies to clearly demonstrate the existence of hydrocarbon- and fluorocarbon-rich domains within the bilayers. In one study the polymerizable lipid was a hydrocarbon-based zwitterionic dienoyl lipid (**11**), which was combined with a nonpolymerizable fluorocarbon lipid (**10**) [1]. The dienoyl lipid domains were then photopolymerized in a manner that should result in cross-linking of the hydrocarbon lipids. Following the polymerization the fluorocarbon lipid **10**, a double chain disulfide, derived from oxidized cysteine, was selectively reduced by the addition of dithiothreitol (DTT). This reduction of the disulfide bond produced a single chain thiol which dissolves in water leaving behind pores in the polymerized bilayer wall. Unilamellar vesicles composed of a 9:1 molar mixture of polymerizable **11** and reducable **10** were prepared in the presence of the self-quenched water soluble fluorophore, eosin, which was then removed from the liposome exterior by GPC. Prior to DTT treatment, the liposomes at pH 9.0 effectively encapsulated the eosin. However, after reduction of the disulfide lipid by DTT the integrity of the bilayer barrier was compromised and the eosin was released. The eosin leakage was nearly quantitative in the case of the polymerized vesicles, but somewhat less than complete if the vesicles were nonpolymerized [1]. This difference does not appear to be due to differences in phase separation

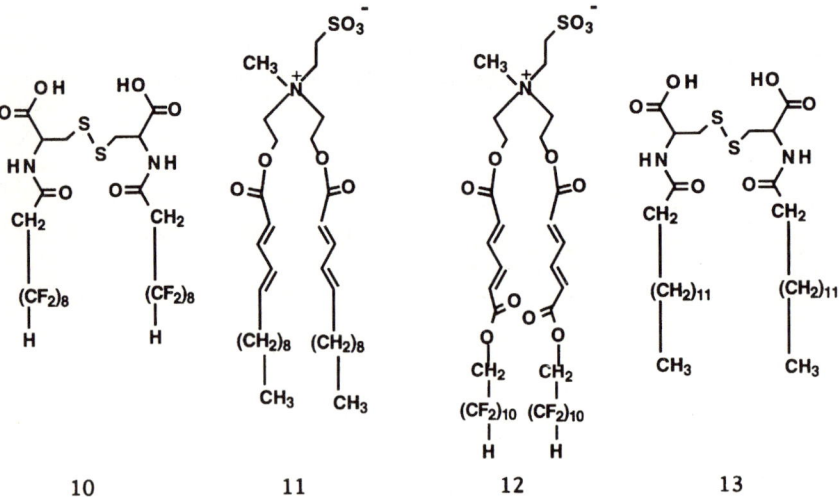

Fig. 8. Two pairs of a polymerizable zwitterionic dienoyl lipid and a cleavable disulfide amphiphile derived from cysteine. In each pair, one amphiphile has a hydrocarbon tail and the other a fluorocarbon tail.

of the polymerized and unpolymerized vesicles, but rather to be a consequence of the lateral diffusion of the monomeric lipids to seal membrane defects. A control experiment with vesicles composed solely of the disulfide lipid showed a more rapid release of eosin upon DTT reduction of the disulfide bond. A similar set of experiments were performed with a partially fluorinated polymerizable muconyl lipid (12) and a hydrocarbon-based cystine (13) [1]. After the fluorinated domains were photopolymerized the vesicles were combined with DTT to reduce the disulfide linkage in 13 and initiate the efficient release of encapsulated eosin.

The reductive cleavage of cystine amphiphiles illustrates the potential utility of phase-separated bilayers for the stimulated release of encapsulated reagents. Ringsdorf coined the phrase "uncorking the liposome (vesicle)" to graphically describe the process [41]. Figure 9 illustrates an idealized end result of such a process. A partially polymerized vesicle with labile domains is uncorked by a chemical or physical process to yield multiple holes in the otherwise stabilized polymeric bilayer. Electron micrographs of such "uncorked liposomes" were obtained using scanning electron microscopy by Folda and Ringsdorf in collaboration with Lando [1]. The review by Ringsdorf et al. shows several of the micrographs obtained for the vesicles composed of 10 and 11 or 12 and 13 [1]. The great interest in the ability to selectively release encapsulated reagents from vesicles led to the design of other methods to trigger the opening of labile domains in an otherwise stabilized vesicle. These include hydrolytic (enzyme-mediated), photochemical, and redox processes. Most of the reported examples of these methods involve polymerization-induced domain formation and are therefore described in Sect. 3.

Ringsdorf, Sackmann, and coworkers characterized the behavior of mixtures of the polymerizable bis-dienoylammonium lipid 14 and DMPC [42]. Evidence for phase separation in these mixtures was obtained from electron microscopy and light scattering. Since the intensity of scattered light is dependent on the physical state of the membrane, plots of scattering intensity versus temperature exhibit inflections at phase transitions. This technique was used in conjunction

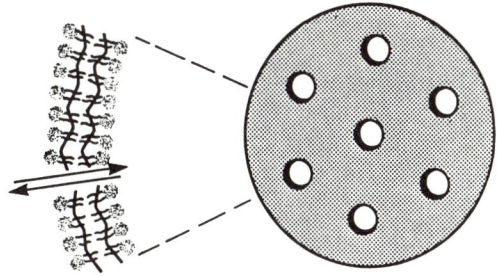

Fig. 9. Schematic representation of a polymerized phase-separated vesicle, i.e. a molecular whiffle ball. The holes are formed by removal of the nonpolymerized lipid domains by the procedures described in the text.

with direct observation of domain formation by electron microscopy to generate a phase diagram for the **14**/DMPC system. The diagram featured a miscibility gap extending over the approximate ranges of 10–22 °C and 40–90 mole % DMPC. Electron microscopy of samples prepared in this region of the phase diagram exhibited large patches of rippled and smooth surface textures. The rippled domains were found to increase in size with increasing DMPC content. The rippled domains were assigned to the P'_b phase of DMPC, the smooth patches to **14**.

Polymerization of lipid **14** was achieved by photoirradiation with UV light. Polymerization of bilayer vesicles composed of 40% **14** and 60% DMPC at 17 °C (a region of the phase diagram associated with phase separation of the two lipids) yielded bilayers which exhibited smooth domains as well as two distinct rippled domains in electron microscopy. The rippled domains could be distinguished by their different periodicities. One of these periodicities was the same as observed in the unpolymerized vesicles. The other rippled domain was attributed to new DMPC domains formed during the polymerization. This interpretation implies that the polymerization induces a reorganization of lateral distribution of lipids in the membrane. Thus this polymerization reaction cannot necessarily be thought of as locking-in or fixing a given lateral distribution of lipids in the bilayer. In fact in some instances it can be used to trigger a reorganization of the original distribution. The phase diagram of **14** and DMPC indicates a composition-dependent fluid (miscible) phase at temperatures above 28 to 34 °C. Polymerization of a 1:1 molar mixture of the lipids in the fluid phase led to the formation of a DMPC rich phase and polymeric lipid domains, that could be observed by freeze fracture electron microscopy [43]. The same type of electron microscopy images were observed whether the polymerization was performed at temperatures that the lipids were miscible or partially immiscible.

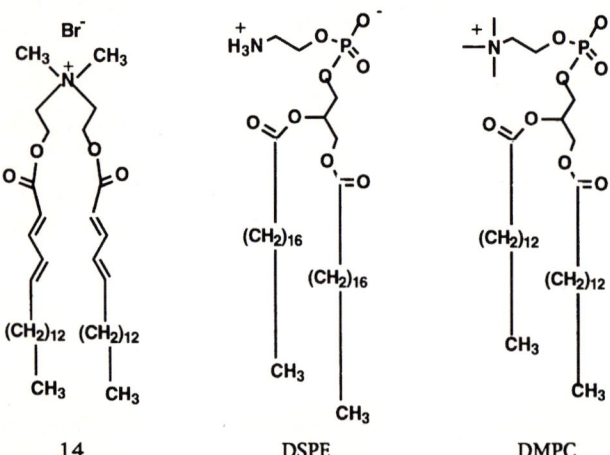

Fig. 10. Structures of polymerizable dienoyl ammonium lipid and phospholipids.

The authors surmised that the crosslinking polymerization of **14** resulted in the formation of only small polymers. They estimated an aggregation number, $N \leq 10$, for the domains of crosslinked **14** [43]. The careful analysis of the phase behavior of these lipids and the effect of polymerization at selected conditions coincided with the realization that polymerization of a lipid that is miscible with a second nonreactive lipid could cause phase separation of the lipids into enriched domains. This polymerization-induced lipid domain formation is considered more completely in the following section of this review.

3 Polymerization-Induced Domain Formation

The polymerization of reactive lipids in two or multi-component membranes can cause the separation of the polymerized lipid and the colipid(s) into enriched domains. Several lines of evidence for polymerization induced lipid domain formation have been reported. Dorn et al. [31] found that mono-methacryloyl dialkylammonium lipids form vesicles that encapsulate hydrophilic molecules. The thermally initiated polymerization of vesicles of these lipids did not alter the internal volume and external diameter even though the polymerization gave linear polymer chains with a degree of polymerization of ca. 500 [44]. DSC of the lipid membranes at intermediate stages of the conversion to polymer showed two endothermic transitions. One occurred at the same temperature observed for membranes composed of the monomer and was ascribed to domains of the monomeric lipid. The polymerization induced transition occurred at a lower temperature and increased in proportion to the percent conversion [41]. These early results show that a linear polymerization of bilayers of a single lipid can proceed with the formation of two phases, one highly enriched in monomer and the other enriched in polymer. Other studies described below have utilized two-component lipid membranes and examined the effect of polymerization of a reactive lipid on the lateral distribution of both lipids. Several factors may be important for the phase separation process including: the nature and size of the polymer chains, the relative location of the polymer chains and the bilayer, the composition of the lipid hydrophilic head group (charged or neutral) and the lipid hydrophobic tails. Most of these variables have not been systematically characterized. Therefore the studies to date are grouped into those that examined (1) bilayers composed of two neutral lipids (neutral-neutral) and (2) bilayers composed of a charged lipid and a neutral lipid (charged-neutral), e.g. bilayers of the cationic lipid **14** and DMPC [43].

3.1 Polymeriztion of Bilayers of Neutral Lipids

Tsuchida and coworkers reported the polymerization behavior of lipid membranes composed of the bis-DenPC (**15**) and dipalmitolPC (DPPC) [45, 46].

Since the fatty acid chains in each lipid were 18 carbons and 16 carbons, respectively, it is reasonable that they could form a mixed lipid phase. Furthermore the bis-dienoyl substitution of **15** favors the formation of crosslinked polymer networks. Ohno et al. showed that the dienoyl group associated with the *sn*-1 chain could be polymerized by lipid soluble initiators, e.g. AIBN, whereas the dienoyl in the *sn*-2 chain was unaffected by AIBN generated radicals. Conversely, radicals from a water-soluble initiator, e.g. azo-bis(2-amidinopropane) dihydrochloride (AAPD), caused the polymerization of the *sn*-2 chain dienoyl group, but not the *sn*-1 chain. These data provide clear evidence for the hypothesis of Lopez et al. that the same reactive group located in similar positions in the *sn*-1 and *sn*-2 chains of polymerizable 1,2-diacyl phospholipids are positionally inequivalent [23].

Vesicles of **15**/DPPC (1:1) were prepared by ultrasonication. The vesicular nature of the resulting structures was shown by the effect of water-soluble NMR shift reagents on the proton spectra. DSC and fluorescence depolarization experiments provided evidence of some phase separation of the lipids, because two-phase transitions attributable to domains of the individual lipids were observed by both techniques. An intermediate transition, which indicates the presence of a mixed lipid phase before the photoreaction, was also observed. The photopolymerization of the diene group was followed by the disappearance of the 255 nm absorption maxima, due to the dienoyl chromophore. The sample irradiations were performed either at room temperature, well below the phase transition for DPPC ($T_m = 41\,°C$) but close to that for **15** ($T_m = 18\,°C$); or at 50 °C which is well above the main phase transitions. The lipids were considered to be homogeneously mixed at the higher temperature, and partially phase-separated at room temperature. Polymerization caused the intermediate phase transition assigned to the mixed lipid phase to disappear. The photoreaction preserved the transitions due to DPPC as well as a transition near the temperature associated with the T_m for **15**, albeit with lower enthalpy. This lower temperature phase transition was assigned to poly**15** on the following basis. A lipid chain melting transition was expected to be retained in poly**15**, because the polymerization occurs fairly close to the glycerol backbone. The motion of the chain termini should be somewhat decoupled from the polymer backbone permitting the phase transition to persist. Ringsdorf and coworkers reported similar phenomena for the polymerization of lipid vesicles with reactive groups in the head groups (see Fig. 1C) [47]. Further evidence for polymerization enhancement of phase separation was obtained by freeze fracture electron microscopy of the polymerized vesicles of **15**/DPPC, which revealed domains of the P'_b (ripple) phase of DPPC and patches of smooth surfaces. The latter were assigned to domains of poly**15**. Similar micrographic images were obtained whether the polymerization was performed on a partially phase-separated sample at room temperature or a homogeneous lipid sample at 50 °C, indicating that a crosslinking polymerization of neutral lipids can cause its phase separation from another neutral lipid.

The formation of domains in two-component vesicles was dramatically demonstrated by electron microscopy of partially solubilized vesicles. Ohno et al. coined the term "skeletonized" vesicles, i.e. porous vesicles, for the resulting supramolecular structure [48]. They utilized mixtures of cholesterol and 1-(9-styrylnonanoyl)-2-octadecyl-*rac*-glycero-3-PC to form giant vesicles by slow hydration with nitrogen-outgassed water. The vesicles were examined by optical microscopy and ranged in size from ca. 1 to 100 µm. After the styryl-lipid was photopolymerized by UV light exposure for 4 h, the polymerized vesicles were washed with chloroform to remove the cholesterol and residual unpolymerized lipid. Subsequent scanning electron microscopy revealed images of spherical structures containing many circular pores that varied in size from 50 to 500 nm diameter. The initial interpretation of the observed results suggested that the polymerizable lipid and the cholesterol were separated into enriched domains after the polymerization. The possible existence of domains prior to polymerization was not discussed in the report. The polymerized lipid was presumably resistant to chloroform solubilization. The apparent lack of solubility of the linear polystyrene lipid is most likely due to the zwitteronic head group of the lipid monomers and/or the length of the polymer chains.

The formation of skeletonized vesicles was also reported for vesicles composed of **15** and DPPC, where the mole fraction of DPPC varied from 5 to 25 mole percent. Takeoka et al. analyzed the release of entrapped water soluble molecules in order to assess the size of the pores formed in the vesicle wall [49]. The ease of release of saccharides, primarily dextrans, of various molecular

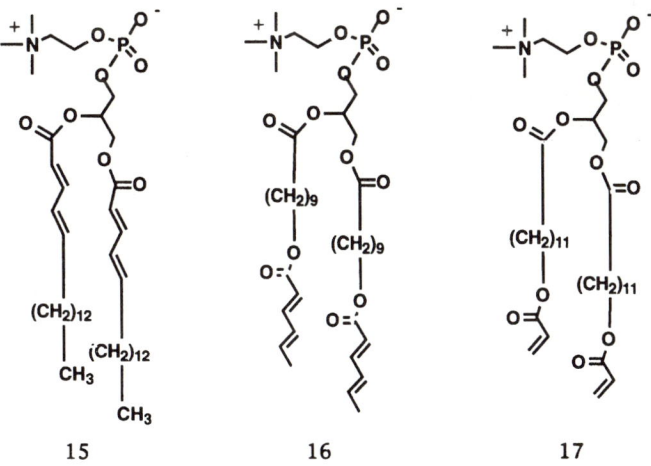

Fig. 11. Structures of crosslinkable phosphatidylcholines which have been usefully employed for polymerization-induced domain formation: left to right the compounds are bis-DenPC$_{18,18}$; bis-SorbPC$_{17,17}$ and bis-AcrylPC$_{16,16}$, where the subscripts indicate the number of atoms in the fatty acid chain.

weights and radius of gyration was determined for the skeletonized vesicles. Thus vesicles of a particular composition were polymerized and the nonpolymerized domains were dissolved with sodium dodecylsulfate. If the entrapped saccharide was not released, then the largest pore sizes were too small to permit facile release of the marker. Larger saccharides were released at greater mole fractions of DPPC in the vesicle, thereby suggesting that increasing the fraction of non-polymerizable lipid results in the formation of larger membrane domains of the DPPC. Based on this analysis the authors were able to estimate the diameter of the largest pores to be 3 nm or 4 nm, when the mole fraction of DPPC was 0.10 or 0.20, respectively. An initial analysis of the effect of the polymerization conditions on the DPPC domain size showed that the temperature of the polymerization relative to the phase transitions of the lipid membrane can influence the ease of release of the aqueous markers. Thus polymerization in the L_α phase appeared to yield smaller lipid domains than when the polymerization was performed at lower temperatures where the two lipids were partially phase-separated prior to polymerization.

Phospholipase A_2 hydrolysis of 1,2-diacyl-L-α-glycerophospholipids occurs specifically at the *sn*-2 chain ester bond to produce the corresponding lysophospholipid and the fatty acid [50]. Büschl observed that the permeability of vesicles composed of polymerizable and natural lipids could be dramatically altered by phospholipase A_2 hydrolysis of the lipids [1, 51]. The studies primarily used bis-DenPC (**15**) and DPPC, the same lipids employed by Tsuchida and coworkers. Büschl performed the experiments at 30 °C and assumed the lipids were homogeneously mixed prior to polymerization. The characterization data from Tsuchida at al. indicate that these lipids were probably partially phase-separated before polymerization at this temperature [52]. Prior to reaction with phospholipase A_2, a population of vesicles composed of a 1:1 molar mixture of **15** and DPPC were prepared in the presence of carboxyfluorescein (CF) and the non-encapsulated CF was removed by gel chromatography or dialysis. The vesicles with encapsulated CF could then be polymerized by photoreaction of **15** via UV exposure for different periods of time. This experimental protocol allowed an indirect examination of the enzymatic hydrolysis of vesicles, by a direct determination of the kinetics of CF release. The release or leakage of self-quenched CF from vesicles can be readily detected by an increase in emission intensity as the CF is diluted upon escaping the vesicles. Little if any release was observed prior to phospholipase hydrolysis of the lipid bilayer vesicles. Interestingly the kinetics of CF release from the vesicles after the addition of phospholipse A_2 depended on the estimated extent of vesicle polymerization. In the case of the unpolymerized vesicles the CF was released slowly as the phospholipase hydrolyzes the lipids. In contrast, as shown in Fig. 12, the enzyme mediated leakage of polymerized vesicles occurred much sooner. The authors proposed that the hydrolysis products could readily dissolve in the unpolymerized **15**/DPPC bilayers and thereby maintain bilayer integrity. However after polymerization of **15** and the more complete formation of domains of poly**15** and DPPC the vesicle had less flexibility and capability to accommodate the

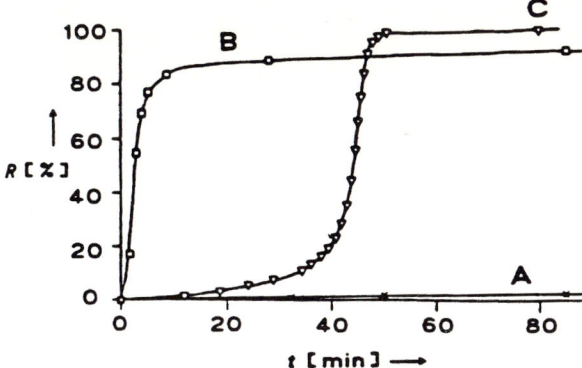

Fig. 12. Phospholipase A_2 mediated release of self-quenched carboxyfluorescein from unpolymerized and polymerized mixed vesicles of **15** and DPPC (1:1) at 30 °C. (*A*) Monomeric and polymerized vesicles in the absence of enzyme. (*B*) Polymerized vesicles after the addition of phospholipase A_2 (30 µg/ml). (*C*) Unpolymerized vesicles after the addition of phospholipase A_2 (30 µg/ml).

hydrolysis products. Therefore the polymerized vesicles reached the point of rapid CF release almost immediately.

A detailed analysis of the effect of mixed monolayers of **15** and DMPC on the activity of phospholipase A_2 was reported by Grainger et al. [53]. Monolayers composed of different ratios of DMPC and either **15** or primarily poly**15** were characterized by Langmuir isotherms and isobars. The phospholipse-A_2-mediated hydrolysis of selected monolayer compositions was usefully employed to ascertain the effectiveness of the enzyme. Both **15** and poly**15** were resistant to hydrolysis. The DMPC hydrolysis was sensitive to its molecular environment in a manner that suggests the phase separation of the poly**15** from DMPC. Phospholipase A_2 activity is known to be sensitive to the concentration of the hydrolytic products, i.e. the fatty acid and lysophospholipid. The effect of these reaction products of the activity of phospholipase A_2 on mixed monolayers of nonpolymerizable lipids is the subject of a series of interesting studies which are beyond the scope of this review. Ahlers et al. reviewed some of this research [54].

In 1985 Tyminski et al. [55, 56] reported that two-component lipid vesicles of a neutral phospholipid, e.g. DOPC, and a neutral polymerizable PC, bis-DenPC (**15**), formed stable homogeneous bilayer vesicles prior to photopolymerization. After photopolymerization of a homogeneous 1:1 molar lipid mixture, the lipid vesicles were titrated with bovine rhodopsin-octyl glucoside micelles in a manner that maintained the octyl glucoside concentration below the surfactant critical micelle concentration. Consequently there was insufficient surfactant to keep the membrane protein, rhodopsin, soluble in the aqueous buffer. These conditions favor the insertion of transmembrane proteins into lipid bilayers. After addition and incubation, the bilayer vesicles were purified on a sucrose gradient by ultracentrifugation, which gave a major lipid/rhodopsin band and a smaller band at a greater lipid/rhodopsin ratio. Analysis of the phosphorus content and rhodopsin absorption intensity of the two purified membrane populations showed that the major and minor bands had lipid-to-rhodopsin ratios of 100 and 390, respectively. Very little if any aggregated rhodopsin was observed at the bottom of the sucrose gradient. Therefore almost all of the rhodopsin was successfully inserted into the vesicles of poly15 and DOPC. A control experiment showed that rhodopsin could not be readily inserted into vesicles composed of only the photopolymerized 15. Thus rhodopsin incorporation into the poly15/DOPC bilayer vesicles appeared to occur preferentially at enriched domains of the DOPC. A large fraction of the rhodopsin inserted into these vesicles was both photochemically and enzymatically functional. Since rhodopsin functionality requires unsaturated phospholipids, the observed functionality provides further evidence that the rhodopsins were inserted into domains of the nonpolymerizable unsaturated PC. The enzymatic functionality of rhodopsin in these partially polymerized bilayer membranes was shown by combining the other purified components of the G-protein coupled enzymatic cascade with the rhodopsin-poly15/DOPC vesicles. These included the rod-outer-segment G protein and phosphodiesterase (PDE), both of which are peripheral proteins that associate with the lipid bilayer surface [57]. The efficiency of binding of these enzymes to poly15 bilayers was comparable to that of nonpolymerizable PC lipids, e.g. DOPC [58]. Thus the biocompatibility of the phosphatidylcholine surface was maintained even after the polymerization of the bis-DenPC.

Polymerization induced domain formation via the separation of neutral phosphatidylcholine lipids into poly-PC domains and unreactive PC domains was successfully demonstrated by detergent lysis or enzymatic hydrolysis of the nonpolymeric domains. These studies showed that polymerization produced a bilayer vesicle that was stable until further reaction, i.e. the polymerized vesicle exhibited a latent instability. O'Brien and coworkers reasoned that the polymerization of vesicles composed in part of polymorphic unreactive lipids could immediately destabilize the bilayer vesicle [59, 60]. This premise was based on the accumulated evidence that lipids such as phosphatidylethanolamine (PE) can form a variety of supramolecular structures. Processes that form enriched domains of PE in bilayers result in destabilization of the bilayer (lamellar) structure with the formation of precursors to nonlamellar phases. The effect of photopolymerization on bilayer stability was experimentally examined with vesicles composed of the photosensitive bis-SorbPC (16) in combination with either dioleolylPE (DOPE), or DOPC. The bis-SorbPC had the same λ_{max} at 258 nm whether it was in organic solvents or hydrated bilayers, which indicates that the chromophores do not strongly interact in the bilayer interior. Photolysis of the vesicles with 254 nm light diminished the monomer absorption in parallel with polymer formation. The photoactivated polymerization of 16 gave a crosslinked polymer, which was ascribed to the unequal depth of penetration of the reactive α and β chains [23, 61]. The polymer structure of the corresponding

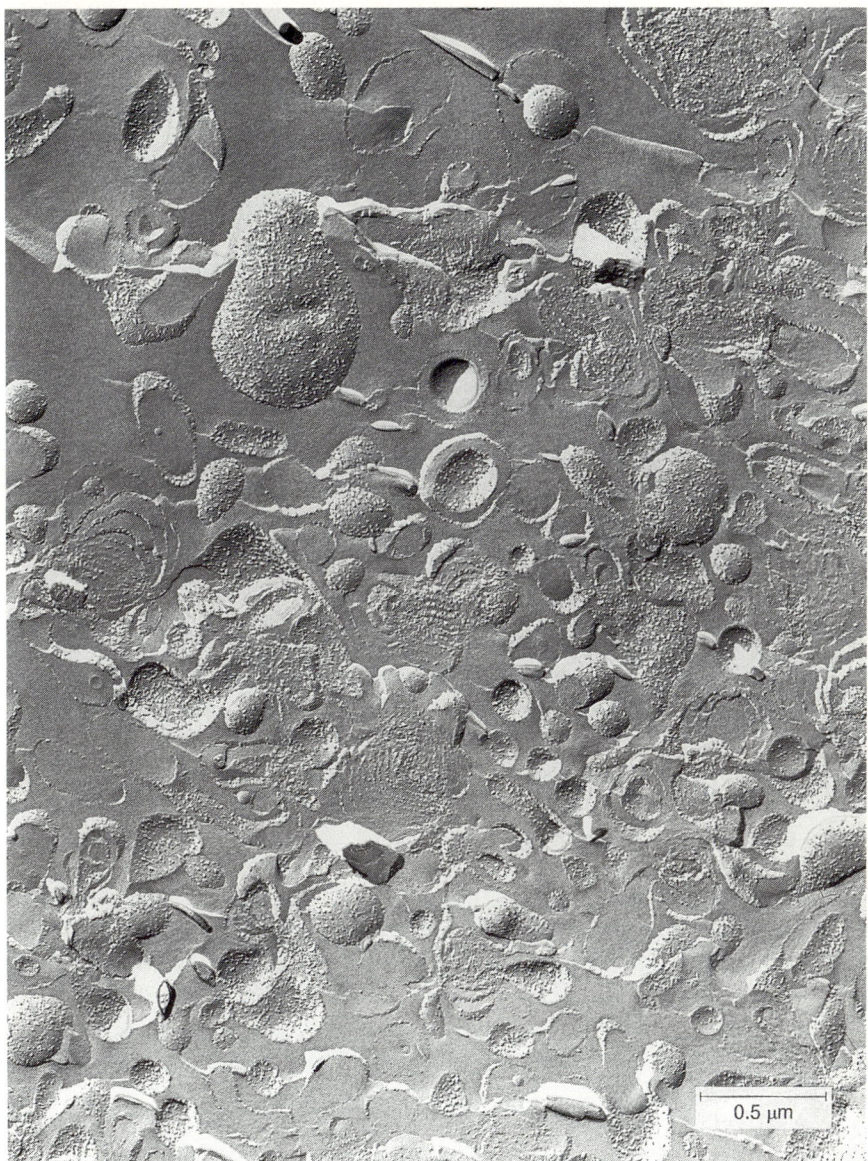

Fig. 13. Freeze-fracture replicas of rhodopsin-poly15/DOPC membranes in 30% glycerol/water frozen from room temperature. The particles are morphological manifestations of the protein rhodopsin. The nonrandom distribution of particles indicates the presence of enriched domains of lipid and of protein. The particle-free domains constitute about 30% of the surface area.

mono-SorbPC, which forms linear polymer chains, was shown to be a 1,4-polymer with a degree of polymerization of ca. 10 [19].

Vesicles composed of DOPE/**16**, or DOPC/**16** in either a 2/1 or 3/1 molar ratio were prepared by a variety of methods, including mild sonication of hydrated bilayers, reverse phase evaporation (REV) [62], and freeze-thaw extrusion [63, 64], with encapsulated calcein, a water-soluble fluorescent marker [60]. The stability of the vesicles was ascertained by determining the fraction of the calcein released upon photolysis of the vesicles. Calcein was released from the DOPE/**16** vesicles after ca. 40% loss of monomeric **16** (1 to 2 min of exposure to 254 nm light with an intensity of $\sim 5 \times 10^{14}$ photons/s). The DOPC/**16** vesicles were stable throughout the photolysis, even though nearly 85% of **16** was polymerized. Surprisingly at the vesicle concentration employed only oligolamellar or multilamellar DOPE/**16** vesicles were efficiently destabilized by the photopolymerization. Bilayer contact of PE-rich lamellae of separate vesicles has previously been shown to be a prerequisite for vesicle fusion and bilayer destabilization [65–67]. Bilayer contact within a vesicle (intravesicular) is possible between lamellae only if the vesicle consists of more than one bilayer, i.e. the vesicle is oligo- or multilamellar. Intervesicular bilayer contact occurs between the exterior monolayers if two or more vesicles aggregate.

The photopolymerization of phospholipid/**16** vesicles changes the monomer composition of the bilayer and the lateral distribution of the lipids. In principle this phase separation might be sufficient to cause the observed dye leakage, if the leakage occurs at the boundaries of the poly**16** and nonpolymerized lamellar lipid domains. However, since the photoinduced phase separation of the DOPC/poly**16** membranes did not release the entraped dye, the phase separation alone was insufficient to cause an increase in membrane permeability. Phase separation in DOPE/**16** bilayer vesicles creates domains enriched in DOPE. Bilayers of randomly distributed PE/PC lipids are repelled from neighboring bilayers by strong hydration forces [68]. Processes which promote phase separation and cause the enrichment of domains of the poorly hydrated PE decrease the repulsive forces in these regions and facilitate bilayer contact. Contact between apposing bilayers enriched in PE in oligolamllellar vesicles can lead to the formation of interlamellar attachment(s) (ILA). It has been proposed that an ILA is a hourglass-shaped bilayer attachment between two original lamellae that effectively fuses these bilayers [69–71]. ILAs exhibit isotropic ^{31}P NMR signals since lipid molecules diffusing along the surfaces of the ILA assume all headgroup orientations [72]. Under appropriate circumstances ILAs can proceed to form long-lived arrays which take on the characteristics of the inverted cubic phase.

The information provided by the ^{31}P-NMR spectra of DOPE/**16** as a function of temperature and the extent of polymerization was critical to the characterization of the nature of the lipid structures responsible for the destabilization of the photolyzed DOPE/**16** vesicles [73]. The progressive appearance of an isotropic NMR signal at the expense of the lamellar signal (Fig. 14) indicated that a lipid structure with isotropic symmetry was associated with the photoinduced leakiness of DOPE/**16** vesicles. The enriched domains of PE facilitates the interaction and formation of intermediate lipid structures between bilayers, with the eventual development of an ILA(s) that connect the bilayers of an olgiolamellar vesicle and provides an aqueous pathway for the leakage of water-soluble encapsulated reagents. ILAs have been proposed as a critical intermediates in the fusion of bilayer vesicles with one another as well as precursors to the formation of nonlamellar phases, e.g. inverted cubic structures. The bilayer reorganization suggests that the photopolymerization of DOPE/**16** unilamellar vesicles (LUV) could usefully initiate vesicle-vesicle fusion.

The rate and extent of vesicle fusion is a function of several factors including temperature, lipid composition, and pH. Bennett and O'Brien (1995) demonstrated that the photopolymerization of **16** in DOPE/**16** vesicles (100 nm diameter LUV) significantly reduced the critical fusion temperature, i.e. the temperature threshold for the onset of rapid fusion [74]. They utilized fluorescent probes to study the interaction of photolyzed LUV of DOPE/**16** with either photolyzed or dark (unpolymerized) DOPE/**16** LUV. Both lipid mixing and aqueous contents mixing between the donor and target LUV was enhanced by the photoactivated polymerization of **16**. Fusion of vesicles has been described by a mass action kinetic model in which the first step is aggregation of two stable vesicles to form a dimer, and the second step is the actual fusion process which produces the fusion product [75]. The latter step includes destabilization of the two vesicles and their communion via the fusion process. The kinetics of the overall fusion process depends on the rates of both aggregation and the fusion event. Polymerization of DOPE/**16** bilayer vesicles is proposed to induce fusion by facilitating both aggregation and subsequent fusion. A model for homo-fusion between (3:1) DOPE/**16** vesicles following polymerization is shown in Fig. 15. In the figure the DOPE lipid headgroups are unfilled and bis-SorbPC headgroups are filled. The vesicles were photopolymerized at pH 9.5 where ca. half of the PE component is negatively charged and where the SorbPC lipids are zwitterionic. The vesicle surface charge

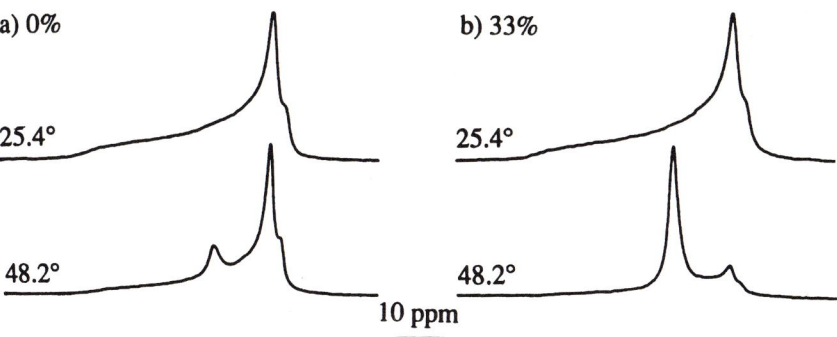

Fig. 14. Representative proton-decoupled ^{31}P NMR spectra of multilamellar dispersions of (3:1) DOPE/**16** in excess water at either 25.4 °C or 48.2 °C. The spectra on the left (*a*) are from an unpolymerized sample, and the spectra on the right (*b*) are for a sample where lipid **16** has been 33% photopolymerized.

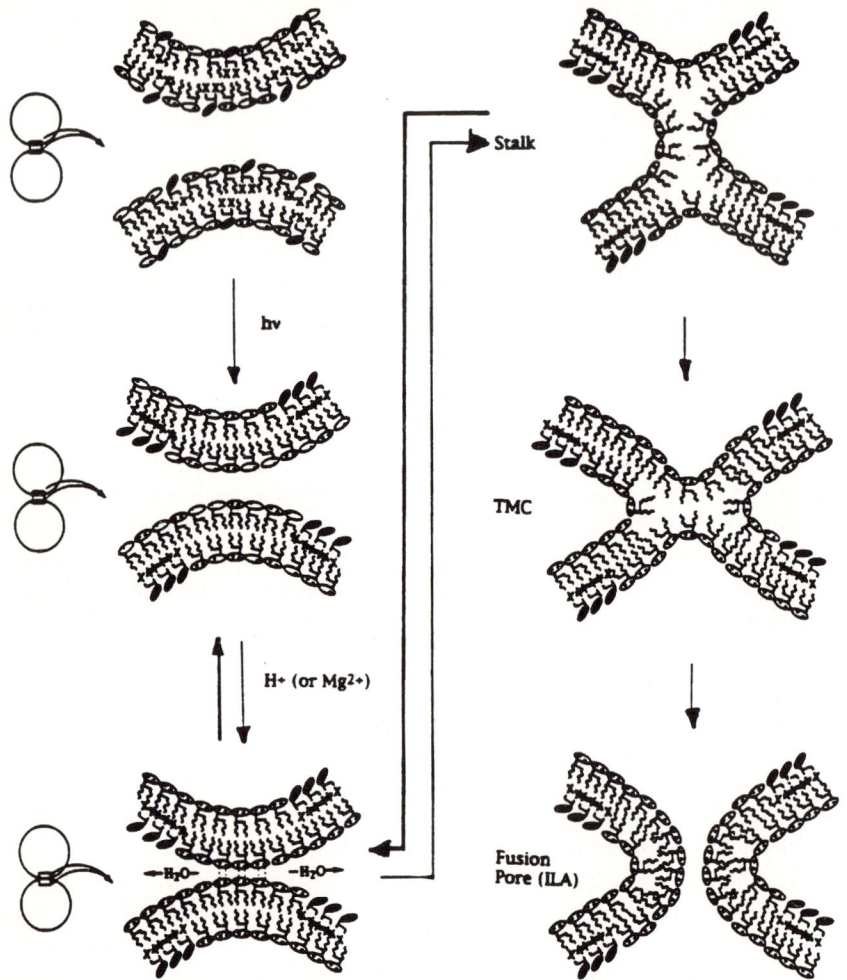

Fig. 15. A schematic representation for the polymerization-induced phase separation of DOPE and poly16 leading to fusion of vesicles due to bilayer contact of the enriched PE domains. See text for description.

inhibited vesicle interaction and allowed the experimentalists to separate in time the photopolymerization of vesicles from their fusion. Following polymerization the vesicles were neutralized by addition of protons or magnesium ions. Adhesion of the neutralized bilayers initiates the sequential formation of previously proposed intermediates termed a stalk, a TMC, and finally an ILA as shown in the figure. Polymerization induces lipid phase separation with consequent decreased hydration of PE-rich domains, and decreased R_0 (intrinsic radius of curvature) [76, 77] of the lipids composing the PE-rich domains. Since the domains of monomeric lipids contain only small fractions of PC, the surface hydration of these regions is diminished. The decreased R_0 of the PE domains is a consequence of both the increased ratio of DOPE to monomeric-PC and the decreased membrane hydration. Attractive hydrophobic forces promote adhesion of the membranes at these sites. After adhesion, formation of fusion intermediates is facilitated by the decreased R_0 of the lipids composing the apposed leaflets of the membranes.

The photoactivated polymerization of vesicles significantly enhanced the fusion of LUV. The results complement the previous demonstration of photo-induced destabilization of oligolamellar and multilamellar vesicles (MLV). Both methods rely on the temporal and spatial characteristics of light to deliver reagents from vesicles to other bilayer-bounded structures in the case of LUV or release of reagents to the aqueous media surrounding MLV. In each case the polymerization-induced phase separation of lipids in a multicomponent bilayer triggers the subsequent reorganization of the lipids to compromise the bilayer boundary. Such processes hold out the promise of new approaches to the delivery of therapeutic agents, if vesicle fusion with cellular bilayers can be enhanced by similar strategies. Studies of these phenomena are currently in progress.

3.2 Polymerization of Bilayers of Neutral and Charged Lipids

Chain polymerizations of homogeneous mixtures of polymerizable and non-polymerizable lipids can produce lipid domains when both lipids are neutral, or even if one of the lipids is charged. The investigation of Gaub et al. [43] into the characteristics of lipid mixtures of **14** and DMPC indicates that polymerization of a 1:1 molar mixture of these lipids in the fluid phase caused the formation of a DMPC rich phase and polymeric lipid domains. These experiments showed that the polymerization of charged lipids could occur despite the electrostatic repulsion of the monomers of **14**. Eggl et al. analyzed the linear polymerization of a charged lipid with a polymerizable head group, **18**, in a binary mixture with the zwitterionic DMPC [78]. A phase diagram for this pair of hydrated lipids, constructed by static light scattering of large vesicles, showed nearly ideal behavior for the monomeric mixture. Radical polymerization of **18**, utilizing the water-soluble ACVA, produced linear polymers with an estimated degree of polymerization of 300. The phase diagram of poly**18** and DMPC suggested the formation of a semidilute 2-D macromolecular solution. Diffusion measurements of the poly**18** indicated the mixture could be represented as densely coiled poly**18** surrounded by DMPC. Thus the formation of relatively large linear polymers from ionized monomers can lead to lipid domain formation.

18

Recently the experiments of Armitage et al. demonstrated that polymerization of neutral lipids could serve to concentrate nonreactive charged colipids into domains [79]. Recall that lipid polymerization in the L_α phase of the bilayer proceeds in the presence of unreactive lipids because of the rapid lateral diffusion of the monomeric lipids. The high lateral mobility of lipid monomers permits their diffusion to the growing polymer chain-ends even if the bilayers are composed in large part of nonreactive lipids. The polymerization-induced formation of enriched domains of ionic lipids was demonstrated by interactions of molecules at the bilayer surface that are electrostatically associated with the nonpolymerizable ionic lipids. Armitage et al. examined the effect of polymerization of vesicles composed of a neutral polymerizable PC, bis-AcrylPC (**17**), and the nonreactive anionic 1,2-dioleoyl phosphatidic acid (DOPA), which can bind cationic dyes to the bilayer surface [79]. The efficiency of electronic energy transfer between a tetra(trimethylammoniumphenyl)porphyrin (TAPP) donor (λ_{max}^{abs} = 417 nm, λ_{max}^{em} = 650 nm) and a tricationic thiadicarbocyanine dye (Cy^{3+}) acceptor (λ_{max}^{abs} = 669 nm, λ_{max}^{em} = 696 nm) is sensitive to the average distance of separation between the dyes. Electronic energy transfer, from the blue light excited TAPP to the Cy^{3+}, quenched the porphyrin emission at 650 nm with a concomitant increase in the cyanine emission at 696 nm. The vesicles (100 nm diameter) were composed of a 9:1 molar ratio of **17** and DOPA, i.e. 800 DOPA ionized molecules/vesicle under the experimental conditions. At the TAPP concentration employed there were (on average) 10 porphyrins per vesicle, and the cyanine dye was added in increments of 20 Cy^{3+} per vesicle until the ratio of TAPP to Cy^{3+} was 1:20. Steady state energy transfer was determined at each ratio of TAPP to Cy^{3+}, and reflected the average distance of separation of the donor and acceptor. Calculation of the energy transfer efficiency revealed that

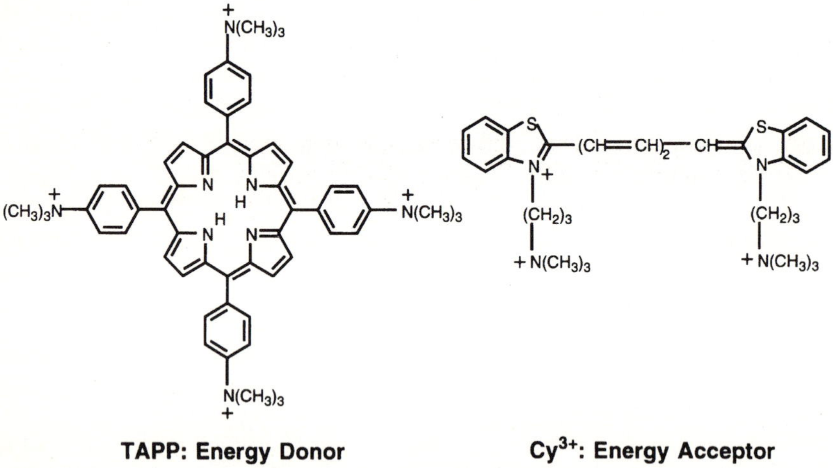

TAPP: Energy Donor **Cy³⁺: Energy Acceptor**

Fig. 16. Structures of the tetracationic porphyrin (TAPP) donor and the tricationic cyanine dye (Cy^{3+}) acceptor used in the polymerization-enhanced energy transfer experiments.

the quantum efficiency was 0.09 for the unpolymerized vesicles, whereas it increased to 0.71 for the completely polymerized vesicles. The bilayer-polymerization-increased efficiency of energy transfer at the vesicle surface was a direct reflection of the increased local concentration, i.e. domain formation, of the bilayer anionic binding sites. These results show that polymerization of bilayers in a manner that renders a proportion of the bilayer inaccessible can be a useful strategy for achieving high effective concentrations of energy-transfer "cofactors". This becomes especially important in the design of synthetic photochemical molecular devices that employ energy-transfer and redox components [80, 81].

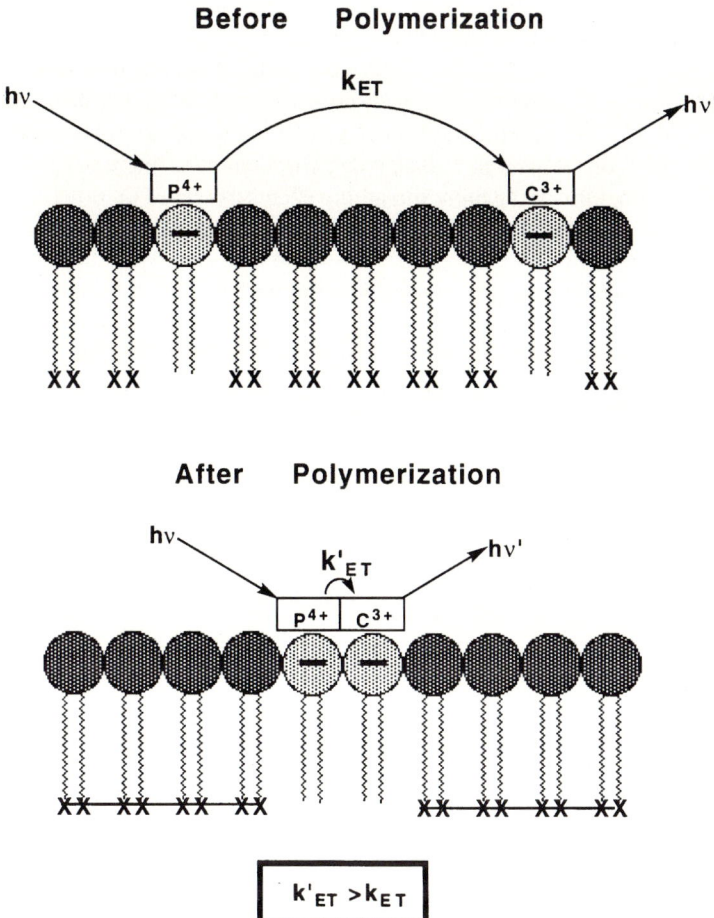

Fig. 17. Schematic representation of the effect of bilayer polymerization on the average distance of separation of nonreactive anionic lipids and their associated cationic dyes, where P^{4+} and C^{3+} are the TAPP and Cy^{3+} shown in Fig. 16.

4 Summary

It is well known that polymerization processes can enhance the stability of materials. The mutual interaction of lipid polymerization and lipid domains described in this review offers other possibilities. The recognition that the polymerization of amphiphilic assemblies can "lock in" preexisting lipid domains or create lipid domains opens the way to new strategies for the use of lipid polymerizations. We have seen that both lipid diacelylenes and fluorinated lipids provide a convenient means to form an immiscible mixture of reactive and nonreactive lipids in monolayers or bilayers, as long as the head groups do not have a net charge. In contrast the polymerization of dienoyl-, sorbyl-, or acryloyl-substituted lipids can be usefully employed to phase-separate unreactive lipids from the newly formed polymeric domains. These polymerization-induced domains endow bilayer vesicles with latent instability sites that can be triggered by the addition of reagents, or by a lamellar to nonlamellar transition. Polymerization-induced domain formation can also serve to concentrate membrane-associated electron or energy transfer cofactors, thereby increasing the efficiency of interaction. One of the more intriguing effects observed to date is the effectiveness of two-dimensional polymerizations in bilayers and to reorganize a rapidly diffusing random mixture of charged and neutral lipids to concentrate ionized amphiphiles into enriched domains despite their electrostatic repulsion.

5 References

1. Ringsdorf H, Schlar B, Venzmer J (1988) Angew Chem Int Ed Engl 27: 113
2. O'Brien DF (1994) Trends Polym Sci 2: 183
3. Lee Y-S, Yang J-Z, Sisson TM, Frankel DA, Gleeson JT, Aksay E, Keller SL, Gruner SM, O'Brien DF (1995) J Am Chem Soc 117: 5573.
4. Singer SJ, Nicolson GL (1972) Science 175: 720
5. Möhwald H (1990) Annu Rev Phys Chem 41: 441
6. McConnell H (1991) Annu Rev Phys Chem 42: 171
7. Chi LF, Anders M, Fuchs H, Johnson RR, Ringsdorf H (1993) Science 259: 213
8. Klausner RD, Kleinfeld AM (1984) In: Perelson AS, DeLisi C, Wiegel FW (eds). Cell Surface Dynamics. Marcel Dekker, New York, p 23
9. McElhaney RN (1982) Chem Phys Lipids 30: 229
10. Krisovitch SM, Regen SL (1992) J Am Chem Soc 114: 9628
11. Silvius JR (1986) Biochim Biophys Acta 857: 217
12. Sankaram MB, Marsh D, Thompson TE (1992) Biophys J 63: 340
13. Singh A, Schnur JM (1993) In: Cevc G (ed). Phospholipids Handbook. Marcel Dekker, NY, p 233
14. O'Brien DF, Ramaswami V (1989) Encyclopedia of Polymer Science and Engineering 17: 108
15. Fahey PF, Webb WW (1978) Biochemistry 17: 3046
16. Sells TD, O'Brien DF (1991) Macromolecules 24: 336
17. Sells TD, O'Brien DF (1994) Macromolecules 27: 226

18. Lei J, O'Brien DF (1994) Macromolecules 27: 1381
19. Lamparski H, O'Brien DF (1995) Macromolecules 28: 1786
20. Hitchcock PB, Mason R, Thomas KM, Shipley GG (1974) Proc Natl Acad Sci USA 71: 3036
21. Haberkorn RA, Griffin RG, Meadows MD, Oldfield E (1977) J Am Chem Soc 99: 7353
22. Hauser H, Pascher I, Pearson RH, Sundell S (1981) Biochim Biophys Acta 650: 21
23. Lopez E, O'Brien DF, Whitesides TH (1982) J Am Chem Soc 104: 305
24. Kölchens S, Lamparski H, O'Brien DF (1993) Macromolecules 26: 398
25. Helenius A, Simons K (1975) Biochim Biophys Acta 415: 29
26. Lichtenberg D, Robson R, Dennis EA (1983) Biochim Biophys Acta 737: 285
27. Regen SL, Singh A, Oehme G, Singh M (1982) J. Am Chem Soc 104: 791
28. Tsuchida E, Hasegawa E, Kimura N, Hatashita M, Makino C (1992) Macromolecules 25: 207
29. Sisson T, Lamparski HG, Kölchens S, Elyadi A, O'Brien DF, (unpublished observations)
30. Stein WD (1986) Transport and Diffusion Across Cell Membranes, Academic Press, San Diego, CA
31. Dorn K, Klingbiel RT, Specht DP, Tyminski PN, Ringsdorf H, O'Brien DF (1984) J Am Chem Soc 106: 1627
32. Büschl R, Ringsdorf H, O'Brien DF (unpublished observations)
33. Hupfer B, Ringsdorf H, Schupp H (1981) Makromol Chem 182: 247
34. Crisp DJ (1967) In: Surface Chemistry. Butterworth's London: Proc Joint Meeting Faraday Soc Chem Phys 1944
35. Wegner G (1972) Makromol Chem 154: 35
36. Büschl R, Hupfer B, Ringsdorf H (1982) Makromol Chem Rapid Commun 3: 589
37. Lopez E, O'Brien DF, Whitesides TH (1982) Biochim Biophys Acta 693: 437
38. Sackmann E, Eggl P, Fahn C, Bader H, Ringsdorf H, Schollmeier M (1985) Ber Bunsenges Phys Chem 89: 1198
39. Elbert R, Folda T, Ringsdorf H (1984) J Am Chem Soc 106: 7687
40. Mimms LT, Zampighi G, Nozaki Y, Tanford C, Reynolds JA (1981) Biochemistry 20: 833
41. Büschl R, Folda T, Ringsdorf H (1984) Makromol Chem 6: 245
42. Gaub H, Büschl R, Ringsdorf H, Sackmann E (1985) Chem Phys Lipids 37: 19
43. Gaub H, Sackmann E, Büschl R, Ringsdorf H (1984) Biophys J 45: 725
44. Dorn K, Patton EV, Klingbiel RT, O'Brien DF, Ringsdorf H (1983) Makro Chem Rapid Commun 4: 513
45. Seki N, Tsuchida E, Ukaji K, Sekiya T, Nozawa Y (1985) Polym Bull 13: 489
46. Tsuchida E, Seki N, Ohno H (1986) Makromol Chem 187: 1351
47. Elbert R, Laschewsky A, Ringsdorf H (1985) J Am Chem Soc 107: 4134
48. Ohno H, Takeoka S, Tsuchida E (1985) Polym Bull 14: 487
49. Takeoka S, Sakai H, Ohno H, Tsuchida E (1991) Macromolecules 24: 1279
50. Dennis ED (1983) In: Boyer PD (ed.) The Enzymes, 3rd Edn. Academic Press, New York 16: 307
51. Büschl R (1984) PhD Dissertation, Universität Mainz
52. Tsuchida E, Seki N, Ohno H (1986) Makromol Chem 187: 1351
53. Grainger DW, Reichert A, Ringsdorf H, Salesse C, Davies DE, Lloyd JB (1990) Biochim Biophys Acta 1022: 146
54. Ahlers M, Muller W, Reichert A, Ringsdorf H, Venzmer J (1990) Angew Chem Int Ed Engl 29: 1269
55. Tyminski PN, Latimer LH, O'Brien DF (1985) J Am Chem Soc 107: 7769
56. Tyminski PN, Latimer LH, O'Brien DF (1988) Biochemistry 27: 2696
57. Kuhn H (1980) Nature (London) 283: 587
58. Tyminski PN, O'Brien DF (1984) Biochemistry 23: 3986
59. Frankel DA, Lamparski H, Liman U, O'Brien DF (1989) J Am Chem Soc 111: 9262
60. Lamparski H, Liman U, Frankel DA, Barry JA, Ramaswami V, Brown MF, O'Brien DF (1992) Biochemistry 31: 685
61. Tyminski PN, Ponticello IS, O'Brien DF (1987) J Am Chem Soc 109: 6541
62. Szoka FC, Papahadjopoulos D (1978) Proc Natl Acad Sci USA 75: 4198
63. Hope MJ, Bally MB, Webb G, Cullis PR (1985) Biochim Biophys Acta 812: 55
64. Mayer LD, Hope MJ, Cullis PR (1986) Biochim Biophys Acta 858: 161
65. Ellens H, Bentz J, Szoka FC (1984) Biochemistry 23: 1532
66. Ellens H, Bentz J, Szoka FC (1986) Biochemistry 25: 4141
67. Ellens H, Bentz J, Szoka FC (1986) Biochemistry 25: 285
68. Parsegian VA, Fuller N, Rand RP (1979) Proc Natl Acad Sci USA 76: 2750

69. Siegel DP (1986) Biophys J 49: 1155
70. Siegel DP (1986) Biophys J 49: 1171
71. Siegel DP (1993) Biophys J 65: 2124
72. Cullis PR, DeKruijff B (1979) Biochim Biophys Acta 559: 399
73. Barry JA, Lamparski H, Shyamsunder E, Osterberg F, Cerne J, Brown MF, O'Brien DF (1992) Biochemistry 31: 10114
74. Bennett DE, O'Brien DF (1995) Biochemistry 34: 3102
75. Bentz J, Nir S, Wilschut J (1983) Colloids Surf 6: 333
76. Gruner SM (1985) Proc Natl Acad Sci 82: 3665
77. Gruner SM (1989) J Phys Chem 93: 7562
78. Eggl P, Pink D, Quinn B, Ringsdorf H, Sackmann E (1990) Macromolecules 23: 3472
79. Armitage BA, Klekotka PA, Oblinger E, O'Brien DF (1993) J Am Chem Soc 115: 7920
80. Balzani V, Scandola F (1991) Supramolecular Photochemistry, Ellis Harwood, New York
81. Clapp PJ, Armitage BA, Roosa P, O'Brien DF (1994) J Am Chem Soc 116: 9166

Editor: Prof. H. Ringsdorf
Received: 1995

Concentrated Solutions of Liquid-Crystalline Polymers

Takahiro Sato and Akio Teramoto
Department of Macromolecular Science, Osaka University, Toyonaka, Osaka 560, Japan

This article reviews the following solution properties of liquid-crystalline stiff-chain polymers: (1) osmotic pressure and osmotic compressibility, (2) phase behavior involving liquid crystal phase(s), (3) orientational order parameter, (4) translational and rotational diffusion coefficients, (5) zero-shear viscosity, and (6) rheological behavior in the liquid crystal state. Among the related theories, the scaled particle theory is chosen to compare with experimental results for properties (1)–(3), the fuzzy cylinder model theory for properties (4) and (5), and Doi's theory for property (6). In most cases the agreement between experiment and theory is satisfactory, enabling one to predict solution properties from basic molecular parameters. Procedures for data analysis are described in detail.

List of Symbols and Abbreviations 87

1 Introduction 90
   1.1 Purposes and Scope 90
   1.2 Liquid-Crystalline Polymers Viewed as the Wormlike Chain 91

2 Scaled Particle Theory for Wormlike Hard Spherocylinders 93
   2.1 Free Energy 93
   2.2 Excluded Volume Contribution 94
   2.3 Orientational Entropy 96
   2.4 Expressions for Thermodynamic Quantities 97
   2.5 Comparison with Other Theories for Hard Rods 100

3 Osmotic Pressure and Osmotic Compressibility 101
   3.1 Osmotic Pressure 101
   3.2 Osmotic Compressibility 103

4 Phase Equilibrium Behavior 105
   4.1 Binary Solutions 105
   4.2 Ternary Systems Containing Two Polymer Species with Different Lengths 110
   4.3 Electrostatic Interaction 113

5 Orientational Order Parameter 116

6 Theory of Stiff-Chain Polymer Dynamics 119
   6.1 Kinetic Equation 119

    6.2   Fuzzy Cylinder Model 121
    6.3   Diffusion Coefficients. 123
          6.3.1 Green Function Method. 123
          6.3.2 Hole Theory. 127
          6.3.3 Reptation Model 128
    6.4   Stress Expression 129

7 **Diffusion Coefficients from Computer Simulation and Experiment** . . . 131
    7.1   Computer Simulations for Rodlike Polymers. 131
    7.2   Translational Self-Diffusion Coefficient 134
    7.3   Rotational Diffusion Coefficient 135

8 **Zero-Shear Viscosity** 136
    8.1   Experimental Results. 136
    8.2   Viscosity Equation 139
    8.3   Comparison Between Experiment and Theory 142

9 **Rheological Properties of Liquid-Crystalline Solutions** 147
    9.1   Experimental Results. 147
    9.2   Theoretical Considerations 148
    9.3   Comparison Between Experiment and Theory 150

10 **Conclusions**. 152

11 **Appendices** 152
    Appendix A: Accuracy of Using Onsager's Trial Function
    for the Determination of the Equilibrium Orientational
    Distribution Function 152
    Appendix B: Frictional and Transport Coefficients at Infinite
    Dilution. 153
    Appendix C: Mean-Field Green Function Method 155

12 **References** 158

List of Symbols and Abbreviations

APC	(acetoxypropyl) cellulose
CTA	cellulose triacetate
DBP	dibutyl phthalate
DCE	1,2-dichloroethane
DCM	dichloromethane
DMAc	dimethylacetamide
DMF	dimethylformamide
HPC	(hydroxypropyl) cellulose
PBLG	poly(γ-benzyl L-glutamate)
PHIC	poly(n-hexyl isocyanate)
PPTA	poly(p-phenylene terephthalamide) (Kevlar)
PYPt	poly(yne)-platinum polymer
SPT	scaled particle theory
TCE	trichloroethane
TFA	trifluoroacetic acid
TMV	tobacco mosaic virus
A	strain tensor defined by Eq. (42 b)
a	unit tangent vector to the polymer contour in Sects. 2–5 or unit vector parallel to the polymer end-to-end vector in Sects. 6–9
B	parameter defined by Eq. (10) or Table 3 in Sect. 2 and by Eq. (78) in Sect. 8
B_2, B_3	second and third virial coefficients
C	parameter defined by Eq. (10) or Table 3
c	(total) polymer mass concentration
c'	number concentration of a polymer
c_I, c_A	phase boundary concentrations between isotropic and biphasic regions and between biphasic and liquid crystal regions
$C_\perp, C_r$	parameters appearing in Eqs. (47) and (51)
C_s	added salt molar concentration
d	polymer diameter
d_e	diameter of the fuzzy cylinder model
$D_r(D_{r0})$	rotational diffusion coefficient (at infinite dilution)
$D_s(D_{s0})$	translational self-diffusion coefficient (at infinite dilution)
$D_\parallel(D_{\parallel 0})$	longitudinal diffusion coefficient (at infinite dilution)
$D_\perp(D_{\perp 0})$	transverse diffusion coefficient (at infinite dilution)
$\bar{f}(\mathbf{a})$	orientational distribution function of the tangent vector **a** averaged along the chain contour
$f(\mathbf{r},\mathbf{a};t)$	time-dependent single-particle distribution function
$F_{\parallel 0}, F_{\perp 0}, F_{r0}$	correction factors defined by Eqs. (B1)–(B3)

$f_\perp, f_r$	factors appearing in Eqs. (46) and (50)
H_r	reduced viscosity defined by Eq. (75)
k_B	Boltzmann constant
L	contour length of a polymer
L_c	length of the cylinder part of a (wormlike) spherocylinder
L_e	length of the fuzzy cylinder model
M	molecular weight of a monodisperse polymer (except in Appendix C)
M_L	molar mass per unit contour length of a polymer
M_n, M_w, M_v	number, weight, and viscosity average molecular weights
n	number of polymer molecules in solution
N	number of Kuhn's statistical segments per polymer chain
N_A	Avogadro constant
q	persistence length
R	gas constant
$S(\mathbf{S})$	orientational order parameter (tensor)
S_{ex}	translational entropy loss due to the intermolecular excluded volume effect
S_{or}	entropy with respect to the orientational degree of freedom of polymer
t	time
T	absolute temperature
$v(v_s)$	molecular volume of a polymer (of polymer species s)
V_{ex}^*	mutual excluded volume between the critical hole and a surrounding chain
$V_{scf}(\mathbf{a})$	self-consistent mean field or a molecular field
v_{sp}	partial specific volume of a polymer
X	parameter defined by Eq. (82)
x_s	mole fraction of polymer species s
α	degree of orientation in the Onsager trial function given by Eq. (17)
$\beta_\perp, \beta_r$	numerical constants given by Eqs. (48) and (52)
γ	hydrodynamic parameter in Eq. (B3)
ΔF	excess Helmholtz free energy of a solution over that of solvent
$[\eta]$	intrinsic viscosity
η_0	zero-shear viscosity
$\eta^{(s)}$	solvent viscosity
λ^*	similarity ratio of the critical hole to the fuzzy cylinder
κ	reciprocal of the Debye screening length (in Sect. 4.3) and shear rate (in Sects. 6, 8 and 9)
$\boldsymbol{\kappa}$	gradient tensor of the solvent flow
μ	chemical potential of a polymer
ν	linear charge density of a polyelectrolyte
ξ	weight fraction of a polymer species in the total polymer
Π	osmotic pressure

ρ	parameter defined by Eq. (11)
$\sigma, \boldsymbol{\sigma}$	stress and stress tensor
$\sigma(N)$	orientational entropy; see Eq. (15)
$\sigma_{N_1}, \sigma_{N_2}$	primary and secondary normal stress differences
χ	hydrodynamic parameter in Eq. (42a)
Ω	vorticity tensor defined by Eq. (42b)

1 Introduction

1.1 Purposes and Scope

Since Robinson [1] discovered cholesteric liquid-crystal phases in concentrated α-helical polypeptide solutions, lyotropic liquid crystallinity has been reported for such polymers as aromatic polyamides, heterocyclic polymers, DNA, cellulose and its derivatives, and some helical polysaccharides. These polymers have a structural feature in common, which is elongated (or asymmetric) shape or chain stiffness characterized by a relatively large persistence length. The minimum persistence length required for lyotropic liquid crystallinity is several nanometers[1].

Liquid-crystalline polymers with stiff backbones have many static and dynamic solution properties markedly distinct from usual flexible polymers. For example, their solutions are transformed from isotropic to liquid crystal state with increasing concentration. While very high in the concentrated isotropic state, their viscosity decreases drastically as the concentration crosses the phase boundary toward the liquid crystal state. The unique rheological properties they exhibit in the liquid crystal state are also remarkable.

The distinct properties of liquid-crystalline polymer solutions arise mainly from extended conformations of the polymers. Thus it is reasonable to start theoretical considerations of liquid-crystalline polymers from those of straight rods. Long ago, Onsager [2] and Flory [3] worked out statistical thermodynamic theories for rodlike polymer solutions, which aimed at explaining the isotropic–liquid crystal phase behavior of liquid-crystalline polymer solutions. Dynamical properties of these systems have often been discussed by using the tube model theory for rodlike polymer solutions due originally to Doi and Edwards [4]. This theory, the counterpart of Doi and Edward's tube model theory for flexible polymers, can intuitively explain the dynamic difference between rodlike and flexible polymers in concentrated systems [4].

However, as accurate experimental data were accumulated, it has become apparent that these earlier theories of rodlike polymers fail to describe quantitatively the behavior of real liquid-crystalline polymers, which are not completely rigid but more or less flexible.

The growing interest in liquid-crystalline polymers has stimulated many theoretical and experimental studies of their solutions, and the results have already been summarized by many authors. For instance, the statistical thermodynamic theories were reviewed by Flory [5], Odijk [6], Semenov and Khokhlov [7], Ciferri et al. [8], and Vroege and Lekkerkerker [9], while the dynamical theories were discussed by Doi and Edwards [4] and Moscicki [10].

---

[1] Thermotropic liquid crystalline polymers, like polyesters containing mesogenic units on the main chain, may not be described by the wormlike chain model (cf. Sect. 1.2). The present article does not consider this type of polymers.

In the present article, we focus on the scaled particle theory as the theoretical basis for interpreting the static solution properties of liquid-crystalline polymers. It is a statistical mechanical theory originally proposed to formulate the equation of state of hard sphere fluids [11], and has been applied to obtain approximate analytical expressions for the thermodynamic quantities of solutions of hard (sphero)cylinders [12–16] or wormlike hard spherocylinders [17, 18]. Its superiority to the Onsager theory lies in that it takes higher virial terms into account, and it is distinctive from the Flory theory in that it uses no artificial lattice model. We survey this theory for wormlike hard spherocylinders in Sect. 2, and compare its predictions with typical data of various static solution properties of liquid-crystalline polymers in Sects. 3–5. As is well known, the wormlike chain (or wormlike cylinder) is a simple yet adequate model for describing dilute solution properties of stiff or semiflexible polymers.

The dynamic behavior of liquid-crystalline polymers in concentrated solution is strongly affected by the collision of polymer chains. We treat the interchain collision effect by modelling the stiff polymer chain by what we refer to as the "fuzzy cylinder" [19]. This model allows the translational and rotational (self-)diffusion coefficients as well as the stress of the solution to be formulated without resort to the hypothetical tube model (Sect. 6). The results of formulation are compared with experimental data in Sects. 7–9.

1.2 Liquid-Crystalline Polymers Viewed as the Wormlike Chain

Before proceeding to a review of both scaled particle theory and fuzzy cylinder model theory, it would be useful to mention briefly the unperturbed wormlike (sphero)cylinder model which is the basis of these theories. Usually the intramolecular excluded volume effect can be ignored in stiff-chain polymers even in good solvents, because the distant segments of such polymers have little chance of collision. Therefore, in the subsequent reference to wormlike chains, we always mean that they are "unperturbed".

The wormlike cylinder is characterized by three parameters, contour length L, persistence length q, and diameter d. The contour length can be calculated from the molecular weight M of the polymer and the molar mass per unit contour length M_L by $L = M/M_L$. The number of Kuhn's statistical segments N, defined by $L/2q$, is a measure of the global shape of a polymer chain; when N is much smaller or larger than unity, the chain is rodlike or random coil-like respectively.

The model parameters q and M_L can be estimated from experimental data for radius of gyration, intrinsic viscosity, sedimentation coefficient, diffusion coefficient and so on in dilute solutions. The typical methods are expounded in several recent articles and books [20–22]. Here we refer only to the results of the application to representative liquid-crystalline polymers (See Table 1).

From the table, we see that polymers having lyotropic liquid crystallinity are not rare exceptions. Their smallest q values are larger than 5 nm, while those of

Table 1. Persistence length q and molar mass per unit contour length M_L for liquid-crystalline stiff-chain polymers

polymer	solvent	q/nm	M_L/nm^{-1}	Ref.
tobacco mosaic virus (TMV)	tris-HCl buffer	∞	133 000	[23]
fd-virus	tris-HCl buffer	2200	21 600	[24]
schizophyllan	water	200	2 150	[25]
poly(γ-benzyl-L-glutamate) (PBLG)	dimethylformamide (DMF)	150	1 450	[21, 26]
xanthan	0.1 mol/l aqueous NaCl	120	1 940	[27]
deoxynucleic acid (DNA)	0.2 mol/l aqueous NaCl[a]	60	1 950	[28]
	0.179 mol/l aqueous NaCl[b]	62.5	1 950	[29]
poly(p-benzamide)	96% sulfuric acid	50	198	[30]
poly(n-hexyl isocyanate)	toluene (25 °C)	37	740	[31]
(PHIC)	dichloromethane (DCM, 20 °C)	21	740	[31]
poly(p-phenylene tere-phthalamide) (PPTA)	96% sulfuric acid	18	198	[30]
poly(yne)-Pt (PYPt)	trichloroethane (TCE)	13	692	[32, 33]
(hydroxypropyl)cellulose (HPC)	dimethylacetamide (DMAc)	6.5	720	[34]
(acetoxypropyl)cellulose (APC)	dibutyl phthalate (DBP)	5.9	821	[35]
cellulose triacetate (CTA)	trifluoroacetic acid (TFA)	5.3	560	[36]

[a] Containing 2 mmol/l EDTA and 2 mmol/l phosphate buffer
[b] Containing 1 mmol/l EDTA and 4 mmol/l phosphate buffer

Table 2. Values of the diameters of liquid-crystalline polymers calculated by different methods

polymer	solvent	d/nm			
		(from v_{sp})[a]	(from Π)[b]	(from HD)[c]	(from CRY)[d]
schizophyllan	water	1.68	1.52	2.6 [25]	1.8 [37]
PBLG	DMF	1.56	1.4	1.7 [26]	1.5 [38]
PHIC	toluene	1.24	1.15 [39]	1.6 [31]	
	DCM	1.25	0.95		
PPTA	sulfuric acid	0.51		0.6 [30]	0.52 [30]
PYPt	TCE	1.08		1.2 [32]	
CTA	DMAc	0.88			
	TFA	0.95			
HPC	water	1.13			1.28 [40]
APC	DBP	1.19			

[a] Calculated from $d = (4v_{sp}M_L/\pi N_A)^{1/2}$ with the Avogadro constant N_A
[b] Calculated from the fitting to osmotic pressure and osmotic compressibility data (cf. Sect. 3)
[c] Calculated from intrinsic viscosity or sedimentation coefficient data
[d] Obtained from crystallographic data

ordinary flexible polymers incapable of forming mesophase are about 1 nm [20]. Therefore the critical persistence length for lyotropic liquid crystallinity may be taken as being between 1 and 6 nm.

The diameter d of a polymer chain can be estimated from (1) hydrodynamic quantities such as intrinsic viscosity and sedimentation coefficient, (2) the partial specific volume v_{sp} of the polymer, and (3) X-ray crystallographic data of the polymer. Table 2 lists the values of d for liquid-crystalline polymers estimated by different methods. Those determined from hydrodynamic data are close to but slightly larger than those from v_{sp} and crystallographic data, though this may not always be the case.

Most liquid-crystalline polymer solutions have a large second virial coefficient ($\gtrsim 10^{-4} \text{cm}^{-3} \text{mol/g}^2$) [41], which means that it is rather difficult to find poor or theta solvents for these polymers and that liquid-crystalline polymers in solution interact repulsively. This fact is essential in formulating their static solution properties (osmotic pressure, phase separation, etc.).

2 Scaled Particle Theory for Wormlike Hard Spherocylinders

2.1 Free Energy

We begin by formulating the free energy of liquid-crystalline polymer solutions using the wormlike hard spherocylinder model, a cylinder with hemispheres at both ends. This model allows the intermolecular excluded volume to be expressed more simply than a hard cylinder. It is characterized by the length of the cylinder part $L_c (\equiv L - d)$, the Kuhn segment number N, and the hard-core diameter d. We assume that the interaction potential between them is given by

$$u_0 = \begin{cases} \infty & \text{when the hard cores of the two molecules overlap} \\ 0 & \text{otherwise.} \end{cases} \quad (1)$$

In general, the excess Helmholtz free energy ΔF of a hard-particle solution over that of the solvent is written in the form

$$\Delta F/n = \mu° + k_B T \ln c' - TS_{ex} - TS_{or} \quad (2)$$

where n is the number of solutes, $\mu°$ and c' the standard chemical potential and the number concentration of the solute, respectively, k_B the Boltzmann constant, and T the absolute temperature. The second term on the right-hand side is the translational (or mixing) entropy per one solute molecule for the ideal solution. The intermolecular excluded volume lowers the translational entropy by the amount expressed by the third term (note that S_{ex} is negative). The term S_{or} is the entropy relating to the orientational degree of freedom of the solute molecule. We remark that Eq. (2) includes no energetic term, i.e., concerns an athermal solution.

2.2 Excluded Volume Contribution

The hard-core repulsion prevents spherocylinders from overlapping. This effect reduces the space available for the cylinders, and gives rise to a loss of their translational entropy ($-S_{ex}$). Many statistical thermodynamic techniques were used to calculate it, as has been extensively reviewed by Vroege and Lekkerkerker [9].

The scaled particle approach is exact at the limit of infinite dilution and makes it possible to formulate static solution properties at finite dilutions in an approximate way. In this approach, we first calculate S_{ex} for one hypothetical scaled particle with a size smaller or larger than the real particle and then find S_{ex} of the real size by interpolation. For the wormlike spherocylinder, the scaled particle is assumed to have a cylinder length λL_c and a hard-core diameter κd where λ and κ are scaling factors. The persistence length q of the scaled particle may be chosen rather arbitrarily. Here we do not scale q of the scaled particle but take it to be the same as the real q [17, 18].

When λ and $\kappa \ll 1$ and thus λL_c is much smaller than q, the scaled particle reduces to a small straight spherocylinder. We represent its orientation in space by a unit vector **a**. Then the probability P that the scaled particle can be placed at some arbitrary point in the solution with no overlap with other (real) spherocylinders is given to a good approximation by

$$P = 1 - c'v_{ex}(\lambda L_c, \kappa d, \mathbf{a}; L_c, d) \quad (\lambda, \kappa \ll 1) \tag{3}$$

Here, $v_{ex}(\lambda L_c, \kappa d, \mathbf{a}; L_c, d)$ is the excluded volume between the scaled particle and the real wormlike spherocylinder, and can be expressed by

$$v_{ex}(\lambda L_c, \kappa d, \mathbf{a}; L_c, d) = \lambda(1 + \kappa) L_c d \int_0^{L_c} ds \int d\mathbf{a}'(s) |\sin \gamma(\mathbf{a}, \mathbf{a}'(s))| f(\mathbf{a}'(s))$$
$$+ \frac{\pi}{4}(1 + \lambda)(1 + \kappa)^2 L_c d^2 + \frac{\pi}{6}(1 + \kappa)^3 d^3 \tag{4}$$

where $\mathbf{a}'(s)$ represents the unit tangent vector to the real wormlike spherocylinder contour at a contour point s, $\gamma(\mathbf{a}, \mathbf{a}'(s))$ the angle between **a** and $\mathbf{a}'(s)$, and $f(\mathbf{a}'(s))$ the orientational distribution function of $\mathbf{a}'(s)$. The entropy S_{ex} is related to P and the osmotic pressure Π by [42]

$$S_{ex} = -k_B [\langle \ln P \rangle - \pi/c' k_B T] \tag{5}$$

where $\langle \cdots \rangle$ represents the average over the orientation of the scaled particle.

On the other hand, when the scaled particle is very large ($\lambda, \kappa \gg 1$), the reversible work W of adding the scaled particle to the solution at an arbitrary position at a constant Π can be calculated from

$$W = \Pi \left[\frac{\pi}{4} \lambda L_c (\kappa d)^2 + \frac{\pi}{6}(\kappa d)^3 \right] \quad (\lambda, \kappa \gg 1) \tag{6}$$

and S_{ex} is related to W by [42]

$$S_{ex} = \frac{W + \Pi/c'}{T} \tag{7}$$

To interpolate S_{ex} at $\lambda = \kappa = 1$, we use a functional form as [16]:

$$S_{ex} = C_{00} + C_{10}\lambda + C_{01}\kappa + C_{11}\lambda\kappa + C_{02}\kappa^2$$
$$+ \frac{\Pi}{T}\left[\frac{\pi}{4}\lambda L_c(\kappa d)^2 + \frac{\pi}{6}(\kappa d)^3 - \frac{1}{c'}\right] \tag{8}$$

and determine the unknown coefficients C_{ij} so that Eq. (8) agrees with Eq. (5) at the limit of vanishingly small λ and κ. By this operation, it can be shown that S_{ex} for the real wormlike spherocylinder system is given by [17, 18]

$$S_{ex} = -k_B\left[\ln\frac{1}{(1-vc')-1} + \frac{Bc'}{2(1-vc')} + \frac{Cc'^2}{3(1-vc')^2}\right] \tag{9}$$

where v is the molecular volume of the spherocylinder ($v = (\pi/4)L_c d^2 + (\pi/6)d^3$) and

$$B = \frac{\pi}{2}L_c^2 d\rho + 6v, \quad C = v'(B - 2v') \tag{10}$$

with

$$v' \equiv v + \frac{\pi}{12}d^3$$

and

$$\rho \equiv \frac{4}{\pi}\iint |\sin\gamma(\mathbf{a}, \mathbf{a}')| \bar{f}(\mathbf{a})\bar{f}(\mathbf{a}')\,d\mathbf{a}d\mathbf{a}' \tag{11}$$

with $\mathbf{a}$ and $\mathbf{a}'$ representing unit tangent vectors of two real wormlike spherocylinders, and $\bar{f}(\mathbf{a})$ ($\bar{f}(\mathbf{a}')$) the orientational distribution function of $\mathbf{a}$ ($\mathbf{a}'$) averaged over the wormlike chain contour. [We have used preaverage approximations for the distribution functions f(**a**(s)) and f(**a**'(s)).] Equation (9) for S_{ex} coincides with that obtained by Cotter [16] for a straight hard spherocylinder system using the same approach. Thus we find that the flexibility of the polymer does not affect S_{ex} in the framework of the present scaled particle theory [18, 41]. In the dilute limit, Eq. (9) reduces to Onsager's result [2] for a hard spherocylinder.

The orientation dependent parameter ρ defined by Eq. (11) becomes unity in the isotropic state, and decreases as the polymers are uniaxially oriented. Therefore, it follows from Eqs. (9) and (10) that the wormlike hard spherocylinder system has a smaller translational entropy loss from the ideal solution in the liquid crystal state than in the isotropic state. This difference drives the system to form a liquid crystal phase. However, in order to determine the equilibrium orientation of the system, the orientation dependence of S_{or} has to be formulated, and this is done in Sect. 2.3.

2.3 Orientational Entropy

When a wormlike spherocylinder is in the liquid crystal phase, its tangent vector **a** at each contour point should align more or less to the preference direction of the phase specified by the director **n**. This alignment induces the orientational entropy decrease $-S_{or}$ from the entropy in the isotropic state. Since the orientation of the tangent vector stretches the wormlike spherocylinder, $-S_{or}$ includes a conformational entropy loss of the spherocylinder.

There are various methods to calculate S_{or} of the wormlike chain whose tangent vector aligns according to the average distribution function $\bar{f}(\mathbf{a})$ [6, 43–47]. Khokhlov and Semenov [43, 44] were the first to apply Lifshitz's method [48, 49], which had been originally proposed to calculate the conformational entropy of a polymer chain with a given inhomogeneous segment distribution. This method first assumes a hypothetical mean-field potential $U(\mathbf{a})$ (per unit contour length) inducing an orientation of the tangent vector **a** of each chain; $U(\mathbf{a})$ represents the interaction of a chain under consideration with surrounding chains in solution. Under this field, the partition function $z(s, \mathbf{a})$ of a partial chain with a contour length s and a tangent vector **a** at the end point should satisfy [6, 44, 47]

$$\frac{\partial z(s, \mathbf{a})}{\partial s} = \left(\Delta_a - \frac{U(\mathbf{a})}{k_B T}\right) z(s, \mathbf{a}) \tag{12}$$

Here we have chosen the Kuhn statistical segment length ($=2q$) as the unit for measuring length. The partition function Z of the total chain is given by

$$Z \equiv \int z(N, \mathbf{a}) \, d\mathbf{a} \tag{13}$$

and the average orientation distribution function $\bar{f}(\mathbf{a})$ is related to $z(s, \mathbf{a})$ by

$$\bar{f}(\mathbf{a}) = (NZ)^{-1} \int_0^N z(s, \mathbf{a}) z(N-s, \mathbf{a}) \, ds \tag{14}$$

Since the free energy of the chain is given by $-k_B T \ln Z$, the orientational entropy loss $-S_{or}$ can be written

$$-S_{or} = k_B \sigma(N) \tag{15}$$

with

$$\sigma(N) \equiv \ln(4\pi/Z) - \int d\mathbf{a} \, \bar{f}(\mathbf{a}) \frac{N U(\mathbf{a})}{k_B T} \tag{16}$$

Khokhlov and Semenov derived analytical expressions for $\sigma(N)$ from Eqs. (12)–(16) in the asymptotic limits of $N \ll 1$ and $N \gg 1$, assuming no special form for $U(\mathbf{a})$. Furthermore, by choosing the Onsager function [2] as the trial function for $\bar{f}(\mathbf{a})$, they expressed $\sigma(N)$ as functions of N and α in these asymptotic limits. Here α is the parameter representing the degree of orientation, which appears in the trial function used, i.e.,

$$\overline{f}(\mathbf{a}) = \left(\frac{\alpha}{4\pi \sinh \alpha}\right) \cosh(\alpha \mathbf{a} \cdot \mathbf{n}) \tag{17}$$

where **n** is the director.

Odijk [6] formulated σ(N) by utilizing the analogy between the partition function z(s, **a**) and the density matrix of a two-dimensional quantum harmonic oscillator, and, with a Gaussian function as the trial function for $\overline{f}(\mathbf{a})$, obtained an analytical expression σ(N) valid over the entire range of N. His σ(N), however, would not be accurate except for the highly oriented state, because the Gaussian trial function becomes less accurate with disorientation; this function does not approach the second Legendre polynomial as the degree of orientation diminishes, differing from the Onsager trial function. It may be worth noting that Odijk's σ(N) does not reduce to Khokhlov–Semenov's in the asymptotic limits of N≪1 and N≫1.

Recently, Hentschke [45] and also DuPré and Yang [46] proposed empirical interpolation formulas on the basis of Khokhlov and Semenov's σ(N). The former used the Padé approximation, while the latter modified Odijk's σ(N) so as to agree with Khokhlov and Semenov's in the asymptotic limits of N≪1 and N≫1 and derived

$$\sigma(N) = \ln \alpha - 1 + \pi e^{-\alpha} + \frac{1}{3} N(\alpha - 1) + \frac{5}{12} \ln \left[\cosh\left(\frac{2}{5} N(\alpha - 1)\right)\right] \tag{18}$$

Although Hentschke's σ(N) differs in form from Eq. (18), both interpolation formulas give almost identical numerical results. In what follows, we use Eq. (18) for σ(N).

2.4 Expressions for Theromodynamic Quantities

Inserting Eqs. (9) and (15) into Eq. (2), we obtain the excess Helmholtz free energy in the form

$$\frac{\Delta F}{nk_B T} = \frac{\mu^\circ}{k_B T} - 1 + \ln\left(\frac{c'}{1 - vc'}\right) + \frac{B}{2}\left(\frac{c'}{1 - vc'}\right) + \frac{C}{3}\left(\frac{c'}{1 - vc'}\right)^2 + \sigma(N) \tag{19}$$

This leads to the following expressions for the osmotic pressure Π of the solution and the chemical potential μ of the wormlike hard spherocylinder:

$$\frac{\Pi}{k_B T} = \frac{c'}{1 - vc'}\left[1 + \frac{B}{2}\left(\frac{c'}{1 - vc'}\right) + \frac{2C}{3}\left(\frac{c'}{1 - vc'}\right)^2\right] \tag{20}$$

and

$$\frac{\mu}{k_B T} = \frac{\mu^\circ}{k_B T} + \ln\left(\frac{c'}{1 - vc'}\right) + B\left(\frac{c'}{1 - vc'}\right) + C\left(\frac{c'}{1 - vc'}\right)^2 + \frac{v\Pi}{k_B T} + \sigma(N) \tag{21}$$

These expressions contain two orientation-dependent parameters ρ and σ(N), which can be calculated from the equilibrium orientational distribution function $\bar{f}(a)$ minimizing ΔF. When the Onsager trial function (Eq. (17)) is used for $\bar{f}(a)$, ρ and σ(N) are given as functions of the unknown parameter α; the functional form of σ(N) has already appeared in Eq. (18), while ρ is expressed asymptotically as [2]

$$\rho = \frac{4}{(\pi\alpha)^{1/2}} \left[1 - \frac{30}{32\alpha} + \frac{210}{(32\alpha)^2} + \frac{1260}{(32\alpha)^3} + \cdots \right] \quad (22)$$

The equilibrium value of α is determined from the minimization condition of ΔF:

$$\frac{\partial}{\partial \alpha}\left(\frac{\Delta F}{nk_B T}\right) = 0 \quad (23)$$

Since α in the equilibrium nematic phase is fairly large ($\alpha \gtrsim 5$ under usual conditions), Eq. (22) may be truncated at the third or fourth term.

Recently, Chen [47] proposed to determine the equilibrium distribution function directly from a numerical analysis. Thus he determined $\bar{f}(a)$ by requiring that ΔF be minimum for all variations of $\bar{f}(a)$ under the normalization condition. Using the second virial approximation for S_{ex} in ΔF, he obtained

$$\frac{U(a)}{k_B T} = \frac{2\bar{c}}{N} \int |\sin\gamma|\ \bar{f}(a')\, da' + \lambda \quad (24)$$

where λ is the Lagrangian multiplier and $\bar{c}$ is a reduced concentration defined by $L^2 dc'$. When the scaled particle theory is used for S_{ex} [i.e., Eq. (19)], $\bar{c}$ should be replaced by

$$\bar{c} = L_c^2 d \frac{c'}{1 - vc'}\left(1 + \frac{2}{3}\frac{v'c'}{1 - vc'}\right) \quad (25)$$

Combining Eq. (24) with Eqs. (12)–(14), Chen obtained $\bar{f}(a)$ for given N and $\bar{c}$ by the following numerical analysis. Firstly, a zero-th approximation to U(a) was calculated by numerical integration of Eq. (24) using a properly chosen zero-th approximation to $\bar{f}(a)$, secondly the calculated U(a) was used to obtain a first-order approximation to $\bar{f}(a)$ by solving the differential equation, Eq. (12), with Eqs. (13) and (14), and finally the process was iterated until the mean-square relative difference between the two successive approximations to $\bar{f}(a)$ became less than a prescribed small value.

Chen's analysis is more accurate than the procedure in which the Onsager trial function is used with σ(N) given by Eq. (18), but it is very involved to carry through. On the other hand, the Onsager trial function procedure is simple enough for practical purposes. As shown in Appendix A, it predicts the isotropic–nematic phase boundary concentrations that can be favorably compared with those by Chen's procedure.

The Gaussian trial function for $\bar{f}(\mathbf{a})$ used by Odijk [6] is mathematically simpler than Onsager's and allows ρ to be expressed by the leading term of Eq. (22) and σ(N) to be derived analytically. However, it becomes less accurate as the orientation gets weaker. As shown in Appendix A, its use leads to the isotropic–nematic phase boundary concentrations largely different from those by Chen's method and hence is not always relevant for quantitative discussion.

The above formulation for a monodisperse polymer system by the scaled particle theory can be readily extended to a polydisperse polymer system, as described in [17]. The result is

$$\frac{\Delta F}{nk_BT} = \sum_{s=1}^{r} x_s \left(\frac{\mu_s^\circ}{k_BT} + \ln x_s + \sigma_s(N_s) \right) - 1 + \ln\left(\frac{c'}{1-\bar{v}c'}\right) + \frac{B}{2}\left(\frac{c'}{1-\bar{v}c'}\right)$$
$$+ \frac{C}{3}\left(\frac{c'}{1-\bar{v}c'}\right)^2 \tag{26}$$

$$\frac{\Pi}{k_BT} = \frac{c'}{1-\bar{v}c'}\left[1 + \frac{B}{2}\left(\frac{c'}{1-\bar{v}c'}\right) + \frac{2C}{3}\left(\frac{c'}{1-\bar{v}c'}\right)^2\right] \tag{27}$$

$$\frac{\mu_s}{k_BT} = \frac{\mu_s^\circ}{k_BT} + \ln\left(\frac{x_s c'}{1-\bar{v}c'}\right) + \sigma_s(N_s) + B_s\left(\frac{c'}{1-\bar{v}c'}\right) + C_s\left(\frac{c'}{1-\bar{v}c'}\right)^2 + \frac{v_s \Pi}{k_BT} \tag{28}$$

where r is the number of polymer species and the subscript s signifies the polymer species ($1 \le s \le r$). The composition of the system is expressed here in terms of the total polymer number concentration c' and the mole fraction x_s, $\bar{v}$ is the number-average molecular volume ($\equiv \sum_{s=1}^{r} x_s v_s$), and the other parameters appearing in Eqs. (26)–(28) are defined in Table 3.

Table 3. Definitions of the parameters appearing in Eqs. (26)–(28) for multicomponent solutions

notation	definition
B	$\dfrac{\pi}{2} d \sum\limits_{s,t=1}^{r} x_s x_t L_{c,s} L_{c,t} \rho_{st} + 6\bar{v}$
C	$\bar{v}'(B - 2\bar{v}')$
B_s	$\dfrac{\pi}{2} dL_{c,s} \sum\limits_{t=1}^{r} x_t L_{c,t} \rho_{st} + 3v_s + 3\bar{v}$
C_s	$\dfrac{1}{3}[v_s'(B - 4\bar{v}') + 2\bar{v}'(B_s - \bar{v}')]$
v_s	$\dfrac{\pi}{4} L_{c,s} d^2 + \dfrac{\pi}{6} d^3$
v_s' ($\bar{v}'$)	$v_s + \dfrac{\pi}{12}d^3 \left(\bar{v} + \dfrac{\pi}{12}d^3\right)$
ρ_{st}	$\dfrac{4}{\pi} \iint |\sin\gamma(\mathbf{a},\mathbf{a}')| \bar{f}_s(\mathbf{a}) \bar{f}_t(\mathbf{a}') d\mathbf{a} d\mathbf{a}'$

2.5 Comparison with Other Theories for Hard Rods

Various statistical thermodynamic methods are available for formulating the thermodynamics of hard rod systems. Onsager [2] used the cluster expansion of ΔF, retaining only the second virial term in S_{ex} (the second virial approximation). Flory [3] formulated rodlike polymer solutions by using the lattice model. Parsons [50] proposed a decoupling approximation to extend the result of S_{ex} for a hard sphere fluid to a hard rod system. Using this approximation, Khokhlov and Semenov [51, 52] obtained an expression of ΔF, which tends to agree with Flory's in highly oriented concentrated solutions. Lee [53] applied Parson's approach to extend the Carnahan–Starling equation of state for a hard sphere system to hard spherocylinder systems.

Table 4 compares different theoretical approaches with respect to the equations of state and the second and third virial coefficients (B_2, B_3) for a hard rod solution in the isotropic state; B_2 and B_3 are the parameters appearing in the expansion

$$\frac{\Pi}{k_B T} = c' + B_2 c'^2 + B_3 c'^3 + \cdots$$

(In Table 4, terms of the order of $b'^2(d/L)^2$ in B_3 are neglected.) The scaled particle theory (SPT), Lee's theory, and Khokhlov–Semenov's theory give the same B_2 as Onsager's, so that they become identical at infinite dilution. Flory's

Table 4. Equations of state for isotropic rodlike polymer solutions derived from various theories

theory	$\Pi/k_B T$	B_2 [a,b]	B_3 [a,c]
SPT	Eq. (20)	b'	$\dfrac{10d}{3L} b'^2$
Onsager [2]	$c'[1 + b'c']$	b'	
Flory [3]	$-\dfrac{L}{dv}\left[\ln(1 - vc') + \left(1 - \dfrac{d}{L}\right)vc'\right]$	$\dfrac{L}{2d}v \approx \dfrac{1}{2}b'$	$\dfrac{d}{3L} b'^2$
Lee [53]	$c'\left[1 + \dfrac{c'(2 - vc')}{2(1 - vc')^3} b'\right]$	b'	$\dfrac{5d}{2L} b'^2$
Khokhlov-Semenov [51, 52]	$c'\left[1 + \dfrac{L}{d}\dfrac{vc'}{(1 - vc')}\right]$	$\dfrac{L}{d}v \approx b'$	$\dfrac{d}{L} b'^2$
Straley [54]			$4.1 \dfrac{d}{L} b'^2$
Kihara [55]			$4 \dfrac{d}{L} b'^2$

[a] $b' \equiv \frac{\pi}{4} L_c^2 d + 4v$
[b] Second virial coefficient on the number density base
[c] Third virial coefficient on the number density base; only the leading term is listed

theory cannot predict the correct B_2 owing to the use of the mean-field approximation.

Straley [54] and Kihara [55] calculated B_3 for a hard cylinder by numerical integration of the ternary cluster integral. The results of their calculations, which are also listed in Table 4, are most favorably compared with B_3 from the SPT. Although not shown in Table 4, B_3 for the nematic state from the SPT also agrees closely with Straley's and Kihara's B_3 for the same state [39]. Figure 1 compares different theories in terms of the equation of state of a hard rod solution in the isotropic state, with the axial ratio L/d chosen to be 50. It is seen that the SPT begins to deviate from Onsager's theory truncated at the second virial term at around vc' (volume fraction) = 0.05. Lee's theory comes closest to the SPT, but some deviation is apparent at high concentrations.

3 Osmotic Pressure and Osmotic Compressibility

3.1 Osmotic Pressure

The relation between the osmotic pressure Π and the polymer concentration, referred to as the equation of state for the solution, is often used for a critical comparison between theory and experiment (or simulation). Kubo and Ogino

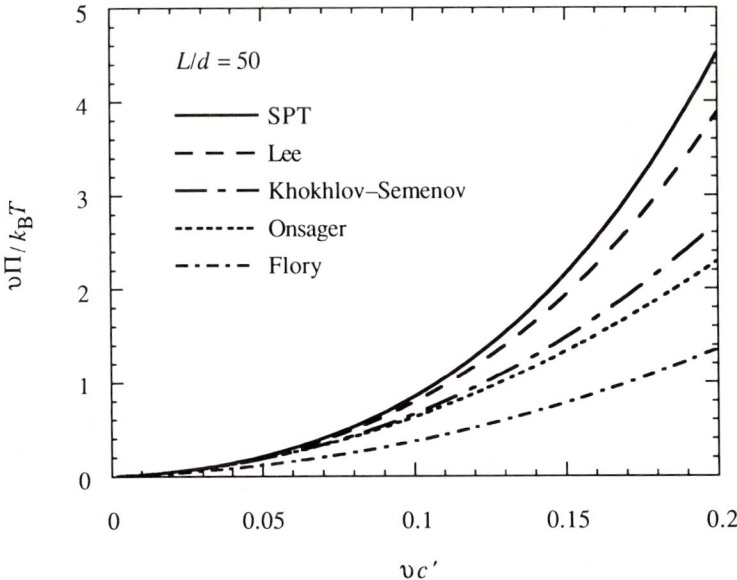

Fig. 1. Equations of state derived from various statistical thermodynamic theories

[56, 57] measured Π for isotropic and cholesteric solutions of poly(γ-benzyl L-glutamate) (PBLG) and dimethylformamide (DMF) over a very wide range of polymer mass concentration c by membrane osmometry and isothermal distillation. Figure 2 shows their results for PBLG with three different molecular weights. Vertical segments in Fig. 2 indicate phase boundary concentrations c_I between isotropic and biphasic regions and c_A between biphasic and cholesteric regions. Although we expect for a binary solution that Π in the biphasic region is constant, the experimental Π in that region monotonically increases with c regardless of the sample's molecular weight. This discrepancy may be due to polydispersity in molecular weight of the PBLG samples used.

Now we compare the above osmotic pressure data with the scaled particle theory. The relevant equation is Eq. (27) for polydisperse polymers. In the isotropic state, it can be shown that Eq. (27) takes the same form as Eq. (20) for the monodisperse system though the parameters (B, C, v, and c') have to be calculated from the number-average molecular weight M_n and the total polymer mass concentration c of a polydisperse system; ρ_{st} in the parameters B and C is unity in the isotropic state. No information is needed for the molecular weight distribution of the sample. On the other hand, in the liquid crystal state[2], Eq. (27) does not necessarily take the same form as Eq. (20), because ρ_{st} depends on the molecular weight distribution.

The three solid curves in Fig. 2 represent the theoretical Π for the three PBLG samples computed from Eq. (20) with M_n, the hard-core diameter d taken to be 1.4 nm, and the wormlike chain parameters of PBLG given in Table 1. Each theoretical curve has a horizontal part which represents the biphasic region, and two break points at its ends which correspond to the phase boundary concentrations c_I and c_A. In the isotropic region, the theoretical curves fit closely the data points for all samples, but, in the cholesteric region, the one for the lowest molecular weight sample deviates downward from the data points, while the agreement between experiment and theory is good for the highest molecular weight sample. Since, as mentioned above, Eq. (20) is not relevant to polydisperse solutions in the cholesteric state, this disagreement is not surprising. Polydispersity may also be responsible for the failure of correct prediction of the experimental phase boundary concentrations c_I and c_A (cf. vertical segments).

Hentschke [45] and DuPré and Yang [46] compared Kubo and Ogino's data with Lee's theory [53] extended to a (monodisperse) wormlike chain system by using $\sigma(N)$ (cf. Sect. 2.3). Hentschke took d to be 1.6 nm, which is consistent with the value estimated from the partial specific volume (cf. Table 2). Though good for the middle and highest molecular weight samples in the

---

[2] Although PBLG solutions form a cholesteric liquid crystal, it is expected that the supramolecular helical arrangement in the cholesterics little contribute to their thermodynamic properties, because cholesteric pitches of PBLG solutions are usually much larger than the average intermolecular distance [1], and then the local structure of the cholesterics is indistinguishable from that of nematics.

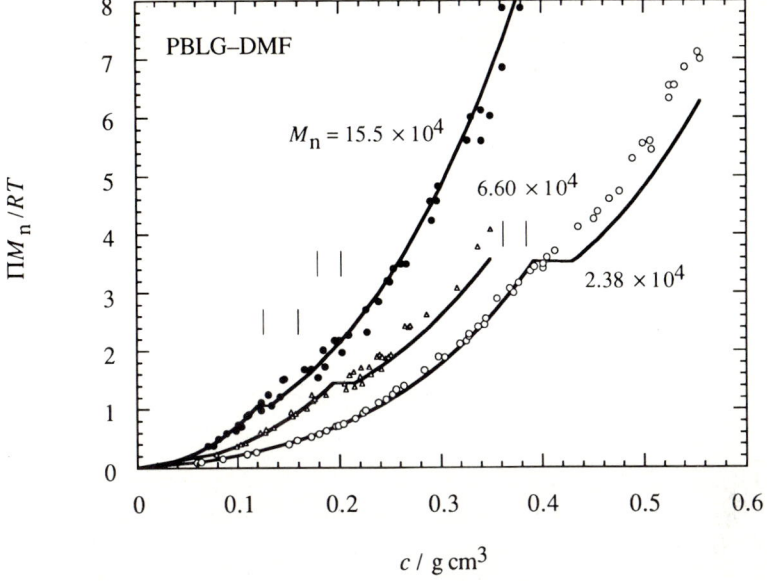

Fig. 2. Comparison between the scaled particle theory (*solid curves*) and experiment (*circles and triangles*) for osmotic pressure Π of PBLG-DMF [56, 57]. For the samples with $M_n = 6.6 \times 10^4$ and 15.5×10^4, the data at T = 15, 30, and 45 °C are plotted with the same symbols

cholesteric region, the agreement between theory and experiment in the isotropic region was not as satisfactory as the one shown in Fig. 2. In calculating Π, DuPré and Yang varied d from 1.96 nm to 2.04 nm depending on the sample to obtain a best agreement.

3.2 Osmotic Compressibility

For a binary polymer solution, the reciprocal of the osmotic compressibility $\partial\Pi/\partial c$ at constant T and the solvent chemical potential μ_0 can be determined by sedimentation equilibrium through the relation [58, 59]:

$$\frac{\partial \Pi}{\partial c} = \frac{c(r)(\partial \rho/\partial c)}{dc(r)/dr} \omega^2 r \tag{29}$$

Here c(r) is the polymer mass concentration at a radial distance r from the axis of rotation, $\partial\rho/\partial c$ the specific density increment of polymer at constant T and μ_0, and ω the angular velocity.

Itou et al. [60] determined $\partial\Pi/\partial c$ for isotropic solutions of poly(n-hexyl isocyanate) (PHIC) and dichloromethane (DCM) from sedimentation equilibrium data on narrow distribution PHIC samples. Figure 3 shows their results, where the ordinate $M_w(\partial\Pi/\partial c)$ represents the osmotic pressure increment on the molar concentration scale.

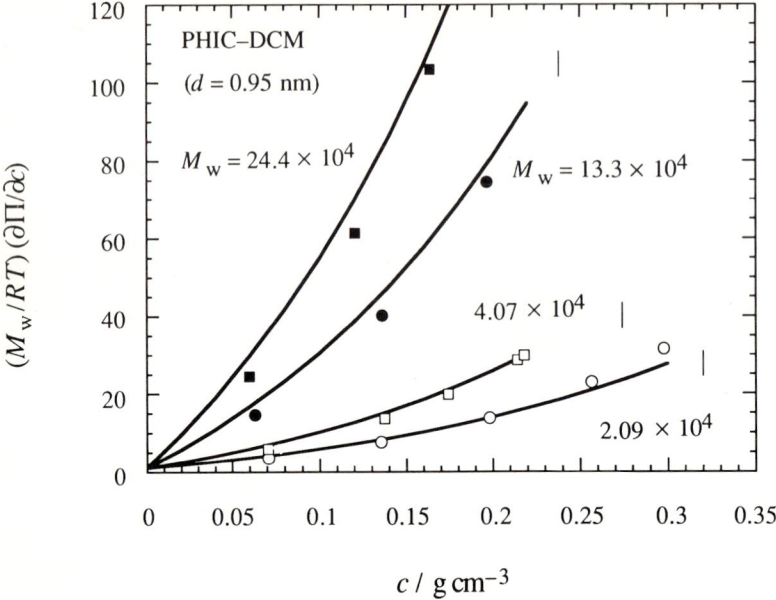

Fig. 3. Comparison between the scaled particle theory and experiment for reciprocal osmotic compressibility of PHIC-DCM [60]. Vertical segments indicate C_I.

It follows from Eq. (20) that

$$\frac{\partial \Pi}{\partial c} = \frac{RT}{M(1-vc')^2}\left[1 + \frac{Bc'}{1-vc'} + \frac{2Cc'^2}{(1-vc')^2}\right] \quad (30)$$

where M is the molecular weight of the monodisperse polymer. When the diameter d is chosen to be 0.95 nm and the molecular parameters other than d are estimated from the wormlike chain parameters for PHIC listed in Table 1, this equation fits the data points closely, as shown by the solid curves in Fig. 3. The chosen d is somewhat smaller than that estimated from the partial specific volume (cf. Table 2). Recently, Jinbo et al. [61] have shown that the $\partial \Pi/\partial c$ data of Itou et al. can be more satisfactorily fitted by the scaled particle theory incorporating a weak attractive interaction.

For a polydisperse polymer, analysis of sedimentation equilibrium data becomes complex, because the molecular weight distribution significantly affects the solute distribution. In 1970, Scholte [62] made a thermodynamic analysis of sedimentation equilibrium for polydisperse flexible polymer solutions on the basis of Flory and Huggins' chemical potential equations. From a similar thermodynamic analysis for stiff polymer solutions with Eqs. (27) for Π and (28) for the polymer chemical potential, we can show that the right-hand side of Eq. (29) for the isotropic solution of a polydisperse polymer is given, in a good approximation, by Eq. (30) if M is replaced by M_w [41].

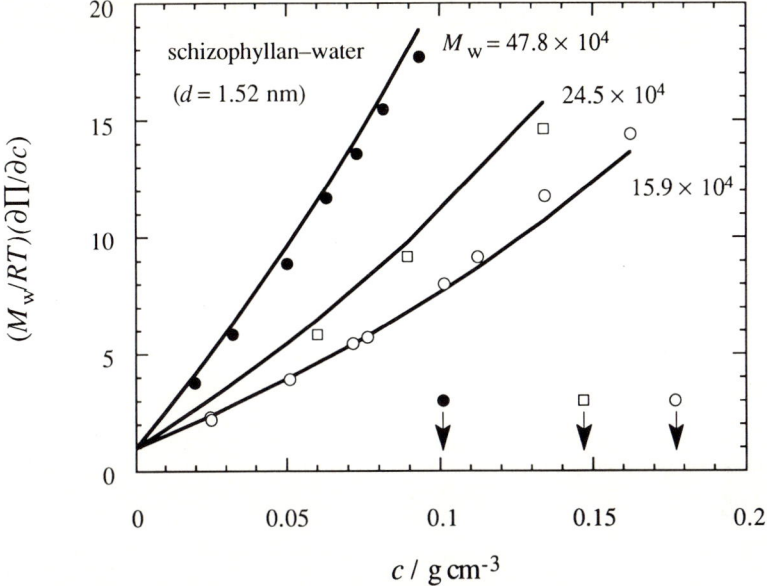

Fig. 4. Comparison between the scaled particle theory and experiment for reciprocal osmotic compressibility of schizophyllan-water [63]. Arrows indicate C_I.

Figure 4 compares osmotic compressibility data for isotropic schizophyllan–water solutions [63] with the scaled particle theory. The ratios of the z-average to the weight-average molecular weights of these schizophyllan samples are ca. 1.2. The solid curves, calculated with d taken to be 1.52 nm and other molecular parameters (L_c, v, and c') estimated from M_W and the wormlike chain parameters in Table 1, are seen to come close to the data points for all samples.

4 Phase Equilibrium Behavior

4.1 Binary Solutions

Many stiff-chain polymer solutions form a liquid crystal phase at high polymer concentrations. Their phase boundary concentrations c_I and c_A usually depend on the polymer molecular weight as well as the temperature, as illustrated in Fig. 5. Here, c_I and c_A are plotted against weight-average molecular weight M_W for the system of poly(n-hexyl isocyanate) (PHIC) and toluene at 25 °C [64]. Both c_I and c_A decrease steeply with increasing M_W in the low M_W region and appear to approach asymptotic values at higher M_W. If M_W is converted to the number N of the Kuhn statistical segments with the wormlike chain parameters

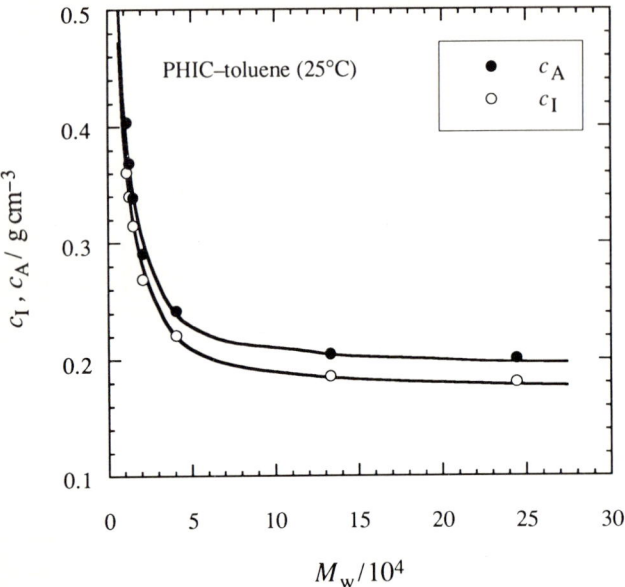

Fig. 5. Concentration–molecular weight phase diagram of the PHIC–toluene system at 25 °C [64]

listed in Table 1, c_I and c_A level off at N of the order of unity (corresponding to $M_W \approx 5.5 \times 10^4$).

The ratios c_A/c_I estimated from Fig. 5 are between 1.08 and 1.12, which means that the biphasic region of this system is very narrow. Such narrow phase gaps are a general feature of lyotropic polymer liquid crystal systems, though they vary with sample polydispersity. In general, the larger the polydispersity, the wider the biphasic region becomes (cf. Sect. 4.2). The M_W/M_n ratios of the PHIC samples in Fig. 5 were less than 1.06 except for the lowest molecular weight sample which had $M_W/M_n = 1.15$.

Figure 6 shows the phase diagrams plotting temperature T vs c for PHIC–toluene systems with different M_W or N [64], indicating c_I and c_A to be insensitive to T, as is generally the case with lyotropic polymer liquid crystal systems. This feature reflects that the phase equilibrium behavior in such systems is mainly governed by the hard-core repulsion of the polymers. The weak temperature dependence in Fig. 6 may be associated with the temperature variation of chain stiffness [64]. We assume in the following theoretical treatment that liquid crystalline polymer chains in solution interact only by hard-core repulsion. The isotropic–liquid crystal phase equilibrium in such a solution is then the balance between S_{ex} and S_{or}, as explained in the last part of Sect. 2.2.

Now we compare the isotropic–liquid crystal phase boundary concentrations for various polymer solution systems with the scaled particle theory for the wormlike spherocylinder. If the equilibrium orientational distribution function $\bar{f}(a)$ in the coexisting liquid crystal phase is approximated by the Onsager trial

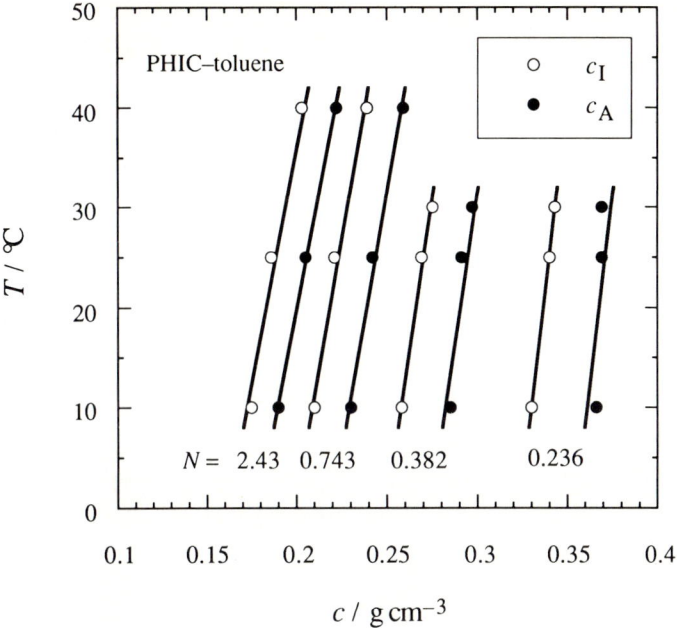

Fig. 6. Temperature–concentration phase diagrams for four PHIC–toluene systems with different molecular weights [64]

function, the phase boundary number densities c'_I and c'_A and the degree of orientation parameter α in $\bar{f}(a)$ can be calculated from Eq. (23) along with the phase equilibrium conditions

$$\Pi_I = \Pi_A, \quad \mu_I = \mu_A \tag{31}$$

where the subscripts I and A identify the isotropic and liquid crystal (anisotropic) phases, respectively. (See Appendix A about the accuracy of the Onsager trial fraction procedure.)

When all lengths associated with polymers are measured in units of the Kuhn statistical segment length $2q$, the thermodynamic functions ΔF, Π, and μ, given by Eqs. (19)–(21), contain two molecular parameters $N \equiv L/2q$ and $\tilde{d} \equiv d/2q$ and two state variables $\tilde{c}' \equiv (2q)^3 c'$ and α. Thus, numerical solution to Eqs. (23) and (31) provides $\tilde{c}'_I$, $\tilde{c}'_A$, and α as functions of N and $\tilde{d}$. The results for the phase boundary concentrations have been found to be represented to a good approximation by the following empirical expressions:

$$\frac{\tilde{v}\tilde{c}'_v}{\tilde{d}a_v(\tilde{d})} = \frac{A_{v,0} + A_{v,1}N + A_{v,2}N^2 + N^3}{N(1 + A_{v,3}N + N^2)} + \Delta_v(N, \tilde{d}) \quad (v = I \text{ and } A) \tag{32}$$

where $\tilde{v} \equiv v/(2q)^3$ and $a_v(\tilde{d})$ and $\Delta_v(N, \tilde{d})$ are defined by

$$a_v(\tilde{d}) = \sum_{i=0}^{4} a_{v,i}(\log \tilde{d})^i \tag{33}$$

and

$$\Delta_v(N, \tilde{d}) = \frac{\beta_v(\tilde{d})}{N} + \frac{\gamma_v(\tilde{d})}{N^2} + \frac{\delta_v(\tilde{d})}{N^3} \tag{34}$$

with

$$\beta_v(\tilde{d}) = \sum_{i=0}^{3} \beta_{v,i}(\log \tilde{d})^i, \quad \gamma_v(\tilde{d}) = \sum_{i=0}^{3} \gamma_{v,i}(\log \tilde{d})^i,$$

$$\delta_v(\tilde{d}) = \sum_{i=0}^{3} \delta_{v,i}(\log \tilde{d})^i \tag{35}$$

The numerical coefficients in these equations as well as the numerical constants $A_{v,i}$ in Eq. (32) are given in Table 5. In fact, Eq. (32) approximates the results of direct numerical analysis to within 3% for $0.0015 \leq \tilde{d} \leq 0.15$, $N \geq 0.05$, and $L/d \geq 5$, the conditions which are fulfilled by most stiff-chain polymer solution systems studied so far. Equation (32) is more accurate at small N than our previous theory [18], in which slightly different empirical equations for $\tilde{c}'_I$ and $\tilde{c}'_A$ were proposed.

Equation (32) has been compared with phase boundary concentration data in the following way. For each solution, N of the polymer sample is estimated from M_W or the viscosity-average molecular weight M_v along with the molecular parameters M_L and q listed in Table 1, and $\tilde{d}$ is calculated with d from Π or $\partial \Pi / \partial c$ data. For systems which lack these data, the values of d from the (partial) specific volume v_{sp} may be substituted. Table 2 lists the resulting values of d from Π, $\partial \Pi / \partial c$, or v_{sp} for various systems. The phase boundary volume fractions vc'_v ($\equiv \tilde{v}\tilde{c}'_v$; $v = I$ and A) are calculated from experimental phase boundary weight fractions (or mass concentrations) with d, M_W (or M_V), and M_L. Finally, with these numerical results, $[vc'_v/\tilde{d}a_v(\tilde{d})] - \Delta_v(N, \tilde{d})$ is computed

Table 5. Numerical parameters appearing in Eqs. (32)–(35) for the phase boundary concentrations c_I and c_A

i	0	1	2	3	4
(1) c_I (v = I)					
$A_{I,i}$	0.2225	1.1651	1.1121	0.9206	
$a_{I,i}$	3.624	8.513	13.97	5.579	0.7111
$\beta_{I,i}$	− 0.04865	0.04060	0.06363	0.01444	
$\gamma_{I,i}$	0.01193	0.03464	0.02033	0.003166	
$\delta_{I,i}$	− 0.0009415	− 0.001639	− 0.0008694	− 0.0001363	
(2) c_A (v = A)					
$A_{A,i}$	0.2232	1.1121	1.1468	0.9887	
$a_{A,i}$	4.631	11.38	16.96	6.551	0.8152
$\beta_{A,i}$	0.03635	0.1696	0.1274	0.02471	
$\gamma_{A,i}$	− 0.02349	− 0.01236	− 0.001490	− 0.0003100	
$\delta_{A,i}$	0.004000	0.005394	0.002430	0.0003731	

by Eqs. (33)–(35), and plotted against N. It is expected that the resulting plot becomes a master curve for experimental data.

Figures 7 and 8 display such plots for various lyotropic liquid-crystalline polymer systems, which range in q from 5.3 to 200 nm. As expected, most data points come close to the theoretical curve. This finding suggests that liquid crystallinity of stiff-chain or semiflexible polymer solutions has its main origin in the hard-core repulsion of the polymers.

Small deviations of the data points from the theory in Figs. 7 and 8 at small N may be attributed to the polydispersity of the samples in molecular weight. As will be mentioned in Sect. 4.2, when N is small, c_I decreases in the slight presence of a larger N component, while when $N \sim 1$, c_A increases in the slight presence of a smaller N component. Thus polydispersity widens the phase gap.

If $N \gg 1$ for all components of a polymer sample, c_I and c_A are not much affected by polydispersity [73], because the molecular weight dependence of c_I and c_A tends to vanish at large N. Thus, small but clear deviations of the data points for cellulose derivative solutions [34, 35, 72] from the calculated curves in Figs. 7 and 8 may not be simply attributed to polydispersity. Furthermore, since the M_w/M_n ratios of the PHIC and PYPt samples were less than 1.1 and 1.2,

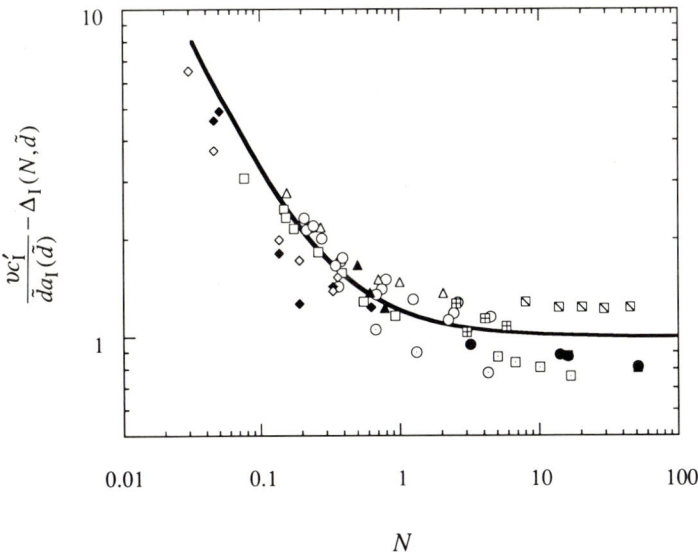

Fig. 7. Comparison of experimental phase boundary concentrations between the isotropic and biphasic regions for various liquid-crystalline polymer solutions with the scaled particle theory for wormlike hard spherocylinders. (□) schizophyllan–water [65]; (△) poly(γ-benzyl L-glutamate) (PBLG)–dimethylformamide (DMF) [66-69]; (▲) PBLG–m-cresol [70]; (♦) PBLG–dioxane [71]; (◇) PBLG–methylene chloride [71]; (o) poly(n-hexyl isocyanate) (PHIC)–toluene at 10, 25, 30, 40 °C [64]; (○) PHIC–dichloromethane (DCM) at 20 °C [64]; (⊞) a poly(yne)-platinum polymer (PYPt)–trichloroethane (TCE) [33]; (●) (hydroxypropyl)-cellulose (HPC)–water [34]; (■) HPC–dimethylacetamide (DMAc) [34]; (◨) (acetoxypropyl) cellulose (APC)–dibutylphthalate (DBP) [35]; (⊡) cellulose triacetate (CTA)–trifluoroacetic acid [72]

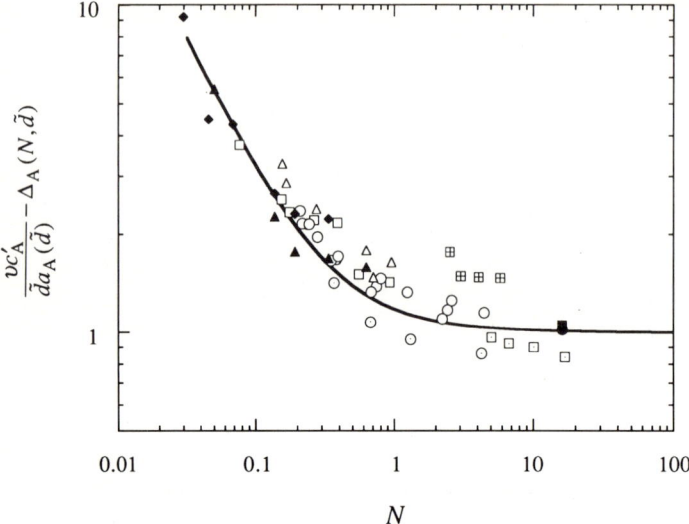

Fig. 8. Comparison of experimental phase boundary concentrations between the biphasic and liquid crystal regions for various liquid crystalline polymer solutions with the scaled particle theory for hard wormlike spherocylinders. The symbols are the same as those in Fig. 7

respectively, the small disagreement of their data with the theory is probably a reflection of weak soft interchain attraction [61] neglected in the present analysis.

4.2 Ternary Systems Containing two Polymer Species with Different Lengths

A ternary system consisting of two polymer species of the same kind having different molecular weights and a solvent is the simplest case of polydisperse polymer solutions. Therefore, it is a prototype for investigating polydispersity effects on polymer solution properties. In 1978, Abe and Flory [74] studied theoretically the phase behavior in ternary solutions of rodlike polymers using the Flory lattice theory [3]. Subsequently, ternary phase diagrams have been measured for several stiff-chain polymer solution systems, and work [6, 17] has been done to improve the Abe–Flory theory.

Figure 9 shows ternary phase diagrams for quasi-ternary systems of PHIC and toluene [73] and schizophyllan and water [75, 76]. Here the composition of each system is expressed in terms of the total polymer mass concentration c and the weight fraction ξ of the lower molecular weight polymer component in the total solute. In either phase diagram, the tie lines represented by the dashed segments indicate that the liquid crystal phase is rich in the higher molecular weight component, in agreement with the fractionation effect first predicted by Abe and Flory.

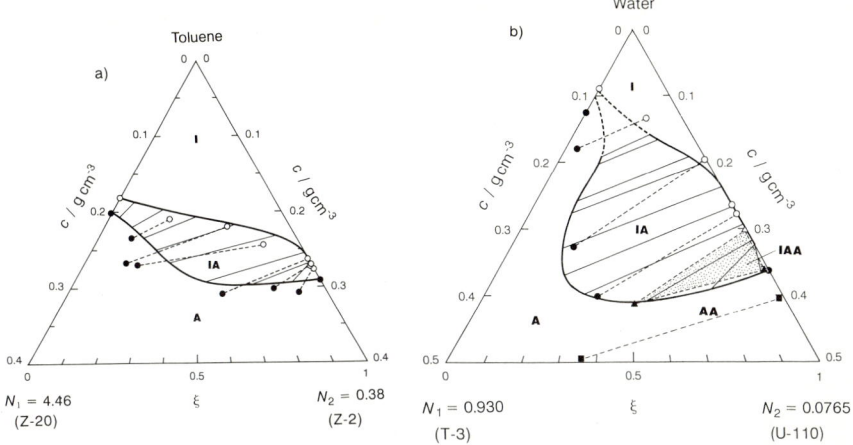

Fig. 9. Phase diagrams of quasi-ternary systems containing two different molecular weight samples: **a** PHIC–toluene with $(N_1, N_2) = (4.46, 0.38)$ [73]; **b** schizophyllan–water system with $(N_1, N_2) = (0.930, 0.0765)$ [75, 76]. ($\bigcirc$, $\triangle$) experimental coexisting isotropic phase; ($\bullet$, $\blacktriangle$, $\blacksquare$) experimental coexisting anisotropic phase; *dashed segments*, experimental tie lines; *the shadowed triangular region*, the IAA triphasic region; *thick full curves*, theoretical binodals; *thin full segments*, theoretical tie lines

By connecting the coexisting phase points in Fig. 9a, we can divide the ternary phase diagram into three regions, the isotropic phase region (I), the anisotropic (nematic) phase region (A), and the isotropic–anisotropic biphasic region (IA). We can observe the following characteristic polydispersity effects on the phase boundary concentrations. At small ξ, as ξ increases, c_A between IA and A increases more rapidly from that of the binary system than does c_I between I and IA, while, at $\xi \approx 1$, as ξ decreases, c_I decreases very sharply from that of the binary system, but c_A remains insensitive to ξ.

The ternary phase diagram shown in Fig. 9b for the schizophyllan–water system has two more phase regions: the isotropic–anisotropic (cholesteric)–anisotropic triphasic region (IAA) and the anisotropic–anisotropic biphasic region (AA). This was the first observation of the complex phase diagram as had been predicted by Abe and Flory, but the diagram quantitatively departs from their prediction in the positions and sizes of the regions [76].

Phase diagrams of ternary systems containing two semiflexible polymer components 1 and 2 can be calculated from thermodynamic functions given by Eqs. 26–28. The thick solid curves and thin solid segments in Fig. 9 are calculated binodals and tie lines, respectively [17][3]. In the calculation, the diameter

---

[3] The IA phase coexistence equations could not be solved in a low ξ region for the aqueous schizophyllan system, where the degree of orientation parameter α_2 of the lower molecular weight polymer component 2 becomes very small. This is mainly due to poor approximations in the asymptotic expansions of ρ_{st} and $\sigma_s(N)$ at small α_2 values. The broken binodals in Panel b of Fig. 9 are not calculated results but interpolation curves.

d was taken to be 1.52 nm for schizophyllan as estimated from $(\partial \Pi / \partial c)$ data, and 1.07 nm for PHIC which was slightly smaller than that from Π data ($= 1.15$ nm). This change in d for PHIC brings c_I and c_A closer to the experimental values at $\xi = 0$ and 1 [41].

The theoretical IA binodals successfully reproduce the experimental binodals for both systems. Furthermore, the theoretical tie lines correctly predict the fractionation effect found by experiment. Thus, the scaled particle theory predicts the IA binodals and tie lines more accurately than the Abe–Flory theory. The success owes much to incorporating chain flexibility into the theory.

Although the scaled particle theory fails to predict correctly the IAA and AA regions observed in the schizophyllan–water system, this may not be taken too seriously, because triphasic equilibria result from too subtle a balance of the chemical potentials (and the osmotic pressures) of the coexisting three phases. In fact, if N_2 is changed from 0.0765 to 0.07, the scaled particle theory yields the IAA and AA regions as depicted in Fig. 10, where the shape and position of the IAA region well resemble those of the triphasic region in the aqueous schizophyllan solutions. We note that no extra interaction other than the hard core repulsion is needed in achieving these calculated results.

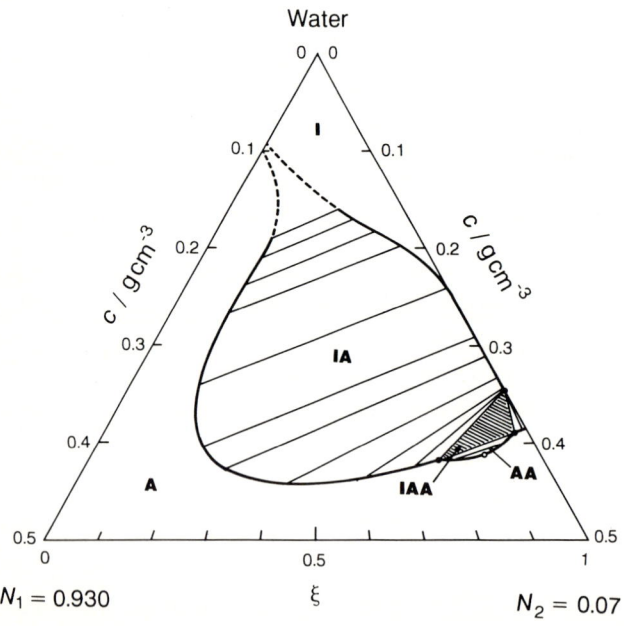

Fig. 10. Theoretical ternary phase diagram calculated from the scaled particle theory for wormlike hard spherocylinders with $(N_1, N_2) = (0.930, 0.070)$, $d = 1.52$ nm, $q = 200$ nm, and $M_L = 2150$ nm^{-1} [17]

4.3 Electrostatic Interaction

More than half a century ago, Bawden and Pirie [77] found that aqueous solutions of tobacco mosaic virus (TMV), a charged rodlike virus, formed a liquid crystal phase at as very low a concentration as 2%. To explain such remarkable liquid crystallinity was one of the central themes in the famous 1949 paper of Onsager [2]. However, systematic experimental studies on the phase behavior in stiff polyelectrolyte solutions have begun only recently. At present, phase equilibrium data on aqueous solutions qualified for quantitative discussion are available for four stiff polyelectrolytes, TMV, DNA, xanthan (a double helical polysaccharide), and fd-virus.

Figure 11 shows the phase boundary concentration data for aqueous Na salt xanthan [78], fd-virus [24], and TMV [23] with added salt. In all these systems, c_I and c_A are very low at low added salt concentration C_s or ionic strength I, and increase with C_s or I. Since such low phase boundary concentrations are not usually observed for neutral liquid-crystalline polymer solutions, it is apparent that polyion electrostatic interactions play an important role in the phase equilibria of these systems.

In Fig. 11a, c_I and c_A for aqueous Na salt xanthan at three different C_s are plotted against N. Both concentrations decrease sharply with N at low N, but almost level off at an N near unity for all C_s. This N dependence is not very different from that for neutral liquid-crystalline polymer solutions (cf. Fig. 5).

In order to calculate the phase boundary concentrations for stiff polyelectrolyte solutions, we express the total intermolecular interaction u for the polyion as the sum of the hard-core interaction u_0 and the electrostatic interaction w_{el}, and assume Eq. (1) for u_0 and the following for w_{el}:

$$w_{el} = \begin{cases} 0 & \text{when the hard cores of the two molecules overlap} \\ w_{el}(r, \mathbf{a}(s_1), \mathbf{a}(s_2)) & \text{otherwise} \end{cases} \quad (36)$$

In $w_{el}(r, \mathbf{a}(s_1), \mathbf{a}(s_2))$, r is the distance between the closest contour points s_1 and s_2 on the two interacting polyions 1 and 2, and $\mathbf{a}(s_1)$ and $\mathbf{a}(s_2)$ are the two tangent vectors at s_1 and s_2 [61]. When the persistence length q of the polyion is much longer than the Debye screening length κ^{-1} and the contour points s_1 and s_2 are not located near the chain ends, $w_{el}(r, \mathbf{a}(s_1), \mathbf{a}(s_2))$ can be approximated by that for an infinitely long straight charged cylinder with the same d, which in turn can be calculated from the electrostatic potential determined by the Poisson–Boltzmann equation. As proposed by Stroobants et al. [79], Philip and Wooding's approximate solution [80] to the nonlinear Poisson–Boltzmann equation is adequate for the calculation of $w_{el}(r, \mathbf{a}(s_1), \mathbf{a}(s_2))$. The chain end effect on $w_{el}(r, \mathbf{a}(s_1), \mathbf{a}(s_2))$ was considered by Odijk [81] and also by Sato and Teramoto [82].

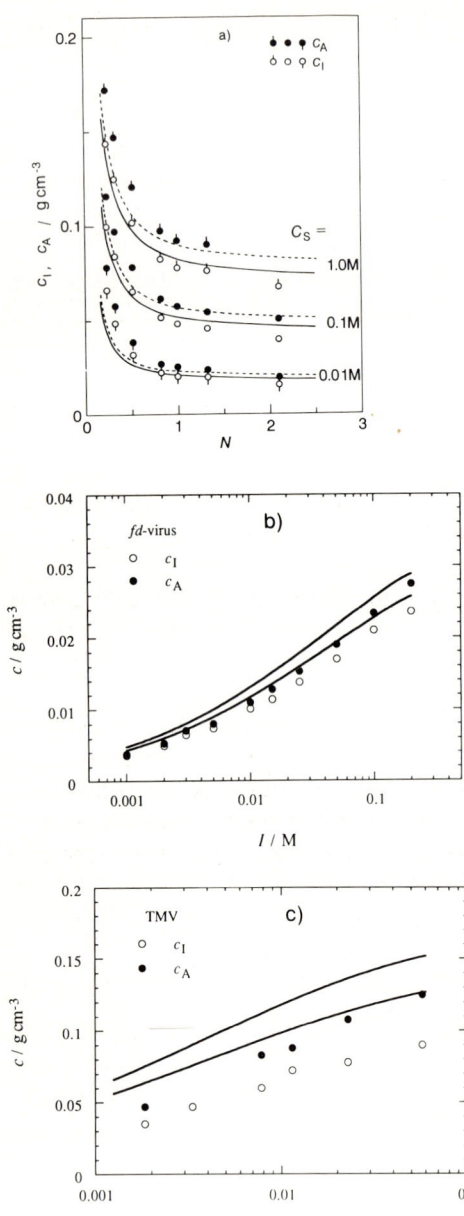

Fig. 11a–c. Phase boundary mass concentrations for aqueous solutions of three stiff polyions: **a** xanthan (a double-helical polysaccharide) [78]; **b** fd-virus [24]; **c** tobacco mosaic virus (TMV) [23]. *Circles*, experimental results; *curves*, predictions by the first-order perturbation theory (see text)

With the wormlike hard spherocylinder system with $u = u_0$ taken as a reference system, and the electrostatic interaction w_{el} regarded as a thermodynamic perturbation, we can derive by perturbation calculation [82]

$$\Delta F = (\Delta F)_{u=u_0} - \tfrac{1}{2} n k_B T \langle \beta_{el} \rangle c' \tag{37}$$

where $(\Delta F)_{u=u_0}$ is given by Eq. (2), and $\langle \beta_{el} \rangle$ is the orientation-averaged binary cluster integral for w_{el}. Equation (37) neglects the second and higher order electrostatic perturbations.

For an actual calculation of the phase boundary concentrations, the following remarks on $\langle \beta_{el} \rangle$ are important. (1) As pointed out by Stroobants et al. [79], $\langle \beta_{el} \rangle$ for the anisotropic phase contains a new orientation-dependent parameter. This parameter concerns the effect that the electrostatic repulsion between two polyions causes them to orient mutually perpendicularly (the twisting effect), and it is given in [79]. (2) At low ionic strengths, the effective diameter of the polyion defined by the second virial coefficient becomes large and the polyion end effect on $\langle \beta_{el} \rangle$ as discussed in [82] cannot be neglected for relatively short polyions. (3) The Debye screening length κ^{-1} appearing in $\langle \beta_{el} \rangle$ changes with the polyion concentration owing to the following two effects. Firstly, small mobile ions are expelled from the vicinity of polyelectrolytes to the "solvent region" by the salt exclusion effect. Hence the effective mobile ion concentration in the solvent region increases with the polyion concentration. Secondly, the redistribution of small mobile ions occurs between the coexisting isotropic and liquid crystal phases. These two effects are treated in [82, 83].

With the above polyion effects taken into account, the phase boundary concentrations c_I and c_A have been computed from the phase coexistence equations and the free energy minimization condition [82, 83]. The curves in Fig. 11a–c represents the results from such calculations with the use of the molecular parameters listed in Table 6. The agreement between theory and experiment is almost quantitative for aqueous xanthan solutions (except at $C_s = 0.01$ M in the low N region) and aqueous fd-virus. The calculated C_s dependence for the medium molecular weight xanthan sample almost quantitatively predicts experiment [82]. However, for TMV, this theory overestimates the phase boundary concentrations, probably owing to the neglect of the higher-order perturbations. Fraden et al. [23] allowed for the higher virial terms in both u_0 and w_{el} by replacing the total pair potential u by the hard-core potential so as to make the second virial coefficient coincide with that of the polyion. This treatment, however, is not justified theoretically.

Rill and Strzelecka [85–88] studied isotropic–cholesteric phase equilibria in aqueous DNA, with the phase boundary concentration results which are compared with those from the perturbation theory in Table 7. Their experimental

Table 6. Molecular parameters of liquid crystalline polyions

polyion	q/nm	d/nm	M_L/nm^{-1}	v[a]	Ref.
TMV	∞	18	133 000	14	[23]
fd-virus	2200	6.6	21 600	10	[24]
xanthan	120	2.2	1 940	3.0	[84]
DNA	50	2.5	1 950	6.1	[28]

[a] Linear charge density in units of (elementary charge)/nm

Table 7. Comparison of experimental phase boundary concentrations for aqueous DNA (N = 0.49) with the perturbation theory

		experiment				perturbation theory	
C_s/M	T/°C	c_I/g cm^{-3}	c_A/g cm^{-3}	c_A/c_I	Ref.	c_I/g cm^{-3}	c_A/g cm^{-3}
2	30	0.180	0.210	1.2	[85]	0.169	0.180
0.21	20	0.123	0.152	1.24	[86]	0.139	0.148
1.0	20	0.17	0.272	1.6	[88]	0.163	0.174
0.1	20	0.136	0.271	2.0	[88]	0.122	0.130
0.01	20	0.131	0.255	1.95	[88]	0.068	0.072

results at $C_s = 2$ M [85] and 0.21 M [86] are in fair agreement with the theoretical values, but more recent data of Strzelecka and Rill [88] at $C_s = 0.01, 0.1$, and 1.0 M are not, especially for c_A. However, since Strzelecka and Rill observed a new phase called "precholesteric phase" at a concentration between c_I and c_A, the disagreement may not be taken too seriously [78].

In concluding this section, we should touch upon phase boundary concentration data for poly(p-benzamide)–dimethylacetamide + 4% LiCl [89], poly(p-phenylene terephthalamide) (PPTA; Kevlar)–sulfuric acid [90], and (hydroxypropyl)cellulose–dichloroacetic acid solutions [91]. Although not included in Figs. 7 and 8, they show appreciable downward deviations from the prediction by the scaled particle theory for the wormlike hard spherocylinder. Arpin and Strazielle [30] found a negative concentration dependence of the reduced viscosity for PPTA in dilute solution of sulfuric acid, as often reported on polyelectrolyte systems. Therefore, the deviation of the c_I data for PPTA in sulfuric acid from the scaled particle theory may be attributed to the electrostatic interaction. For the other two systems too, the low c_I values may be due to the protonation of the polymer, because the solvents of these systems are very polar.

5 Orientational Order Parameter

The order parameter S is used to characterize the degree of orientation in the liquid crystal phase. Experimentally it can be determined from NMR quadrapolar splitting, X-ray diffraction, or IR dichroism, but the measurement has to make the director in the liquid crystal sample align in a specific direction. For a liquid-crystalline polymer with a high dissymmetry of magnetic susceptibility, the alignment can be made by a strong external magnetic field. A special treatment of the inner glass surface of a thin sample cell can also cause the director to align. In general, polymer liquid crystals are more difficult to align than small molecular liquid crystal samples, probably because of their high viscosity.

Recently, experimental data of S for various liquid-crystalline polymer solutions have become available. Figure 12 illustrates the concentration dependence of S for three polymer liquid crystal systems: (a) PBLG–DMF [92, 93], (b) PHIC–toluene [94], and (c) poly(yne)–platinum polymer (PYPt)–trichloroethane (TCE) [33]. For systems (a) and (c), with the alignment made by

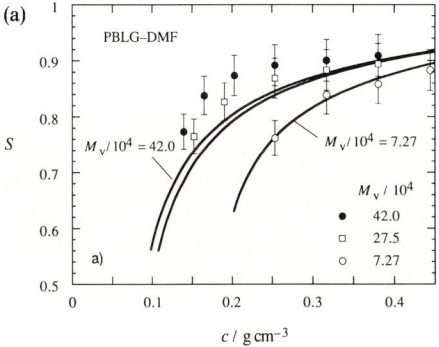

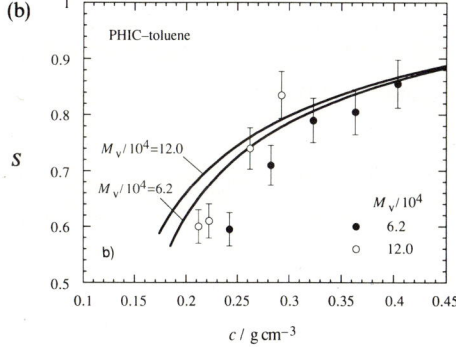

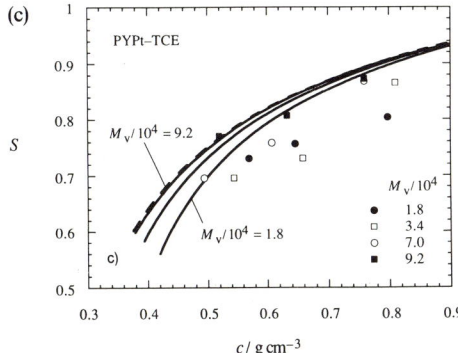

Fig. 12a–c. Polymer concentration dependence of the orientational order parameters S for three liquid-crystalline polymer systems: **a** PBLG–DMF [92, 93]; **b** PHIC–toluene [94]; **c** PYPt–TCE [33]. *Marks* experimental data; *solid curves*, theoretical values calculated from the scaled particle theory. The *left end* of each curve gives the phase boundary concentration c_A

application of an external magnetic field, S was determined by a deuterium NMR technique. On the other hand, the alignment of system (b) was made by rubbing the inner surfaces of NaCl cell windows, and S was determined by IR dichroism. The calculation of S from the NMR quadrapolar splitting needs the component of the quadrapolar interaction tensor along the molecular axis. Chapman et al. [95] and Abe et al. [96, 97] proposed different values for the constant of N–D bond in the deuterated polypeptide chain. The error bars in Fig. 12a indicate the variance of S due to this difference. The IR dichroism method needs the angle between the chromophor transition moment and the polymer axis. The error bars in Fig. 12b represent the uncertainty in the angle between the transition moment of the C–N vibration and the molecular axis of PHIC.

For all systems in Fig. 12, S is an increasing function of the polymer concentration. Furthermore, S increases with increasing polymer molecular weight, markedly for PBLG and PHIC, but not very much for PYPt.

The order parameter S is the orientational average of the second-order Legendre polynomial $P_2(\mathbf{a} \cdot \mathbf{n})$ ($\mathbf{n}$ = the director), and if the orientational distribution function is approximated by the Onsager trial function, it can be related to the degree of orientation parameter α by

$$S = 1 - \frac{3}{\alpha} \coth \alpha + \frac{3}{\alpha^2} \tag{38}$$

The equilibrium value of α in the nematic phase can be determined by minimizing ΔF. With Eq. (19) for ΔF from the scaled particle theory, S has been computed as a function of c, and the results are shown by the curves in Fig. 12. Here, the molecular parameters L_c and N were estimated from the viscosity average molecular weight M_v along with M_L and q listed in Table 1, and d was chosen to be 1.40 nm (PBLG), 1.15 nm (PHIC), and 1.08 nm (PYPt), as in the comparison of the experimental phase boundary concentrations with the scaled particle theory (cf. Table 2).

It can be seen that the theory moderately well describes experimental results for all systems investigated. DuPré and Yang [46] showed Lee's theory [53] to agree equally well with these data for PBLG and PHIC if polymer flexibility effect was taken into account. The disagreement between calculated and experimental results in Fig. 12 may again be ascribed to the polydispersity effect and/or a weak attractive interaction of polymers, as invoked for similar finding regarding the phase boundary concentrations in Sect. 4.1.

Abe et al. pointed out that the experimental S of PBLG [93] and PYPt [33] solutions at the phase boundary concentration c_A largely departed from the prediction of Khokhlov and Semenov's second virial approximation theory [7, 44]. Similar deviations of the scaled particle theory from experiment are seen in Fig. 12a, c, where the left ends of the theoretical curves and the experimental data points at the lowest c correspond to c_A. Sato et al. [17] showed theoretically that ternary solutions containing two polymer species with different lengths have considerably higher S at c_A than the binary solutions containing either of these polymer species (cf. Fig. 3 of [17]). Therefore, it is likely that sample polydispersity causes the disagreement in S at c_A in Fig. 12.

Wang and DuPré [94] found that for the PHIC-chloroform system the S values were much larger than predicted by modified Lee's theory with q = 21–37 nm (cf. Table 1). Although DePré and Yang [46] fitted their data to the theory incorporating the polymer flexibility effect, the value of q = 89 nm used is too large when compared with that estimated from the intrinsic viscosity data [98].

6 Theory of Stiff-Chain Polymer Dynamics

6.1 Kinetic Equation

In the second half of this article, we discuss dynamic properties of stiff-chain liquid-crystalline polymers in solution. If the position and orientation of a stiff or semiflexible chain in a solution is specified by its center of mass and end-to-end vector, respectively, the translational and rotational motions of the whole chain can be described in terms of the time-dependent single-particle distribution function $f(\mathbf{r}, \mathbf{a}; t)$, where $\mathbf{r}$ and $\mathbf{a}$ are the position vector of the center of mass and the unit vector parallel to the end-to-end vector of the chain, respectively, and t is time. ($\mathbf{a}$ should be distinguished from the unit tangent vector to the chain contour appearing in the previous sections, except for rodlike polymers.) Since this distribution function cannot describe internal motions of the chain, our discussion below is restricted to such global chain dynamics as translational and rotational diffusion and zero-shear viscosity.

In general, the motion of a polymer chain in solution is governed by intermolecular interaction, hydrodynamic interaction, Brownian random force, and external field. The hydrodynamic interaction consists of the intra- and intermolecular ones. The intramolecular hydrodynamic interaction and Brownian force play dominant roles in dilute solution, while the intermolecular interaction and the intermolecular hydrodynamic interaction become important as the concentration increases.

The kinetic equation for the distribution function $f(\mathbf{r}, \mathbf{a}; t)$ must include all these effects. Doi and Edwards [4, 99] proposed for it the generalized Smoluchowski equation

$$\frac{\partial}{\partial t} f = T[f] + R[f] \qquad (39)$$

where the first and second terms on the right-hand side relate to the translational and rotational motions of the chain under consideration, respectively, and are explicitly written

$$T[f] \equiv \frac{\partial}{\partial \mathbf{r}} \left\{ [D_\parallel \mathbf{aa} + D_\perp (\mathbf{I} - \mathbf{aa})] \left(\frac{\partial f}{\partial \mathbf{r}} + \frac{f}{k_B T} \frac{\partial h}{\partial \mathbf{r}} \right) - \mathbf{v}f \right\} \quad (40a)$$

$$R[f] \equiv \Re \cdot \left[D_r \left(\Re f + \frac{f}{k_B T} \Re h \right) - \omega f \right] \quad (40b)$$

Here $D_\parallel$, $D_\perp$, and D_r are, respectively, the longitudinal, transverse, and rotational diffusion coefficients of the chain averaged over the internal degree of freedom, h an external field, and $\mathbf{v}$ and ω the macroscopic velocity and angular velocity of the chain induced by a flow field in the solution. Furthermore, $\mathbf{I}$ is the unit tensor and $\Re$ is the rotational operator defined by

$$\Re \equiv \mathbf{a} \times \frac{\partial}{\partial \mathbf{a}}$$

Both $T[f]$ and $R[f]$ consist of three terms associated with the diffusion and convection induced by the external field and the macroscopic flow field.

When the solution is dilute, the three diffusion coefficients in Eq. (40a, b) may be calculated only by taking the intramolecular hydrodynamic interaction into account. In what follows, the diffusion coefficients at infinite dilution are signified by the subscript 0 (i.e., $D_{\parallel 0}$, $D_{\perp 0}$, and D_{r0}). As the polymer concentration increases, the intermolecular interaction starts to become important to polymer dynamics. The chain incrossability or topological interaction hinders the translational and rotational motions of chains, and slows down the three diffusion processes. These are usually called the entanglement effect on the rotational and transverse diffusions and the jamming effect on the longitudinal diffusion. In solving Eq. (39), these effects are taken into account by use of *effective* diffusion coefficients as will be discussed in Sect. 6.3.

In addition to the above effects, the intermolecular interaction may affect polymer dynamics through the *thermodynamic force*. This force makes chains align parallel with each other, and retards the chain rotational diffusion. This slowing down in the isotropic solution is referred to as the pretransition effect. The thermodynamic force also governs the unique rheological behavior of liquid-crystalline solutions as will be explained in Sect. 9. For rodlike polymer solutions, Doi [100] treated the thermodynamic force effects by adding a *self-consistent mean field* or a *molecular field* $V_{scf}(\mathbf{a})$ to the external field potential h in Eq. (40b). Using the second virial approximation (cf. Sect. 2), he formulated $V_{scf}(\mathbf{a})$, as follows [4]:

$$V_{scf}(\mathbf{a}) = 2k_B T L^2 dc' \int |\sin\gamma(\mathbf{a}, \mathbf{a}')| f(\mathbf{r}, \mathbf{a}'; t) \, d\mathbf{a}' \quad (41)$$

where L and d are the length and diameter of the rod, respectively, and $\gamma(\mathbf{a}, \mathbf{a}')$ the angle between $\mathbf{a}$ and $\mathbf{a}'$.

At finite polymer concentrations, the intermolecular hydrodynamic interaction may also alter polymer dynamics. Except for spherical particles, the hydrodynamic calculations of the effective diffusion coefficients including this effect have not yet reached a satisfactory state. However, for highly anisotropic molecules, the topological interaction so strongly retards the transverse and rotational motions that the intermolecular hydrodynamic interaction (or the hydrodynamic screening effect) may be ignored. Muthukumar and Edwards [101] have concluded by an effective medium argument that the screening effect in rodlike polymer solutions is less important than in flexible polymer solutions. For these reasons, we neglect for the intermolecular hydrodynamic interaction in the following formulation of stiff chain polymer solutions. Then the velocity **v** and the angular velocity ω in Eq. (40a,b) can be expressed in terms of the unperturbed velocity gradient tensor κ of the solvent as follows:

$$\begin{cases} \mathbf{v} = \mathbf{\kappa} \cdot \mathbf{r} \\ \omega = \chi[\mathbf{a} \times (\mathbf{A} \cdot \mathbf{a})] - [\mathbf{a} \times (\mathbf{\Omega} \cdot \mathbf{a})] \end{cases} \tag{42a}$$

where

$$\mathbf{A} \equiv \tfrac{1}{2}(\mathbf{\kappa}^\dagger + \mathbf{\kappa}), \quad \mathbf{\Omega} \equiv \tfrac{1}{2}(\mathbf{\kappa}^\dagger - \mathbf{\kappa}) \tag{42b}$$

with the superscript † denoting the transpose. The coefficient χ is a factor concerning the intramolecular hydrodynamic interaction and must be calculated by hydrodynamics (cf. Appendix B and Table B1).

6.2 Fuzzy Cylinder Model

If **r** (the center-of-mass position) and **a** (the end-to-end axis direction) of a semiflexible chain are fixed in a solution, the segments of the chain distribute on average inside a cylindrical domain, whose length and diameter may be equated to $\langle R^2 \rangle^{1/2}$ and $(\langle H^2 \rangle + d^2)^{1/2}$, respectively. Here, $\langle R^2 \rangle$ is the mean-square end-to-end distance of the chain, $\langle H^2 \rangle$ the mean square of the distance between the midpoint and the end-to-end axis of the chain, and d the chain diameter[4]. We view a semiflexible polymer chain as a cylindrical domain in which chain segments are dispersed as illustrated in Fig. 13, and refer to it as the *fuzzy cylinder* [19]. If the fluctuation of the segment distribution is neglected, its length L_e and diameter d_e may be expressed by

$$L_e = \langle R^2 \rangle^{1/2}, \quad d_e = [\langle H^2 \rangle + d^2]^{1/2} \tag{43}$$

In the rod limit, $L_e = L$, and $d_e = d$. On the other hand, in the coil limit, $L_e = \sqrt{6} d_e$, if we assume the chain to be in the unperturbed state. It is to be noted that the axial ratio of the fuzzy cylinder is greater than unity even in the coil limit. At intermediate N, L_e and the axial ratio L_e/d_e may be calculated, respectively, from the Kratky–Porod equation [102, 103] for the (unperturbed)

---

[4] If polymer chains extensively interwind in the solution, the change in the internal conformation of each chain is severely restricted by entanglements with other chains. In such a case the segment distribution may not be cylindrical. We consider highly entangled solutions in Sect. 6.3.3.

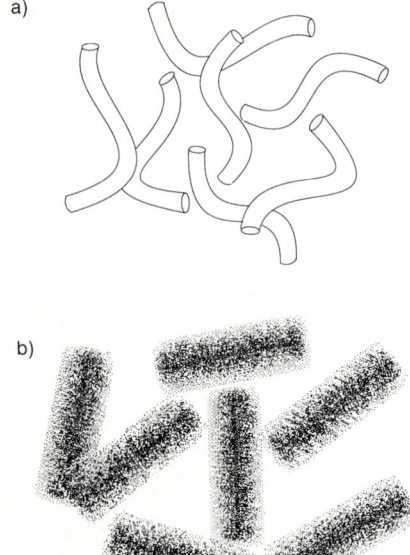

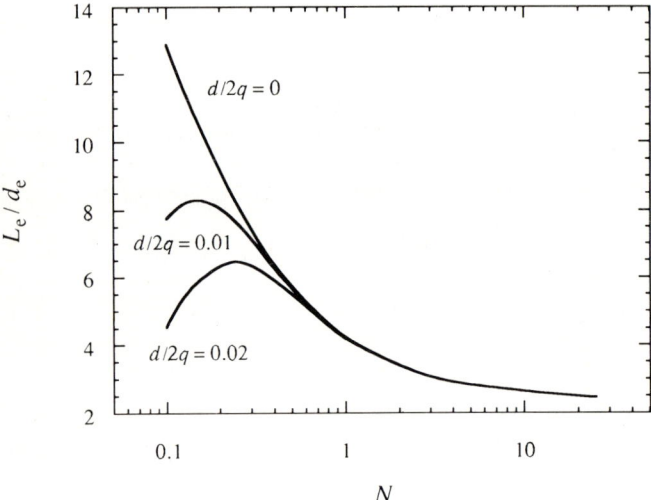

Fig. 13a, b. Solutions containing: **a** wormlike cylinders; **b** fuzzy cylinders

Fig. 14. The Kuhn segment number N dependence of the axial ratio L_e/d_e for fuzzy cylinders with different d [104]

wormlike chain and Hoshikawa et al.'s results [104] for Tagami's model [105]. No formulation of $\langle H^2 \rangle$ for the wormlike chain model is yet made. Although Tagami's model allows the chain contour to stretch in contrast to the wormlike chain, both models give identical expressions for $\langle R^2 \rangle$ and the mean-square radius of gyration, so that Hoshikawa et al.'s $\langle H^2 \rangle$ is expected to be a good approximation to the wormlike chain. The N dependence of L_e/d_e for a chain is shown in Fig. 14 for different d.

We may have to consider that the segment distribution fluctuates in the cylindrical domain in order to formulate the effects of entanglement and jamming in a solution as illustrated in Fig. 13b. In other words, we may no longer be permitted to consider the fuzzy cylinder a "hard-core cylinder" of the geometry specified by Eq. (43), but have to make its periphery fluctuate.

The diffusion coefficients at infinite dilution ($D_{\parallel 0}$, $D_{\perp 0}$, and D_{r0}) for the fuzzy cylinder reduce to those for the wormlike cylinder, which can be calculated as explained in Appendix B. On the other hand, these diffusion coefficients, $D_{\parallel}$, $D_{\perp}$, and D_r, for the fuzzy cylinder at finite concentrations can be formulated by use of the mean-field Green function method and the hole theory, as detailed below.

6.3 Diffusion Coefficients

6.3.1 Green Function Method

The mean-field Green function method was first applied by Edwards and Evans [106] to formulate the jamming effect on the longitudinal diffusion of rodlike polymers. Subsequently it was used by Teraoka and Hayakawa [107, 108] to derive the transverse and rotational diffusion coefficients of a rod entangling with the surrounding rods. They treated in a refined way the multiple perturbations to the Green function, but neglected the jamming effect. Recently we [109] have formulated D_r by incorporating the jamming effect.

To explain the Green function method for the formulation of $D_{\perp}$, D_r, and $D_{\parallel}$ of the fuzzy cylinder [19], we first consider the transverse diffusion process of a test fuzzy cylinder in the solution. As in the case of rodlike polymers [107], we imagine two hypothetical planes which are perpendicular to the axis of the cylinder and touch the bases of the cylinder (see Fig. 15a). The two planes move and rotate as the cylinder moves longitudinally and rotationally. Thus, we can consider the motion of the cylinder to be restricted to transverse diffusion inside the laminar region between the two planes. When some other fuzzy cylinders enter this laminar region, they may hinder the transverse diffusion of the test cylinder. When the test fuzzy cylinder and the portions of such other cylinders are projected onto one of the hypothetical planes, the transverse diffusion process of the test cylinder appears as a two-dimensional translational diffusion of a circle (the projection of the test cylinder) hindered by ribbon-like obstacles (cf. Fig. 15a).

This two-dimensional diffusion can be treated by the Green function method, which is expounded in detail in Appendix C, and shows that $D_{\perp}$ can be

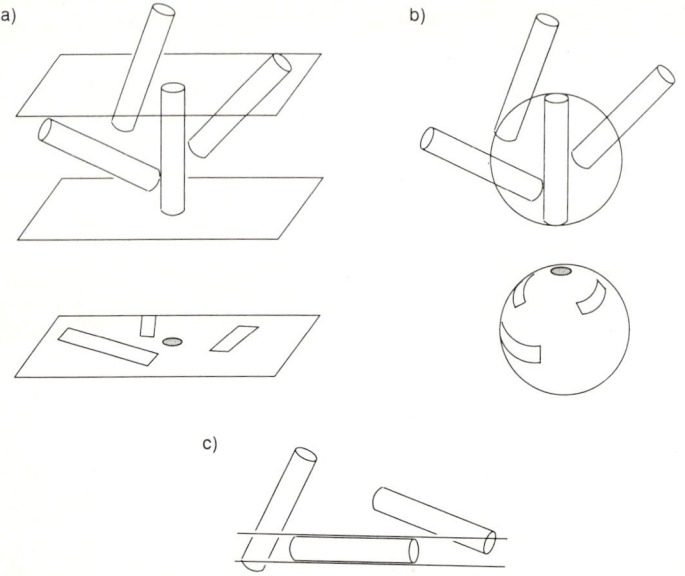

Fig. 15a–c. Projection procedures for the transverse, rotational and longitudinal diffusion processes of a fuzzy cylinder in concentrated solution

identified with $D^{(2)}$ given by Eq. (C17). Thus we have

$$\frac{D_\perp}{D_{\perp 0}} = \left(1 + \frac{2}{3\sqrt{\pi}} \rho_{B\perp} b_\perp \tau^{3/2} D_{\perp 0}^{1/2}\right)^{-2} \tag{44}$$

where $\rho_{B\perp}$ is the number of the obstacles appearing per unit time and per unit area of the projection plane in Fig. 15a, $b_\perp$ the average length of an obstacle on the projection plane, and τ the mean lifetime of the obstacle.

Since the fuzzy cylinders act as obstacles when entering the laminar region in Fig. 15a, the probability of their appearance (and disappearance) on the projection plane should be governed essentially by the longitudinal diffusion of the hindering or test fuzzy cylinder, and the lifetime τ should obey the relation

$$\tau \propto L_e^2/D_\parallel \tag{45}$$

The quantity $\rho_{B\perp}$ can be calculated from $v_{ex,\perp} c'/\tau$, where $v_{ex,\perp}$ is the excluded volume between one hindering fuzzy cylinder and the laminar region of unit surface area, and is approximately equal to $\frac{3}{2} L_e [1 + \frac{1}{3}(d_e/L_e)]$. Furthermore, the average length $b_\perp$ is proportional to L_e for the longer sides of ribbon-like obstacles and to d_e for the shorter sides of the obstacles.

Before inserting τ, $\rho_{B\perp}$, and $b_\perp$ obtained above into Eq. (44), we have to mention the fluctuation effect of the fuzzy cylinder. When a hindering fuzzy cylinder is entering the periphery of the laminar region, the hindrance that it exerts may be released by fluctuation of the segment distributions in the hindering and/or test fuzzy cylinders. The lifetime of such a hindering fuzzy cylinder may be much shorter than that expected from Eq. (45), and the fluctuation effect may be formulated by subtracting some fuzzy cylinders contacting the laminar region on its periphery from the count of obstacles. This manipulation is equivalent to reducing the (d_e/L_e) term in $v_{ex,\perp}$. Similar release by segment fluctuation may also occur when the test fuzzy cylinder collides with the shorter sides of the ribbon-like obstacles, and its effect can be taken into account by reducing the d_e term in $b_\perp$.

From Eq. (44) and the above considerations, we have

$$\frac{D_\perp}{D_{\perp 0}} = \left[1 + \beta_\perp^{-1/2} L_e^3 c' f_\perp (d_e/L_e) \left(\frac{F_{\|0} D_{\|0}}{F_{\perp 0} D_\|}\right)^{1/2}\right]^{-2} \tag{46}$$

where $\beta_\perp$ is a numerical constant, $F_{\|0}$ and $F_{\perp 0}$ the hydrodynamic factors relating to the effects of finite thickness and flexibility of the polymer on $D_{\|0}$ and $D_{\perp 0}$, respectively, which are described in Appendix B, and $f_\perp(d_e/L_e)$ the correction factor for the segment fluctuation defined by

$$f_\perp(x) = (1 + \tfrac{1}{3} C_\perp x)(1 + C_\perp x) \tag{47}$$

with $C_\perp$ being a parameter which takes a value between 0 and 1 depending on the effectiveness of the hindrance release by segment fluctuation. Although $C_\perp$ remains as an adjustable parameter at present, it contributes only to the correction term in d_e/L_e. Since d_e/L_e is always smaller than $6^{-1/2}$, $f_\perp(d_e/L_e)$ changes only from 1 to 1.6.

The factor $D_{\|0}/D_\|$ in Eq. (46) represents the contribution of the jamming effect to the transverse diffusion, and becomes more important at higher polymer concentrations. It is formulated later in this subsection and Sect. 6.3.2.

For an infinitely thin rodlike polymer for which $d/L = d_e/L_e = 0$, we have $f_\perp = F_{\|0} = F_{\perp 0} = D_{\|0}/D_\| = 1$, and Eq. (46) reduces to Teraoka and Hayakawa's original expression [107] of $D_\perp$ for rodlike polymers. At high concentrations, the results from the Green function method approach the one from the cage model [107]. Teraoka [110] calculated stochastic geometry and probability of the entanglement for infinitely thin rods by use of the cage model, and evaluated $\beta_\perp$ to be

$$\beta_\perp = 561 \tag{48}$$

We use this value for $\beta_\perp$ in the subsequent analysis.

The rotational diffusion coefficient of the fuzzy cylinder can be formulated in a similar way. For the rotational diffusion process, it is convenient to imagine a hypothetical sphere which has the diameter equal to L_e, just encloses the test fuzzy cylinder, and moves with the translation of the fuzzy cylinder. If the test cylinder and the portions of surrounding fuzzy cylinders entering the sphere are projected onto the spherical surface as depicted in Fig. 15b (cf. [108]), the rotational diffusion process of the test cylinder can be treated as the translational diffusion process of a circle on the hypothetical spherical surface with ribbon-like obstacles.

When the diffusion time is short enough, the translation on the spherical surface is approximately identical with that on the tangent plane to the spherical surface. The latter is the two-dimensional diffusion process treated by the Green function method in Appendix C, and we can use Eq. (C17) again. Since the rotational diffusion coefficient D_r is related to the translational diffusion coefficient $D^{(2)}$ in Eq. (C17) by $D_r = D^{(2)}/(L_e/2)^2$, we have

$$\frac{D_r}{D_{r0}} = \left(1 + \frac{2}{3\sqrt{\pi}} \rho_{Br} \theta_B \tau^{3/2} D_{r0}^{1/2}\right)^{-2} \tag{49}$$

where $\rho_{Br}[\equiv \rho_B(L_e/2)^2]$ is the number of ribbon-like obstacles per unit solid angle and per unit time appearing on the spherical surface, and $\theta_B[\equiv b/(L_e/2)]$ is the average central angle subtended by one obstacle. Teraoka and Hayakawa [108] showed that Eq. (49) can be applied even to long-time diffusion in a good approximation.

The quantity ρ_{Br} can be calculated from $v_{ex,r} c'/\tau$ with the excluded volume $v_{ex,r}$ between one hindering fuzzy cylinder and the sphere depicted in Fig. 15b, which is approximately equal to $\frac{5}{12}\pi L_e^3[1 + (d_e/L_e)]^2[1 - \frac{1}{5}(d_e/L_e)]$. The average central angle θ_B of the obstacle is proportional to $1 + d_e/L_e$, when both longer and shorter sides of the ribbon-like obstacles are considered as hindering obstacles. The lifetime τ is given by Eq. (45). The hindrance release by fluctuation of the segment distribution must also be taken into account for the rotational motion. These considerations allow Eq. (49) to give [19]

$$\frac{D_r}{D_{r0}} = \left[1 + \beta_r^{-1/2} \frac{L_e^4}{L} c' f_r(d_e/L_e) \left(\frac{F_{\parallel 0} D_{\parallel 0}}{F_{r0} D_\parallel}\right)^{1/2}\right]^{-2} \tag{50}$$

where β_r is a numerical constant, F_{r0} the hydrodynamic factor relating to the effects of finite-thickness and flexibility of the polymer on D_{r0} (see Appendix B), and $f_r(d_e/L_e)$ the correction factor due to the fluctuation effect defined by

$$f_r(x) = (1 + C_r x)^3 (1 - \tfrac{1}{5} C_r x) \tag{51}$$

The parameter C_r expresses the effectiveness of the hindrance release by segment fluctuation and varies between 0 and 1. An empirical expression for it as a function of the Kuhn statistical segment number N is given in Sect. 8. Although C_r contributes only to the correction terms in d_e/L_e, $f_r(d_e/L_e)$ changes from 1 to 2.56 in the range of allowable values of d_e/L_e. Thus, the factor $f_r(d_e/L_e)$ is more important than the factor $f_\perp(d_e/L_e)$ in $D_\perp$.

In the infinitely thin rod limit, Eq. (50) reduces to Teraoka and Hayakawa's original expression of D_r for rodlike polymers [108]. The latter approaches the equation of D_r derived on the cage model [108, 111] at high concentrations. Teraoka et al. [111] estimated β_r from calculations of stochastic geometry and probability of the entanglement for infinitely thin rods with the cage model, and obtained

$$\beta_r = 1350 \tag{52}$$

This β_r value is used in the following discussion.

The longitudinal diffusion in fuzzy cylinder systems may be treated as follows. We imagine a hypothetical infinite tube just enclosing the test fuzzy cylinder, and project the test cylinder and the portions of other fuzzy cylinders which enter the hypothetical tube onto the centerline of the tube. Note that unlike Doi and Edwards' tube model, the present hypothetical tube moves and rotates as the test cylinder translates and rotates. This operation reduces the longitudinal diffusion problem of the test cylinder to a one-dimensional diffusion problem of a segment on a line containing hindering segments [106] (cf. Fig. 15c). If the diffusing segment can be replaced by a point, the longitudinal diffusion under consideration is approximated by one-dimensional diffusion of a point particle in the presence of point obstacles.

Using Eq. (C16) in Appendix C and a similar procedure to those mentioned above for $D_\perp$ and D_r, we can derive for the longitudinal diffusion coefficient $D_\parallel$ [19, 109]

$$\frac{D_\parallel}{D_{\parallel 0}} = [1 - \beta_\parallel^{-1/2} L_e^3 c' f_\parallel (d_e/L_e)]^2 \tag{53}$$

where $\beta_\parallel$ is a numerical constant and $f_\parallel(d_e/L_e)$ is the correction due to the fluctuation of the segment distribution, defined by

$$f_\parallel(x) = C_\parallel x \left[1 + \left(\frac{8}{\pi} - 1\right) C_\parallel x\right] \tag{54}$$

The parameter $C_\parallel$ appears in the leading term unlike $C_\perp$ and C_r in Eqs. (47) and (51). At present, it is impossible to evaluate the dependence of $C_\parallel$ on N and c'.

6.3.2 Hole Theory

The jamming effect, i.e., the slowing down of the longitudinal diffusion of a polymer chain by the head-on collision with other chains, can be treated by a model similar to that proposed by Cohen and Turnbull [112] for self diffusion of small molecules in a fluid. This model assumes that if at least one surrounding polymer chain exists within the *critical hole* ahead of a test chain, both collide, and this prevents the test chain from diffusing longitudinally. With this assumption, we express the longitudinal diffusion coefficient $D_\parallel$ of the test chain as

$$D_\parallel = D_{\parallel 0} P_h \tag{55}$$

where P_h is the probability of finding no surrounding chains entering the critical hole.

In this model, the geometry of the critical hole must be specified. Here we simply assume that the critical hole for a semiflexible polymer chain is similar in shape to the fuzzy cylinder [19]. The similarity ratio λ^* between the critical hole and the fuzzy cylinder is left as an adjustable parameter. The mutual excluded volume V_{ex}^* between the critical hole and one surrounding semiflexible chain can be written in the form [2]

$$V_{ex}^* = F(\lambda^* L_e, \lambda^* d_e; L, d) \tag{56}$$

with

$$F(L_1, d_1; L_2, d_2) \equiv \frac{\pi}{4}\Big[L_1 L_2(d_1 + d_2) + (L_1 d_1^2 + L_2 d_2^2)$$
$$+ \frac{1}{2}(L_1 d_2^2 + L_2 d_1^2) + \frac{\pi}{2}(L_1 + L_2) d_1 d_2$$
$$+ \frac{\pi}{4} d_1 d_2 (d_1 + d_2)\Big]$$

The probability that one surrounding chain enters the critical hole is given by V_{ex}^*/V where V is the volume of the solution. The probability $w_n(x)$ of finding x chains from the total n chains that enter the critical hole is expected to follow the Poisson distribution:

$$w_n(x) = \frac{\langle x \rangle^x}{x!} \exp(-\langle x \rangle)$$

with the mean value $\langle x \rangle = nV_{ex}^*/V = V_{ex}^* c'$. The probability P_h is equal to $w_n(0)$, and hence is written

$$P_h = \exp(-V_{ex}^* c') \tag{57}$$

Combination of Eqs. (55) and (57) gives

$$D_{\parallel}/D_{\parallel 0} = \exp(-V_{ex}^* c') \tag{58}$$

which contains only a single adjustable parameter λ^* that depends on the system as shown later.

6.3.3 Reptation Model

As chain flexibility and polymer concentration increase, the internal motion of the polymer chain is so severely restricted by entanglement with other chains that the chain would behave as if it were trapped in a fixed winding tube. In such a situation, the fuzzy cylinder model may not be relevant. Odijk [113], Doi [170], and Semenov [115] discussed independently the reptative translational and rotational motion of a semiflexible polymer, with the assumption that the chain can move only along its contour and the lateral fluctuation of the chain is negligible. This model is an extreme opposite to the fuzzy cylinder model, and may be useful for highly entangled systems, though no clear evidence for it is as yet available. If the longitudinal diffusion along the chain contour is independent of the polymer concentration, the reptation model leads to the rotational and translational diffusion coefficients which do not vary with concentration, sharply different from what the fuzzy cylinder model predicts.

With increasing polymer concentration, we may expect that the polymer global motion changes from the fuzzy cylinder model mechanism to the reptation model mechanism. The onset of the crossover should depend on the degree in which the lateral motion of a polymer chain is suppressed by entanglement with its surrounding chains, but it is difficult to estimate this degree. There are some disputes over it in the case of flexible polymers [20].

6.4 Stress Expression

In order to discuss the rheological properties of stiff-chain polymer solutions, we need an expression for stress. The stress $\boldsymbol{\sigma}$ induced in a homogeneous isotropic or nematic solution by a macroscopic flow was formulated by Doi [114], who used the Kirkwood general theory [116] to show

$$\boldsymbol{\sigma} = \boldsymbol{\sigma}^{(E)} + \boldsymbol{\sigma}^{(V)} + \boldsymbol{\sigma}^{(S)} \tag{59}$$

Here $\boldsymbol{\sigma}^{(E)}$ is the elastic stress which arises from the change in the (dynamic) free energy in the macroscopic flow, while $\boldsymbol{\sigma}^{(V)}$ and $\boldsymbol{\sigma}^{(S)}$ are the viscous stresses produced by the polymer–solvent friction and the solvent–solvent friction, respectively. In concentrated isotropic polymer solutions, the elastic stress overwhelms the viscous stresses, so the latter are often neglected. However, it should be noticed that the viscous stresses may become significant in more dilute solutions as well as in nematic solutions where the elastic stress diminishes.

Doi [4, 100, 114, 117] formulated $\boldsymbol{\sigma}^{(E)}$ for rodlike polymer solutions by noticing that $\boldsymbol{\sigma}^{(E)}$ is related to the change $\delta(\Delta F)$ in the dynamic Helmholtz free energy of the system due to a virtual deformation $\boldsymbol{\kappa}\,\delta t$ in a short time δt by

$$\delta(\Delta F) = \boldsymbol{\sigma}^{(E)}: \boldsymbol{\kappa}^{\dagger} \delta t \tag{60}$$

Here the dynamic free energy ΔF is calculated from the static free energy by replacing the equilibrium orientational distribution function by the time-dependent distribution function. The calculation of $\delta(\Delta F)$ for a rodlike polymer solution goes through the following three steps.

(1) The virtual deformation rotates the rod by an angle $\boldsymbol{\omega}\delta t$ which is related to $\boldsymbol{\kappa}\delta t$ by Eq. (42).

(2) This rotation modifies the orientational distribution function $f(\mathbf{a};t)$ to $f(\mathbf{a};t) + \delta f(\mathbf{a};t)$ through Eq. (39) with Eq. (40b), the last term on the right hand side dominating over the others if the deformation is very rapid or the time δt is very short. The argument $\mathbf{r}$ in $f(\mathbf{r},\mathbf{a};t)$ is omitted because we are concerned only with uniform solutions.

(3) Finally, the modification of $f(\mathbf{a};t)$ induces a change in the dynamic free energy, which can be calculated by varying the free energy with respect to $f(\mathbf{a};t)$ and by substituting $\delta f(\mathbf{a};t)$ and $\boldsymbol{\omega}$ obtained in the steps (2) and (1).

Doi's final expression of $\sigma^{(E)}$ is written as [117]

$$\sigma^{(E)} = -\frac{1}{2}\frac{c'k_BT}{D_r}\chi \mathbf{F} \tag{61}$$

where $\mathbf{F}$ is defined by

$$\mathbf{F} = -6D_r\left[\mathbf{S} - \frac{1}{6k_BT}(\langle \mathbf{a}(\mathbf{a}\times\mathfrak{R}V_{scf}(\mathbf{a}))\rangle + \langle \mathbf{a}(\mathbf{a}\times\mathfrak{R}V_{scf}(\mathbf{a}))\rangle^\dagger)\right] \tag{62}$$

with the tensor order parameter $\mathbf{S}$ defined by

$$\mathbf{S} \equiv \langle \mathbf{aa} - \tfrac{1}{3}\mathbf{I}\rangle \tag{63}$$

Here $\langle \cdots \rangle$ means the average with respect to $f(\mathbf{a};t)$, and $\mathbf{I}$ is the unit tensor.

If we neglect the distortion of the segment distribution in the fuzzy cylinder by the shear flow, we can apply Doi's stress expression, Eq. (61), to fuzzy cylinder systems as it stands. The neglect of the distortion may be justified when the shear-rate is low. Equation (61) expresses the contribution of the end-over-end rotation of the chain to $\sigma^{(E)}$. If the segment distribution is not distorted, the orientational entropy term S_{or} in the static free energy expression contains only the orientational entropy loss of the entire chain, but not the conformational entropy loss; cf. Sect. 2.3.

In contrast to the case of rodlike polymers, no adequate expression of $V_{scf}(\mathbf{a})$ is available for semiflexible polymers. Thus, at present, we cannot directly calculate $\mathbf{F}$ for semiflexible polymer solutions by Eq. (62). However, as will be shown in Sect. 8.2, we need no direct calculation of $\mathbf{F}$ to obtain $\sigma^{(E)}$ in a *steady-state flow*. So, in the following sections, we will be concerned only with the case of steady-state flow.

The kinetic equation for $\mathbf{S}$ for calculating $\sigma^{(E)}$ is obtained by multiplying the both sides of Eq. (39) by $\mathbf{aa} - \mathbf{I}/3$ and integrating the resulting equation over $\mathbf{a}$ (and $\mathbf{r}$). For a system homogeneous with respect to the translational degree of freedom and subject to no external field, the result is [4, 117]

$$\frac{\partial \mathbf{S}}{\partial t} = \mathbf{F} + \mathbf{G} \tag{64}$$

where

$$\mathbf{G} = \chi(\mathbf{A}\cdot\langle \mathbf{aa}\rangle + \langle \mathbf{aa}\rangle\cdot\mathbf{A}^\dagger - 2\langle \mathbf{aaaa}\rangle:\mathbf{A}) - \Omega\cdot\langle \mathbf{aa}\rangle + \langle \mathbf{aa}\rangle\cdot\Omega^\dagger \tag{65}$$

provided that D_r does not depend on $\mathbf{a}$.

Actual calculations of $\sigma^{(E)}$ for isotropic and nematic solutions will be described in Sects. 8 and 9 respectively. Furthermore, $\sigma^{(V)}$ will be derived approximately for isotropic solutions in a steady shear flow in Sect. 8, but it will be neglected for nematic solutions in Sect. 9.

Ferry and coworkers [118] extensively studied viscoelasticity of dilute solutions of stiff-chain polymers. Their results made clear that the stress or the storage and loss moduli for the solutions are sensitive to chain internal motions at very higher frequencies. Equation (61) does not contain contributions of such motions to $\sigma^{(E)}$, so it cannot be applied at high frequencies or in fast transient flows.

7 Diffusion Coefficients from Computer Simulation and Experiment

7.1 Computer Simulations for Rodlike Polymers

In recent years, several computer simulations have been performed for the dynamics of rodlike polymers in concentrated solutions [119–123], using various models and methods. Although the models used are not necessarily realistic, the simulation gives us information about the quantities of theoretical importance but not experimentally measurable (e.g., $D_{\parallel}$ and $D_{\perp}$). Furthermore, the comparison between simulation and experimental results may reveal the factors mainly responsible for the dynamics under study.

Bitsanis et al. [122, 123] simulated Brownian motion of rodlike polymers over the concentration range $5 \leq L^3 c' \leq 150$, where L and c' are the length and number concentration of the rod, respectively, with the intermolecular potential u given by

$$u = \frac{50}{3} k_B T \exp[-(r/0.03855L)^2]/|\sin \gamma(\mathbf{a},\mathbf{a}')| \qquad (66)$$

where r and $\gamma(\mathbf{a},\mathbf{a}')$ are the nearest distance and the angle between the interacting rod axes, respectively. This potential almost prevents two rodlike polymers from passing through each other. Neglecting both intra- and intermolecular hydrodynamic interactions, they solved numerically the Langevin equation of motion for the systems. The resulting $D_{\parallel}$ and $D_{\perp}$ are presented by unfilled circles in Fig. 16a. With increasing concentration, $D_{\perp}$ sharply decreases from the value $D_{\perp 0}$ at infinite dilution. This decrease is due to the entanglement effect. On the other hand, $D_{\parallel}$ depends on concentration much less than $D_{\perp}$. Bitsanis et al. estimated the effective axial ratio of the rod to be 50 from Eq. (66)

The translational self-diffusion coefficient D_s is related to $D_{\perp}$ and $D_{\parallel}$ by [4]

$$\frac{D_s}{D_{s0}} = \frac{D_{\parallel} + 2D_{\perp}}{D_{\parallel 0} + 2D_{\perp 0}} = \frac{(D_{\parallel}/D_{\parallel 0}) + 2(D_{\perp 0}/D_{\parallel 0})(D_{\perp}/D_{\perp 0})}{1 + 2(D_{\perp 0}/D_{\parallel 0})} \qquad (67)$$

where D_{s0} is D_s at infinite dilution. Since Bitsanis et al. neglected the intramolecular hydrodynamic interaction, their simulation sets $D_{\perp 0}$ equal to $D_{\parallel 0}$ so that the D_s/D_{s0} values they reported are those of $[(D_{\parallel}/D_{\parallel 0}) + 2(D_{\perp}/D_{\perp 0})]/3$. However, the intramolecular hydrodynamic interaction makes $D_{\perp 0}$ smaller than $D_{\parallel 0}$. For example, $D_{\perp 0}/D_{\parallel 0}$ for a spherocylinder with an axial ratio of 50 is

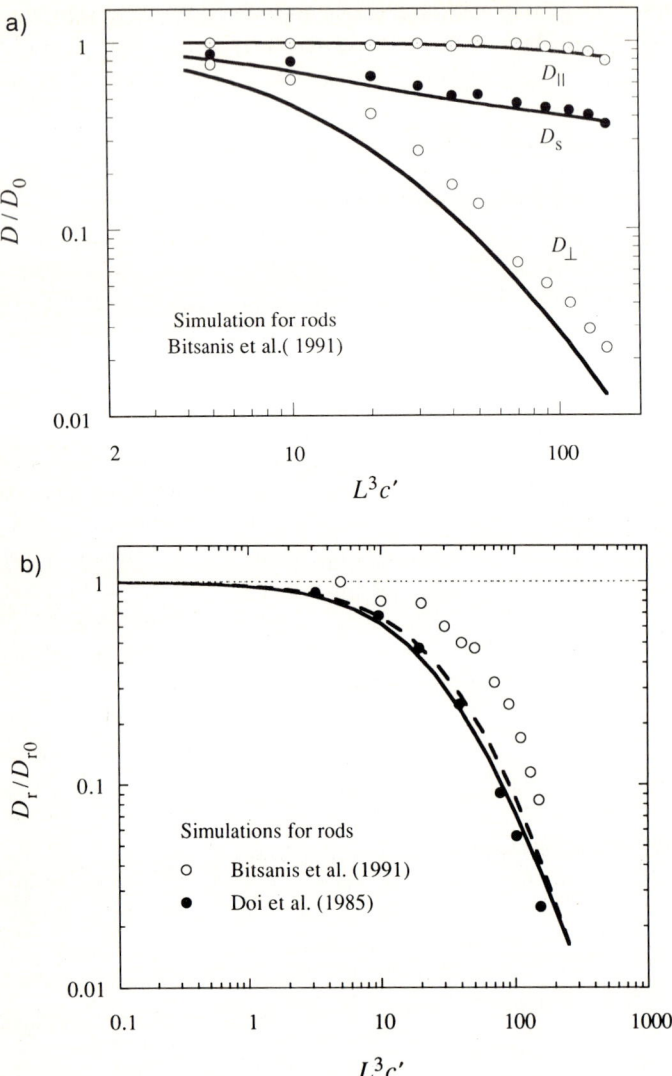

Fig. 16a, b. Computer simulation results for rodlike polymers in solution: **a** the translational diffusion coefficients [122, 123]; **b** the rotational diffusion coefficient [119, 122, 123]

found to be 0.622 from Table B1 in Appendix B. Thus, we have recalculated D_s/D_{s0} from Eq. (67) using Bitsanis et al.'s data and $D_{\perp 0}/D_{\parallel 0} = 0.622$, and obtained the filled circles in Fig. 16a. The contribution of the longitudinal diffusion is enhanced in the recalculated D_s. However, it is noted that this treatment does not fully account for the intramolecular hydrodynamic interaction, because the different diffusivities in the longitudinal and transverse directions also affect $D_\perp$ and $D_\parallel$ themselves (see below).

Bitsanis et al. [122, 123] also calculated the rotational diffusion coefficient D_r, and obtained the concentration dependence of D_r shown by unfilled circles in Fig. 16b. On the other hand, Doi et al. [119] made Monte Carlo simulations for the Brownian motion of infinitely thin rodlike polymers in solution by taking into account the intramolecular hydrodynamic interaction but not the intermolecular one and assuming $D_{\perp 0}/D_{\parallel 0} = 1/2$. The filled circles in Fig. 16b show their results for D_r. Both filled and unfilled circles exhibit a c'^{-2} or slightly stronger concentration dependence of D_r in the high concentration region. However, when compared at the same reduced concentration $L^3 c'$, the entanglement effect on D_r is more enhanced in Doi et al.'s simulation than in Bitsanis et al.'s. This difference may be attributed to the following facts.

(1) As pointed out by Bitsanis et al. [122], Doi et al.'s simulation used a long time interval which would diminish the chances of the test rod to escape from a partially open cage, and thus overestimate the entanglement effect.

(2) Differing from Bitsanis et al. who neglected the intramolecular hydrodynamic interaction and equated $D_{\perp 0}$ to $D_{\parallel 0}$, Doi et al. incorporated this interaction and set $D_{\perp 0}$ equal to $D_{\parallel 0}/2$. The increase of the transverse diffusivity should enhance the disengagement of the rotation of the test rod from hindrances and tend to diminish the entanglement effect.

The longitudinal diffusion coefficient $D_\parallel$ has been formulated by the hole theory in Sect. 6.3.2. If the similarity ratio λ^* in this theory is chosen to be 0.025 for the rod with the axial ratio 50, Eq. (58) with Eq. (56) gives the solid curve in Fig. 16a. Though it fits closely the simulation data, the chosen λ^* is not definitive because the change in $D_\parallel$ is small and the definition of the effective axial ratio is ambiguous. Though not shown here, Eq. (53) for $D_\parallel$ by the Green function method describes the simulation data equally well if $\beta_\parallel$ and $C_\parallel$ are chosen to be 1000 and 1, respectively.

Equations (46)–(48) lead to an expression of $D_\perp/D_{\perp 0}$, in which $L_e = L$ and $d_e = d$ for rodlike polymers. Since $L_e/d_e = L/d = 50$, $f_\perp$ can be equated to unity in a good approximation. Estimating $D_{\parallel 0}/D_\parallel$ from Eq. (58) with $\lambda^* = 0.025$ and $F_{\parallel 0}/F_{\perp 0}$ from Eqs. (B4) and (B5) in Appendix B with $L_e/d_e = L/d = 50$, we have calculated $D_\perp/D_{\perp 0}$ as a function of the reduced concentration $L^3 c'$. The results are compared with Bitsanis et al.'s simulation data in Fig. 16a. It can be seen that the theoretical solid curve for $D_\perp/D_{\perp 0}$ deviates downward slightly from the simulation data points, implying that the Green function method for $D_\perp$ overestimates the entanglement effect compared to Bitsanis et al.'s simulation.

The disagreement between theory and simulation illustrated in Fig. 16a is due partly to the chosen value for $D_{\perp 0}/D_{\parallel 0}$. With this ratio increasing, the transverse motion accelerates the disengagement of the test rod from the hindrance of its surrounding rods. Since Bitsanis et al. took this ratio to be unity, their simulation probably exaggerates the disengagement and thus underestimates the entanglement effect. It is expected that more realistic simulations allowing for hydrodynamic interaction give smaller $D_\perp/D_{\perp 0}$ than Bitsanis et al.'s results. On the other hand, the Green function method using Eq. (45) neglects the disengagement due to the transverse motion of rods.

The theoretical curve for D_s/D_{s0} in Fig. 16a has been drawn using Eq. (67) with $D_{\perp 0}/D_{\parallel 0} = 0.622$, and agrees fairly well with the simulation data. The longitudinal diffusion relative to the total translational diffusion becomes more important at higher concentrations.

The rotational diffusion coefficient can be calculated from Eqs. (50)–(52) formulated by the Green function method. The D_r/D_{r0} values obtained by Doi et al. for infinitely thin rods should be compared with the theory in which $L_e = L$ and $D_{\parallel 0}/D_{\parallel} = F_{\parallel 0}/F_{\perp 0} = f_r = 1$. The theoretical solid curve in Fig. 16b shows a favorable comparison with the simulation data.

Bitsanis et al.'s data must be compared with Eqs. (50)–(52), in which $D_{\parallel 0}/D_{\parallel}$ is calculated by Eq. (58) with $\lambda^* = 0.025$, $F_{\parallel 0}/F_{r0} = p^2/(p^2 + 1)$ [124] is calculated for $p \equiv L/d = 50$, and f_r is taken to be unity. The dashed curve in Fig. 16b is the calculated result, whose agreement with Bitsanis et al.'s data is not as good, however. This finding may be attributed to the disengagement mechanism discussed above regarding the transverse diffusion.

7.2 Translational Self-Diffusion Coefficient

The translational self-diffusion coefficient D_s of a polymer in concentrated solution can be measured by various techniques; e.g., forced Rayleigh scattering, fringe pattern recovery after photobleaching, and NMR [125]. Although in recent years extensive D_s data have been accumulated for flexible polymers in concentrated solutions and melts, those for stiff liquid-crystalline polymers are still meager. Only recently, D_s of a stiff double-helical polysaccharide, xanthan, in isotropic aqueous sodium chloride has been measured by two French groups [126, 127] using fringe pattern recovery after photobleaching.

Figure 17 shows Tinland et al.'s data [127] for D_s relative to D_{s0} for five xanthan samples having different molecular weights in 0.1 mol/l aqueous NaCl. With increasing c, D_s decreases from the infinite dilution value, following a sigmoidal curve except for the highest molecular weight sample. The decrease in D_s is more rapid for higher molecular weight.

As mentioned in Sect. 7.1, D_s/D_{s0} can be calculated from Eq. (67), with Eqs. (56) and (58) for $D_{\parallel}/D_{\parallel 0}$, Eqs. (46)–(48) for $D_{\perp}/D_{\perp 0}$, and $(2D_{\perp 0}/D_{\parallel 0})_{[19]}$ in Table B1 for $2D_{\perp 0}/D_{\parallel 0}$. There are two adjustable parameters: λ^* in Eq. (56) and $C_{\perp}$ in Eq. (47). The former was estimated for aqueous xanthan containing 0.1 mol/l NaCl to be 0.11 [19] from Takada et al.'s zero-shear viscosity data [128] regardless of xanthan's molecular weight (see Sect. 8), but the latter has not been evaluated so far. However, since $C_{\perp}$ does not sensitively affect $D_{\perp}/D_{\perp 0}$ as mentioned in Sect. 6.3.1, we choose $C_{\perp}$ to be zero here. The calculated values of D_s/D_{s0} are shown by the solid curves in Fig. 17. Their agreement with experiment is semi-quantitative except for the three middle fractions at high concentrations.

Tinland et al. [127] interpreted the D_s behavior of xanthan in the high concentration range by the reptation mechanism. As mentioned in Sect. 6.3.3,

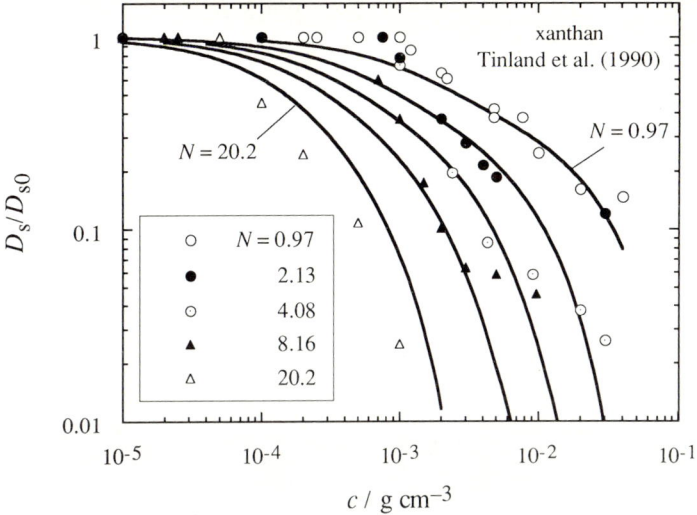

Fig. 17. Translational self-diffusion coefficients for aqueous solutions of five xanthan samples [127]. *Solid curves*, the theoretical values calculated from Eq. (67) along with Eqs. (46) and (58)

the reptation model predicts the concentration independence for the translational and rotational diffusivities, but, as will be shown in Sect. 8, zero-shear viscosities η_0 of aqueous xanthan in a wide molecular weight range strongly depends on concentration ($c^3 - c^7$ dependence) [128] up to a concentration higher than that studied by Tinland et al. This finding implies that the rotational diffusivity exhibits no sign of the reptation mechanism, because η_0 is approximately proportional to c/D_r (cf. Eq. (74)). Thus, Tinland et al.'s interpretation seems to require a reconsideration, but it is apparent that more experimental data must be accumulated to elucidate the mechanism that translational diffusion of stiff-chain polymers in concentrated solution obeys.

7.3 Rotational Diffusion Coefficient

The rotational diffusion coefficient D_r of a rodlike polymer in isotropic solutions can be measured by electric, flow, and magnetic birefringence, dynamic light scattering, and dielectric dispersion. However, if the polymer has some flexibility, its internal motion makes it difficult to extract D_r for the end-over-end rotation of the chain from data of these measurements. In other words, D_r can be measured only for nearly rodlike polymers.

Data of D_r were obtained for low molecular weight poly (γ-benzyl-L-glutamate) (PBLG) samples in isotropic solutions by two methods. Mori et al. [129] used dynamic electric birefringence, while Zero and Pecora [130] and Kubota and Chu [131] applied depolarized dynamic light scattering. Figure 18 shows

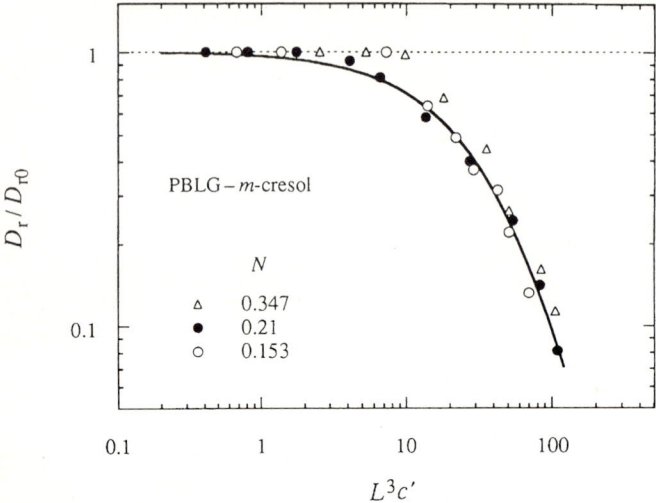

Fig. 18. Double logarithmic plot of D_r/D_{r0} vs L^3c' from dynamic electric birefringence data [129]. *Curves*, eye guide

Mori et al.'s results for three PBLG samples in m-cresol. For every sample, D_r decreases monotonically with increasing c, in a way resembling the simulation results shown in Fig. 16b. These D_r data, as well as the dynamic light scattering data, will be compared with Eqs. (50)–(52) in Sect. 8 together with zero-shear viscosity data.

8 Zero-Shear Viscosity

8.1 Experimental Results

The zero-shear viscosity η_0 has been measured for isotropic solutions of various liquid-crystalline polymers over wide ranges of polymer concentration and molecular weight [70, 128, 132–139]. This quantity is convenient for studying the stiff-chain dynamics in concentrated solution, because its measurement is relatively easy and it is less sensitive to the molecular weight distribution (see below). Here we deal with four stiff-chain polymers well characterized molecularly: schizophyllan (a triple-helical polysaccharide), xanthan (double-helical ionic polysaccharide), PBLG, and poly (p-phenylene terephthalamide) (PPTA; Kevlar). The wormlike chain parameters of these polymers are listed in Tables 1, 2 and 6.

We begin by describing some features of η_0 of semiflexible polymer solutions with data for Na salt xanthan in aqueous NaCl. Xanthan is a semiflexible polyelectrolyte, and its dynamic properties, as its static properties, may be controlled by its elongated shape and the electrostatic interactions among charged sites. The latter can be varied by changing the concentration C_s of added NaCl. Figure 19 shows η_0 plotted against polymer mass concentration c for a Na xanthan sample with $M_v = 5.9 \times 10^5$ in aqueous NaCl of C_s between 0.006 and 1.0 mol/l [140]. It can be seen that all the data points fall on a single composite curve, indicating η_0 to be independent of C_s. This means that the electrostatic interactions play no role in η_0, sharply contrasting with the phase behavior of the same systems shown in Fig. 11 a (cf. Sect. 4).

The viscosity of xanthan solutions is also distinct from that of flexible polyelectrolyte solutions which generally shows a strong C_s dependence [141]. In this connection, we refer to Sho et al. [142] and Liu et al. [143], who measured the intrinsic viscosity and radius of gyration of Na salt xanthan at infinite dilution which were quite insensitive to C_s (≥ 0.005 mol/l). Their finding can be attributed to the xanthan double helix which is so stiff that its conformation is hardly perturbed by the intramolecular electrostatic interactions. In fact, it has been shown that the electrostatic persistence length contributes only 10% to the total persistence length even at as low a C_s as 0.005 mol/l [142]. Therefore, the difference in viscosity behavior between xanthan and flexible polyelectrolyte

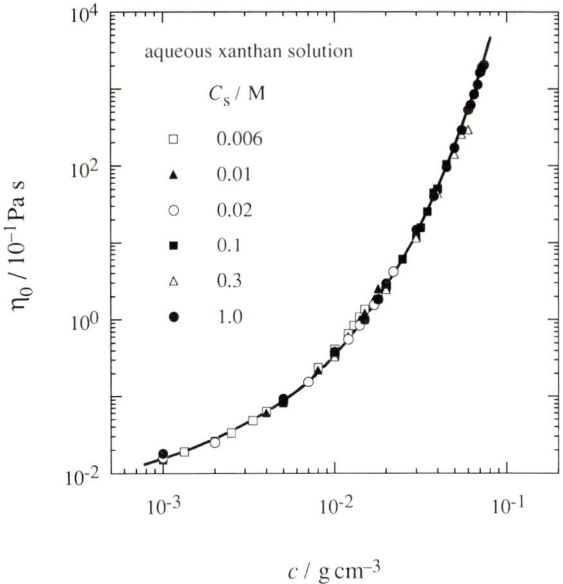

Fig. 19. Zero-shear viscosity of aqueous xanthan solutions at different sodium chloride concentration C_s [140]

solutions may be ascribed to ways that their conformation depends on C_s. Thus, what mainly controls the (steady state) viscosity of stiff polyelectrolyte solutions should be the hard-core interaction or the chain entanglement.

Figure 20 shows double-logarithmic plots on η_0 vs c and of η_0 vs viscosity-average molecular weight M_v for isotropic solutions of eight xanthan samples in 0.1 mol/l aqueous NaCl at 25 °C [128]. We see that η_0 depends very strongly on c and M_v. In particular, the molecular weight dependence far exceeds the 3.4 power law well known for concentrated solutions of flexible polymers. Furthermore, in both plots, the data points follow curves that are concave upward, making it impossible to fit them by simple power laws. The similar viscosity behavior has been observed for other stiff-chain polymers [132, 137].

Aqueous xanthan has also been used to examine the effect of the molecular weight distribution on η_0. Sato et al. [144] measured η_0 of "quasi-ternary solutions" containing two fractionated xanthan samples with different (viscosity-average) molecular weights M_1 and M_2 of 2.5×10^5 and 2.1×10^6 in 0.1 mol/l aqueous NaCl as a function of the total polymer mass concentration c and the weight fraction ξ_2 of the higher molecular weight component in the total polymer. In Fig. 21, the circles show the measured η_0 for this quasi-ternary system at four c plotted against the "weight-average molecular weight" M_w' of

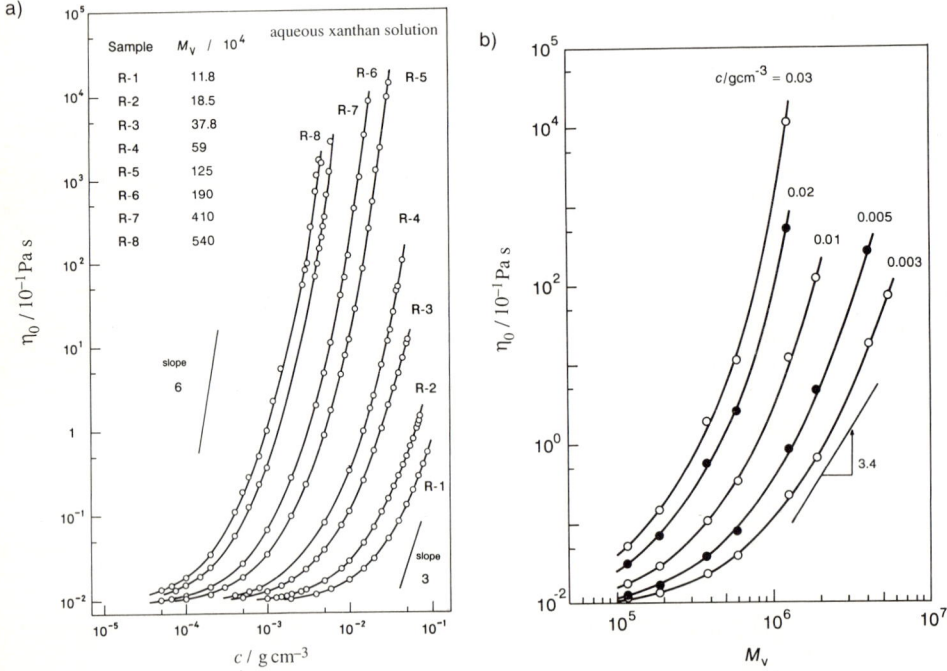

Fig. 20a, b. Double logarithmic plots for aqueous solutions of eight xanthan samples at $C_s = 0.1$ mol/l and 25 °C [128]: **a** η_0 vs c; **b** η_0 vs M_v. *Curves*, eye guide

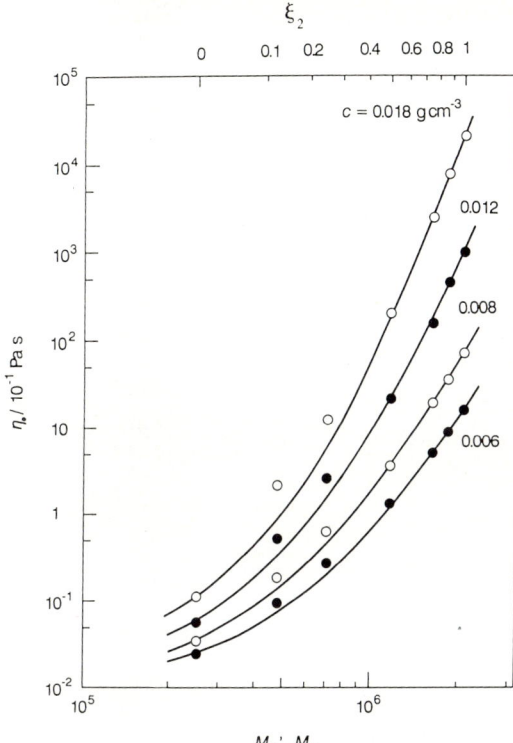

Fig. 21. Average molecular weight M'_w dependence of η_0 for the quasi-ternary system containing two xanthan samples with $(N_1, N_2) = (0.54, 4.5)$ in 0.1 mol/l NaCl [144]. Solid curves represent the M_v dependence of η_0 for quasi-binary solutions of xanthan at 0.1 mol/l NaCl

the polymer mixture, i.e.,

$$M'_w = \xi_1 M_1 + \xi_2 M_2$$

where ξ_s ($s = 1, 2$) is the weight fraction of the polymer component s in the total polymer.

The solid curves in the figure represent the molecular weight dependence of η_0 for quasi-binary system consisting of a fractionated xanthan sample and 0.1 mol/l aqueous NaCl. The circles for quasi-ternary solutions almost follow them at the same c, except at small ξ_2. Thus, to a first approximation, η_0 of stiff polymer solutions is independent of molecular weight distribution, and may be treated as a function of M_w or M_v and c.

8.2 Viscosity Equation

If a solution is in a steady state, $\partial S/\partial t$ in Eq. (64) is zero, so that we have the relation $\mathbf{F} = -\mathbf{G}$. Moreover, if the solution is isotropic and the velocity gradient is weak, we may neglect the higher order terms in κ from $\mathbf{G}$ given by Eq. (65), and obtain

$$\mathbf{G} = \tfrac{1}{5}\chi(\mathbf{\kappa} + \mathbf{\kappa}^\dagger) \quad \text{(linearized approximation)} \tag{68}$$

Substitution of these relations into Eq. (61) allows the elastic stress $\sigma^{(E)}$ to be written

$$\sigma^{(E)} = \frac{c'k_BT}{2D_r}\chi\mathbf{G} = \frac{1}{10}\frac{c'k_BT}{D_r}\chi^2(\mathbf{\kappa} + \mathbf{\kappa}^\dagger). \tag{69}$$

For the steady shear flow given by

$$\kappa_{\alpha\beta} = \begin{cases} \kappa & \text{(for } \alpha = x, \beta = y) \\ 0 & \text{(otherwise)} \end{cases} \tag{70}$$

the elastic term $\eta_0^{(E)}$ of the zero-shear viscosity is given by

$$\eta_0^{(E)} \equiv \frac{\sigma_{xy}^{(E)}}{\kappa} = \chi^2\frac{c'k_BT}{10D_r} \tag{71}$$

Equation (69) or (71) does not contain the self-consistent mean field potential $V_{\text{scf}}(\mathbf{a})$, indicating that the thermodynamic force does not contribute to the steady-state stress or viscosity and thus explaining why η_0 for aqueous xanthan solutions shown in Fig. 19 is independent of C_s. However, this force may play a role in the stress in a *non-steady-state flow* through $V_{\text{scf}}(\mathbf{a})$, as can be seen from Eqs. (61) and (62).

At infinite dilution, the total zero-shear viscosity can be written

$$\eta_0 = \eta^{(S)} + \eta_0^{(V)} + \eta_0^{(E)}$$
$$= \eta^{(S)} + \eta^{(S)}[\eta]c \tag{72}$$

where $\eta^{(S)}$, $\eta_0^{(V)}$, and $[\eta]$ are the solvent viscosity, the viscous term of the zero-shear viscosity ($\equiv \sigma_{xy}^{(V)}/\kappa$), and the intrinsic viscosity, respectively. Combining this equation with Eqs. (71) and (B3) in Appendix B, we obtain

$$\eta_0^{(V)} = (4\gamma^{-1} - 3\chi^2)\frac{c'k_BT}{30D_{r0}} \tag{73}$$

in which the hydrodynamic factors γ and χ may be calculated from the equations given in Table B1 in Appendix B.

In Sect. 6.3, we have neglected the intermolecular hydrodynamic interaction in formulating the diffusion coefficients of stiff-chain polymers. Here we use the same approximation by neglecting the concentration dependence of $\eta_0^{(V)}$, and apply Eq. (73) even at finite concentrations. Then, the total zero-shear viscosity η_0 is represented by [19]

$$\eta_0 = \eta^{(S)} + (4\gamma^{-1} - 3\chi^2)\frac{c'k_BT}{30D_{r0}} + \chi^2\frac{c'k_BT}{10D_r} \tag{74}$$

We introduce the reduced viscosity H_r defined by

$$H_r \equiv \frac{4}{3\gamma\chi^2}\left(\frac{\eta_0 - \eta^{(S)}}{\eta^{(S)}[\eta]c} - 1\right) + 1 \tag{75}$$

Here η_0, $[\eta]$, and $\eta^{(S)}$ are experimentally measurable, while $\gamma\chi^2$ can be calculated from the equations listed in Table B1 in Appendix B. With Eqs. (74) and

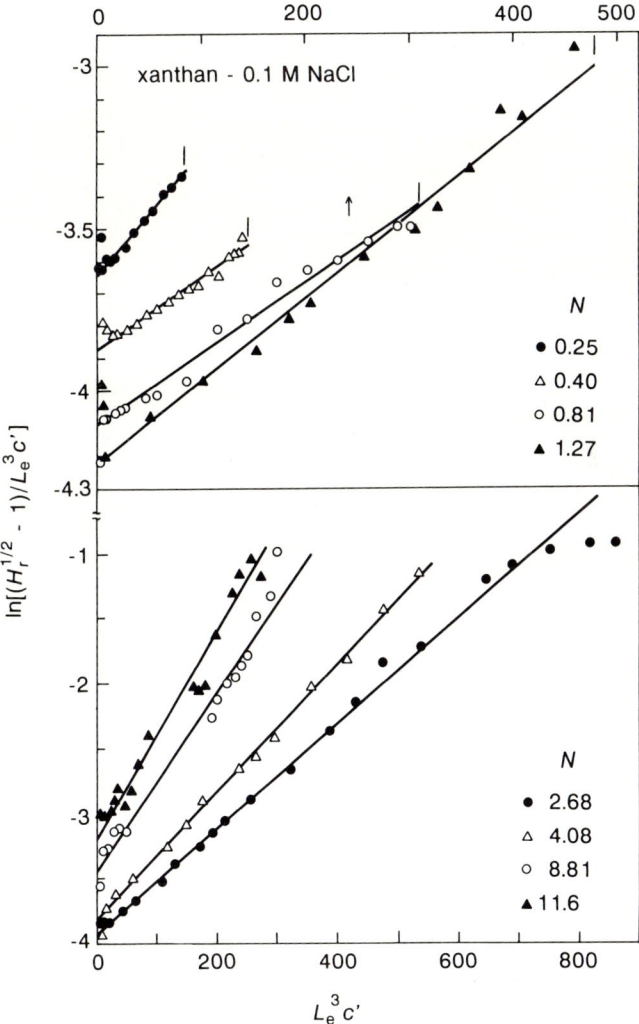

Fig. 22. Plot of $\ln[(H_r^{1/2} - 1)/L_e^3 c']$ vs $L_e^3 c'$ constructed from the η_0 data of aqueous xanthan presented in Fig. 20a. The *vertical segments* in the *upper panel* indicate the phase boundary concentration c_I between the isotropic and biphasic regions

(B3), Eq. (75) gives

$$H_r = D_{r0}/D_r \tag{76}$$

Substitution of Eq. (50) for D_r and Eq. (58) for $D_\parallel$ into Eq. (76) yields

$$H_r = [1 + B(F_{\parallel 0}/F_{r0})^{1/2} c' \exp(\tfrac{1}{2} V_{ex}^* c')]^2 \tag{77}$$

with

$$B \equiv \beta_r^{-1/2} \frac{L_e^4}{L} f_r(d_e/L_e). \tag{78}$$

8.3 Comparison Between Experiment and Theory

Equation (77) can be transformed into

$$\ln\left(\frac{H_r^{1/2} - 1}{c'}\right) = \ln[B(F_{\parallel 0}/F_{r0})^{1/2}] + \frac{1}{2} V_{ex}^* c' \tag{79}$$

Thus, if plotted against c', experimental values of $\ln[(H_r^{1/2} - 1)/c']$ for a given polymer + solvent should follow a straight line whose intercept and slope are equal to $\ln[B(F_{\parallel 0}/F_{r0})^{1/2}]$ and $V_{ex}^*/2$, respectively.

Figure 22 illustrates such plots with the η_0 data for aqueous xanthan solutions [128], where a reduced concentration $L_e^3 c'$ instead of c' is used for the abscissa. In agreement with the expectation, the data points for each sample almost follow a straight line. The vertical segments in the upper panel indicate the (reduced) phase boundary concentration $L_e^3 c_I'$ between the isotropic and biphasic regions. Similar linear plots have been obtained from η_0 data for aqueous schizophyllan solutions [19, 136, 137].

Values of B calculated from the ordinate intercepts are shown in Fig. 23 as a plot of $B/(2q)^3$ against the number of the Kuhn segments N. For $N \lesssim 4$, the data points for the indicated systems almost fall on the solid curve which is calculated by Eq. (78) along with Eqs. (43), (51), (52), and $C_r = 0$. A few points around $N \sim 1$ slightly deviate downward from the curve. Marked deviations of data points from the dotted lines for the thin rod limit, obtained from Eq. (78) with $L_e = L$ and $d_e = 0$, are due to chain flexibility; the effect is appreciable even at N as small as 0.5. The good fit of the solid curve to the data points (at $N \lesssim 4$) proves that the effect of chain flexibility on η_0 has been properly taken into account by the fuzzy cylinder model.

In Fig. 23, the three data points at $N \gtrsim 4$ define a transition region from the solid curve to the dot-dash curve calculated from Eq. (78) with $C_r = 1$[5]. Since C_r represents the effectiveness of hindrance release by the segment fluctuation in

---

[5] Although the dot-dash curve in Fig. 23 is drawn with $d/2q = 0.0092$ for xanthan with $q = 120$ nm and $d = 2.2$ nm (cf. Table 6), the corresponding curve for schizophyllan is indistinguishable from it.

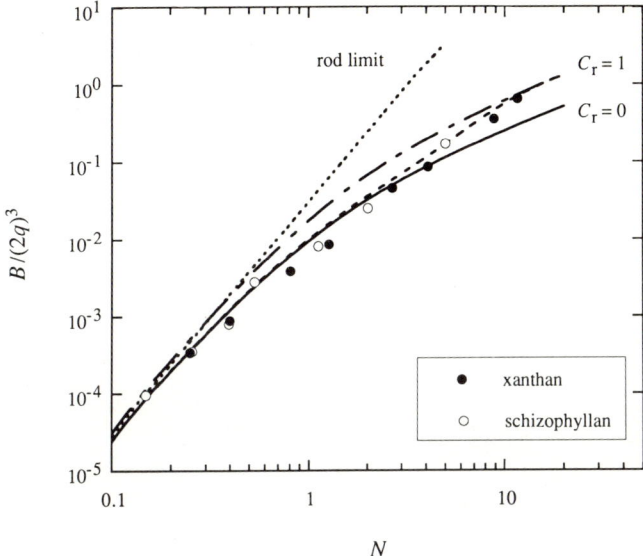

Fig. 23. Plot of $B/(2q)^3$ vs N for aqueous xanthan (*filled circles*) and schizophyllan (*unfilled circles*). *Solid curve*, calculated from Eqs. (78), (43), (51), (52) along with $C_r = 0$; *dot-dash curve*, from the same equation with $C_r = 1$ and $d/2q = 0.0092$; *dashed curve*, from the same equations with the interpolation formula (Eq. (80)) for C_r; *dotted curve*, from Eq. (78) with L_e and d_e replaced by L and d (rod limit)

fuzzy cylinders (cf. Sect. 6.3.1), the crossover of the data points to the dot-dash curve implies that the hindrance release becomes less effective with increasing N. To quantify it, we express the N dependence of C_r by an empirical equation

$$C_r = \tfrac{1}{2}\{\tanh[(N - N^*)/\Delta] + 1\} \qquad (80)$$

The dashed curve in Fig. 23 is the result of calculation from Eqs. (78) and (80), with N^* and Δ chosen to be 6 and 4, respectively.

Figure 24 shows the N dependence of $V_{ex}^*/(2q)^3$ for aqueous xanthan and schizophyllan. For both systems, V_{ex}^* increases monotonically with N. As mentioned in Sect. 6.3.2, if the critical hole for longtitudinal diffusion has similarity ratio λ^* to the fuzzy cylinder, V_{ex}^* is given by Eq. (56), and if λ^* and d are taken to be 0.11 and 2.2 nm for xanthan and 0.13 and 2.6 nm for schizophyllan (cf. Tables 2 and 6), this equation gives the solid curves shown in Fig. 24. They fit closely the data points for the two systems over the entire range of N examined.

Figure 25 shows experimental η_0 (circles) for aqueous xanthan compared with Eq. (74) (solid curves) corresponding to $N^* = 6$, $\Delta = 4$, and $\lambda^* = 0.11$. An excellent fit of the solid curve for each sample to the data points is obtained except in the region $N \sim 1$.

We compare Eq. (74) with the experimental results for two more stiff-chain polymers, PBLG and poly(p-phenylene terephthalamide) (PPTA). Since avail-

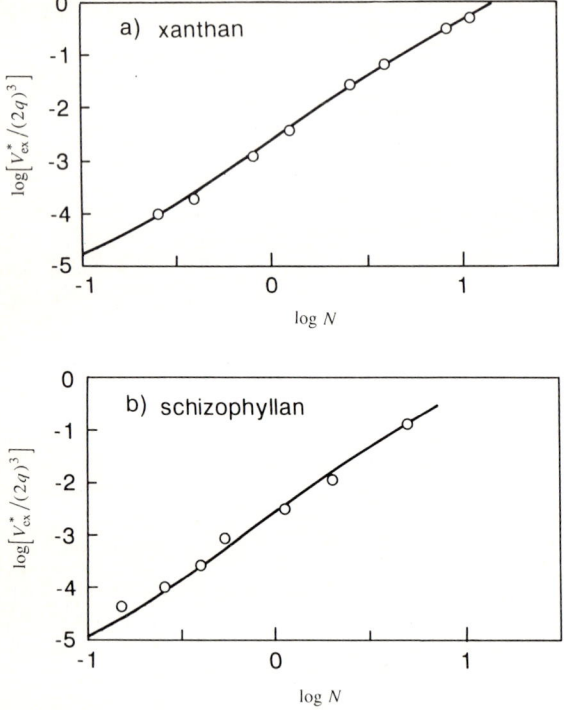

Fig. 24a, b. Plot of $V_{ex}^*/(2q)^3$ vs N for aqueous solutions of: **a** xanthan; **b** schizophyllan. *Solid curves*, calculated from Eq. (56) with $\lambda^* = 0.11$ for xanthan and $\lambda^* = 0.13$ for schizophyllan

able data of η_0 as a function of c and N or the molecular weight are not sufficient for these polymers, we make the comparison in the following way. We have from Eq. (77)

$$H_r = (1 + \beta^{-1/2}X)^2 \qquad (81)$$

where X is defined by

$$X \equiv \frac{L_e^4}{L} c' f_r (d_e/L_e) (F_{\|0}/F_{r0})^{1/2} \exp(\tfrac{1}{2} V_{ex}^* c') \qquad (82)$$

In this equation, $f_r(d_e/L_e)$ can be calculated from Eqs. (51) and (80) (with $N^* = 6$ and $\Delta = 4$), $F_{\|0}/F_{r0}$ from Eqs. (B4) and (B6), and V_{ex}^* from Eq. (56). The last quantity contains the adjustable parameter λ^*. Figure 26 shows the reduced viscosity H_r for PBLG and PPTA solutions plotted against X, calculated with $\lambda^* = 0.05$ and d = 1.7 nm for PBLG and $\lambda^* = 0.03$ and d = 0.6 nm for PPTA (cf. Tables 1 and 2 for the molecular parameters of PBLG and PPTA). Interestingly, all the data points expect for three come fairly close to the theoretical curve.

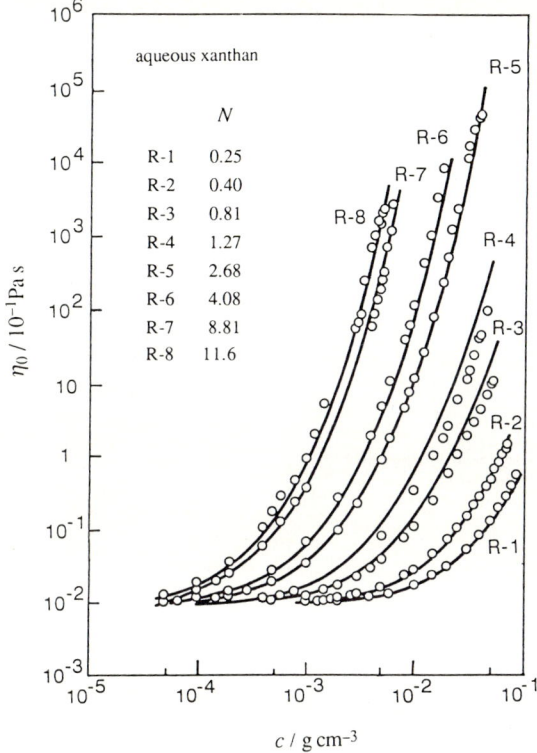

Fig. 25. Comparison of the η_0 data for aqueous xanthan presented in Fig. 19 with the viscosity equation Eq. (74)

Recently Sato et al. [144, 145] have extended the viscosity equation, Eq. (74), to multicomponent solution containing stiff-chain polymer species with different lengths. They showed a favorable comparison of the extended theory with the viscosity data for the quasi-ternary xanthan solutions presented in Fig. 21.

Finally, we make a comparison of the D_r data for PBLG solutions with the fuzzy cylinder model theory. It can be made absolutely, because the values of all parameters appearing in Eq. (50) for D_r are available for PBLG (see above). Figure 27a shows the plot of D_r/D_{r0} vs X for relatively low molecular weight PBLG in m-cresol, with D_r data obtained from dynamic electric birefringence by Mori et al. [129] and the molecular weights of the polymer samples reevaluated by using the [η]-molecular weight relation of Itou et al. [26]. The agreement between theory and experiment is fairly satisfactory, but slight upward deviations of the data points are not to be overlooked.

The same plot constructed with the D_r/D_{r0} data from depolarized dynamic light scattering is shown in Fig. 27b. The data points of Kubota and Chu [131] for dimethylformamide solutions come closer to the theoretical curve than those

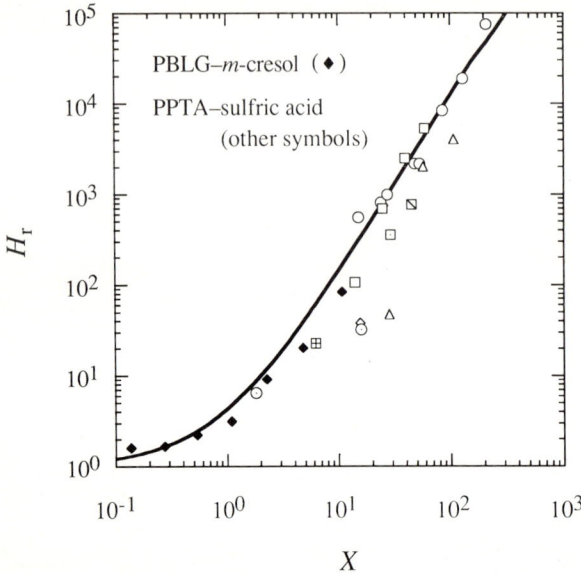

Fig. 26. Comparison of the viscosity equation Eq. (77) with η_0 data for poly(1,4-phenylene terephthalamide) (PPTA)-sulfuric acid solutions and poly(γ-benyl L-glutamate) (PBLG)-m-cresol solutions, in the plot of H_r vs X; see Eq. (82) for the definition of X. PPTA: ($\bigcirc$) N = 5.9 [134]; ($\square$) N = 5.1 [133]; ($\triangle$) N = 4.0 [133]; ($\boxtimes$) N = 3.5 [133]. ($\boxplus$) N = 3.0 [133]; ($\boxdot$) N = 2.6 [133]; ($\diamondsuit$) N = 2.0 [134]; ($\odot$) N = 1.6 [133]. PBLG: ($\blacklozenge$) N = 0.525 [139]

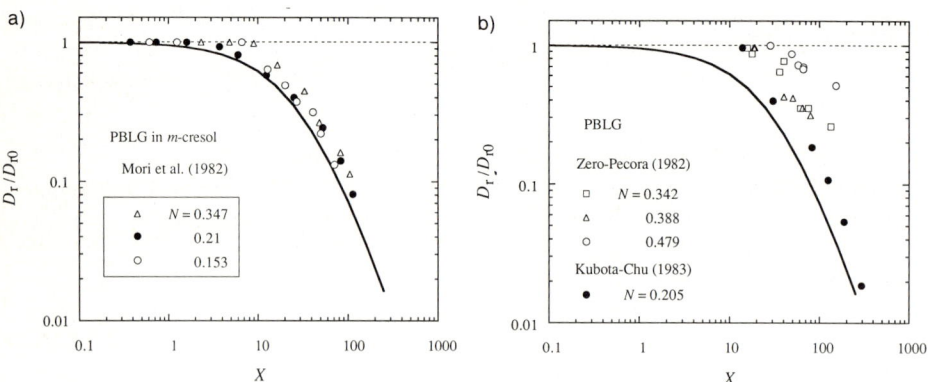

Fig. 27a, b. Comparison of Eq. (50) with the D_r data for isotropic solutions of relatively short PBLG samples in the plot of D_r/D_{r0} vs X: **a** data from dynamic electric birefringence by Mori et al. [129] (the same data as shown in Fig. 18); **b** data from dynamic light scattering by Kubota and Chu [131] and Zero and Pecora [130]. The *solid curve*, theoretical values predicted by Eq. (50)

of Zero and Pecora [130] for 1,2-dichloroethane solutions, the latter showing more pronounced deviations at higher N.

The above three groups of workers evaluated D_r by analyzing the birefringence and light scattering data using the theories for rods. Since PBLG is not a rigid rod, it is likely that such analyses have overestimated D_r. However, the chain flexibility effect on dynamic electric birefringence and light scattering at finite concentrations has not yet been investigated theoretically.

9 Rheological Properties of Liquid Crystalline Solutions

9.1 Experimental Results

So far we have concerned ourselves with the dynamical properties of isotropic solutions of liquid-crystalline polymers. As the polymer concentration is increased, these solutions are converted to a liquid crystal state (nematic or cholesteric) and exhibit some unusual rheological properties. For example, as first found by Kiss and Porter [146, 147], the primary normal stress difference σ_{N1} of liquid crystal solutions of α-helical polypeptides changes its sign with increasing shear rate κ. Moreover, recently, Magda et al. [148] found for liquid crystalline solutions of PBLG that the secondary normal stress difference σ_{N2} also changed its sign with κ.

Figure 28a shows Magda et al.'s data [148] of σ_{N1} and σ_{N2} as functions of the shear rate κ for a liquid crystal (cholesteric at rest) solution of a PBLG sample with the viscocity average molecular weight 23.5×10^4 (L = 162 nm and N = 0.54) dissolved in m-cresol at a polymer concentration of 12.5 wt% (or 0.132 g/cm^3). This concentration is very close to the reported phase boundary concentration (c_A) between the cholesteric and biphasic regions (12.3 wt%) [148]. The value of σ_{N1} is positive at lower κ ($\lesssim 10\,\text{s}^{-1}$), changes to negative at intermediate κ ($10\,\text{s}^{-1} \lesssim \kappa \lesssim 100\,\text{s}^{-1}$), and comes back to positive at higher κ ($\gtrsim 100\,\text{s}^{-1}$). Similar changes in the sign of σ_{N1} with κ were observed in nematic m-cresol solutions of a racemic mixture of poly(γ-benzyl glutamate) [146] and also in cholesteric aqueous hydroxypropyl cellulose [149, 150]. On the other hand, the sign of σ_{N2} is negative at lower and higher κ, but positive at intermediate κ where $\sigma_{N1} < 0$. Furthermore, σ_{N2} is comparable in magnitude to σ_{N1} at almost all κ. Magda et al. [148] also reported a shear rate dependence of viscosity which exhibits a unique change in convexity in the region where $\sigma_{N1} < 0$. These rheological findings indicate some interesting interplay exists between the external flow field and the mean-field (or molecular field) potential in liquid crystal solutions. Theoretical considerations of this indication were made by Marrucci and Maffettone [151–153], Larson [154] and Larson and Mead [155], which will be described in Sect. 9.2.

Polymer liquid crystal solutions respond in an interesting way to a transient flow field. For example, a damped oscillatory response was observed upon a flow inception or flow reversal [150, 156, 157], and the strain was recovered to some extent after a creeping flow [158]. These phenomena suggest that liquid crystal solutions have some textured structure, i.e., the spatial variation in the

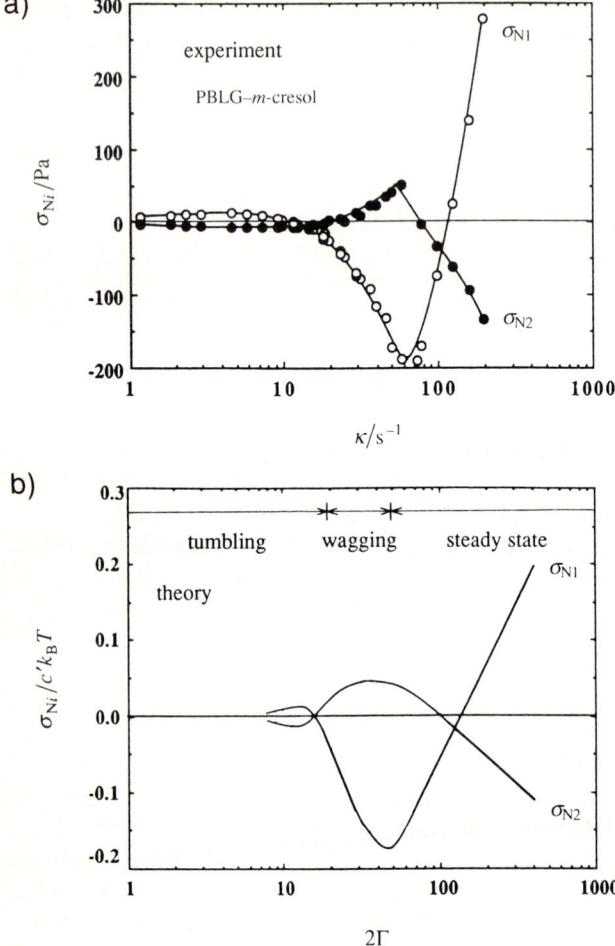

Fig. 28a, b. Shear rate dependence of the primary and secondary normal stress differneces (σ_{N1}, σ_{N2}): **a** Magda et al.'s experimental results [148] for a liquid crsytal solution of PBLG with $M_v = 23.5 \times 10^4$ (N = 0.54) at 25 °C; **b** Larson's theoretical results [154]

director. Larson and Doi [159] discussed the rheological behavior of liquid crystal solutions by applying a (semi-phenomenological) constitutive equation to a mesoscopic domain which is large compared with a monodomain, but small compared with the bulk dimensions.

9.2 Theoretical Considerations

Following Larson [154] and Larson and Mead [155], we discuss the rheological behavior of nematic solutions of a rodlike polymer on the basis of Doi's theory

[100]. We assume the system under consideration to be a single domain. Then the orientational state of the system can be specified by the order parameter tensor **S** defined by Eq. (63). The time evolution of **S** is governed by the kinetic equation, Eq. (64), along with Eqs. (62) and (65). This kinetic equation tells us that the orientational state in the rodlike polymer system in an external flow field is determined by the term **F** related to the mean-field potential V_{scf} and by the term **G** arising from the external flow field. These two terms control the orientation state in a complex manner as explained below.

In a weak flow field, Eq. (64) can be rewritten in a form similar to that for the direct **n** appearing in Leslie and Ericksen's phenomenological theory [160–163] for nematic systems. Thus, we have

$$\frac{\partial}{\partial t}\mathbf{n} = \mathbf{\Omega}^\dagger \cdot \mathbf{n} + \lambda(\mathbf{A}\cdot\mathbf{n} - \mathbf{A}:\mathbf{nnn}) \tag{83}$$

Here λ is an important parameter which determine whether the time evolution of **n** is dominated by the strain tensor **A** or by the vorticity tensor $\mathbf{\Omega}$ (cf. Eq. (42)). If λ is larger than unity, the former dominates and **n** tends to assume a steady-state angle θ relative to the flow direction which is determined by the equation

$$\tan^2\theta = (\lambda - 1)/(\lambda + 1) \tag{84}$$

On the other hand, if λ is smaller than unity, the latter dominates, so that the director tumbles continuously without any steady-state orientation. The period P per rotation is given by

$$P = 4\pi/[\kappa(1 - \lambda^2)^{1/2}] \tag{85}$$

Originally Doi [100] applied the decoupling approximation to the term **A**:$\langle$aaaa$\rangle$ in **G** given by Eq. (65) to calculate λ from Eq. (64). His result was that λ always exceeds unity, so the director tumbling does not occur. However, several authors [151, 154, 164, 165] have suggested this conclusion to be incorrect. For example, Kuzuu and Doi [165] calculated λ by using the term **A**:$\langle$aaaa$\rangle$ directly evaluated from the equilibrium distribution function f(**a**) determined by the Onsager theory [2]. Their λ is close to but definitely smaller than unity in nematic solutions. Therefore, Eq. (64) or (83) predicts that no steady-state orientational state occurs in the nematic solution subject to a weak flow field, but the director **n** is continuously tumbling.

With increasing flow rate, the orientational state in the nematic solution should change. Larson [154] solved numerically Eqs. (39) and (40b) with $V_{scf}(\mathbf{a})$ given by Eq. (41) for a homogeneous system (T[f] = 0) in the simple shear flow to obtain the time-dependent orientational distribution function f(**a**; t) as a function of κ. The non-steady orientational state in the nematic solution can be described in terms of the time-dependent (dynamic) scalar order parameter $S^{(dy)}$ defined by

$$S^{(dy)} = \sqrt{\tfrac{3}{2}\mathbf{S}:\mathbf{S}} \tag{86}$$

and the time-dependent extinction angle θ defined by

$$\theta = \tfrac{1}{2}\tan^{-1}\left(\frac{2S_{xz}}{S_{xx} - S_{zz}}\right) \tag{87}$$

where the x- and z-axes have been chosen parallel to the flow and gradient directions, respectively. Larson calculated these quantities from $f(\mathbf{a};t)$ and Eq. (63).

Larson's results [154] are divided into the three shear rate regimes – tumbling, wagging, and steady-state – as explained below. He chose the strength of mean-field potential $2L^2dc'$ in Eq. (41) to be 10.67, which corresponds to the concentration c_A of the nematic phase coexisting with the isotropic phase (in the second virial approximation), and expressed the shear rate in terms of Γ defined by

$$\Gamma \equiv \kappa/D_r^{(I)} \tag{88}$$

where $D_r^{(I)}$ is the hypothetical rotational diffusion coefficient of the rod in the metastable isotropic solution of the same concentration as the nematic solution under consideration.

(1) *Tumbling regime* At very low shear rates, the birefringence axis (or the director) of the nematic solution tumbles continuously up to a reduced shear rate $\Gamma \lesssim 9.5$. While the time for complete rotation stays approximately equal to that calculated from Eq. (85), the scalar order parameter $S^{(dy)}$ oscillates around its equilibrium value S. Maximum positive departures of $S^{(dy)}$ from S occur at $\theta \approx \pi/4$ and $-3\pi/4$, and maximum negative departures at $\theta \approx -\pi/4$ and $-5\pi/4$, while the amplitude of oscillation increases with increasing Γ.

(2) *Wagging Regime* When Γ exceeds 10, tumbling gives way to oscillation or wagging of the birefringence axis between $\theta = \pm \theta(\text{wag})$, where $\theta(\text{wag})$ decreases with increasing Γ. Synchronizing with θ, $S^{(dy)}$ also oscillates around S.

(3) *Steady-State Regime* When Γ becomes larger than 25, the continuous wagging of the birefringence axis is taken over by a damped oscillation that eventually leads to a steady-state extinction angle θ and also a steady-state scalar order parameter $S^{(dy)}$. In this regime, the steady-state values of θ and $S^{(dy)}$, as well as the damping rate, increase with increasing Γ. It is to be noted that $S^{(dy)}$ exceeds S only when Γ reaches a very large value (≈ 800).

9.3 Comparison Between Experiment and Theory

The elastic shear stress $\sigma_{xz}^{(E)}$ and the elastic normal stress differences $\sigma_{N1}^{(E)}(\equiv \sigma_{xx}^{(E)} - \sigma_{zz}^{(E)})$ and $\sigma_{N2}^{(E)}(\equiv \sigma_{zz}^{(E)} - \sigma_{yy}^{(E)})$ can be calculated from Eq. (61) with Eq. (62). The calculated stresses are for a monodomain nematic specimen whose director is oriented in the z-direction at time $t = 0$, and they should depend on t in the tumbling and wagging regimes. However, the samples used for rheological measurements are usually polydomain specimens at rest, i.e., consist of many domains oriented in different directions at $t = 0$. Even if the specimen was initially oriented in some special direction, it would not necessarily remain at the state of monodomain in a shearing flow, because the molecules near the walls of the shearing device are usually pinned in a fixed (parallel or perpendicular) direction, so that their directions do not change in the same way as those in the bulk.

As a simplest way to account for the effect of the polydomain, Larson [154] averaged the stresses for single domain over one period of a cycle in the tumbling and wagging regimes. This averaging corresponds to assuming that each domain behaves as if it were part of a homogeneous sample in a shear flow with the same shear rate κ, and its phase of tumbling or oscillation shifts randomly. It neglects the influences of the gradient terms of the molecular orientation on both the evolution of the distribution function $f(\mathbf{a};t)$ and the stresses.

Figure 28b shows Larson's theoretical σ_{N1} and σ_{N2} [148, 154] for a nematic solution at the phase boundary concentration c_A and the strength of mean-field potential $2L^2 dc'$ in Eq. (41) chosen to be 10.67. In the figure, the stress differences and the shear rate are reduced by $c'k_B T$ and $D_r^{(l)}$, respectively, and viscous terms in σ_{N1} an σ_{N2} are neglected. Compared with Magda et al.'s experimental results [148] shown in Fig. 28a, Larson's theory successfully describes the changes in the signs of σ_{N1} and σ_{N2}. To facilitate the comparison, the tumbling, wagging, and steady-state flow regimes predicted by Larson are given in the upper part of Fig. 28b. The crossover form the tumbling to wagging regime corresponds to the first changes in the signs of the σ_{N1} and σ_{N2}. By making the theoretical peak heights of σ_{N1} and σ_{N2} coincide with the experimental data, we can estimate $c'k_B T$ to be ca. 1000 Pa, which may be compared favorably with the experimental value 1390 Pa ($= cRT/M_v$). From a similar operation with the peak positions of σ_{N1} and σ_{N2}, we can evaluate $D_r^{(l)}$ to be ca. 3 s^{-1}. In comparing this value with D_r evaluated by Eq. (50), the following attention has to be paid. (1) The PBLG molecule in the liquid crystal solution at c_A should take a more extended conformation than in the unperturbed state. The value of $\langle R^2 \rangle$ of this extended conformation was calculated by Chen [166], who used the method described in Sect. 2.3. From his results, L_e ($= \langle R^2 \rangle^{1/2}$) is estimated to be 142 nm. (In the unperturbed state, $\langle R^2 \rangle^{1/2} = 137$ nm.) (2) The orientation and extension of PBLG create free space in the solution, so that the jamming effect on $D_\parallel$ may be weakened in the liquid crystal solution. To a first approximation, we may neglect the jamming effect by assuming $D_{\parallel 0}/D_\parallel = 1$. (3) D_{r0} can be estimated from Eqs. (B3) and (B6) in Appendix B. Taking these points into account, we obtain $D_r^{(l)} = 0.53$ s^{-1}, which is far smaller than the above estimate 3 s^{-1}.

This disagreement in $D_r^{(l)}$ may be attributed to the choice of the strength U ($\equiv 2L^2 dc'$) of the molecular field $V_{scf}(\mathbf{a})$. In fact, the U value of 10.67 was used to calculate the theoretical curves in Fig. 28b, which is the value at the theoretical phase boundary concentration c_A, while U at the experimental c_A is 24.5. Larson [154] showed that the peak position of the σ_{N1} curve for U = 12.0 appears at a Γ larger than that for U = 10.57, but that the peak heights for these two U are almost the same (cf. Figs. 11 and 15 of [154]). Therefore, it is expected that the choice of larger U decreases the value of $D_r^{(l)}$. However, the strength of $V_{scf}(\mathbf{a})$ is not easy to estimate accurately, because we have to incorporate the effect of chain flexibility into Eq. (41).

10 Conclusions

In this article, we have surveyed typical properties of isotropic and liquid crystal solutions of liquid-crystalline stiff-chain polymers. It had already been shown that dilute solution properties of these polymers can be successfully described by the wormlike chain (or wormlike cylinder) model. We have here concerned ourselves with the properties of their concentrated solutions, with the main interest in the applicability of two molecular theories to them. They are the scaled particle theory for static properties and the fuzzy cylinder model theory for dynamical properties, both formulated on the wormlike cylinder model. In most cases, the calculated results were shown to describe representative experimental data successfully in terms of the parameters equal or close to those derived from dilute solution data.

These two theories neglect the intramolecular excluded volume effect, the multiple contacts between a pair of chains [41], and the reptation-like motion so that their applicability is essentially limited to stiff-polymers. Many favorable comparisons of their predictions with experimental results seem to justify these approximations to stiff polymer systems.

We have omitted discussing such interesting properties of liquid-crystal solutions as the Frank elastic constants, the Leslie viscosity coefficients, cholesteric pitch, textured structure (or defects), and rheo-optics. Some of them are reviewed in recent literature [8, 167], but the level of their experimental and theoretical studies still remains largely qualitative.

11 Appendices

Appendix A: Accuracy of Using Onsager's Trial Function for the Determination of the Equilibrium Orientational Distribution Function

Chen [47] calculated the isotropic–nematic phase boundary concentrations using not any trial function but $\bar{f}(\mathbf{a})$ determined by an numerical–iteration procedure. This method is more rigorous than the one resorting to a trial

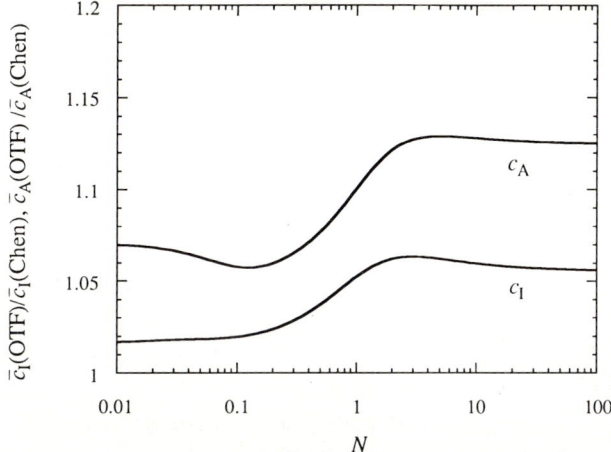

Fig. A1. Comparison of the reduced phase boundary concentrations ($\bar{c}_I$, $\bar{c}_A$) obtained by Chen's procedure [47] and the Onsager trial function procedures for $\bar{f}(a)$

function. In this appendix, we examine the accuracy of the Onsager trial function (OTF) procedure in regard to the phase boundary concentrations.

Since Chen used the second virial approximation in his phase boundary calculation, it is adequate to compare his results with the phase boundary concentrations calculated from Onsager's expression of S_{ex} (cf. Sect. 2.5) and $\sigma(N)$ of Eq. (18) together with the OTF. Chen expressed the phase boundary concentrations in terms of reduced quantities $\bar{c}_I \equiv L^2 dc'_I$ and $\bar{c}_A \equiv L^2 dc'_A$, which depend only on N. As shown in Fig. A1, the relative differences in both $\bar{c}_I$ and $\bar{c}_A$ between the two procedures are less than 13% over the whole range of N. This confirms the relevance of the OTF for semiflexible polymer systems.

Odijk [6] calculated $\sigma(N)$ by using a Gaussian trial function for $\bar{f}(a)$. This trial function, as well as the second virial approximation for use in calculating the phase boundary concentrations, however, leads to $\bar{c}_I$ and $\bar{c}_A$ which differ more than 70% at $N \gtrsim 1$ from Chen's calculations. In addition, c'_I from the Gaussian trial function appears beyond the critical concentration at which the isotropic phase becomes unstable for large N [9]. Therefore, the Gaussian trial function for $\bar{f}(a)$ is inadequate.

Appendix B: Frictional and Transport Coefficients at Infinite Dilution

As explained in Sect. 6, the expressions for $D_\parallel$, $D_\perp$, D_r and η_0 contain several hydrodynamic factors related to frictional coefficients at infinite dilution. We briefly describe the method to calculate these hydrodynamic factors in this appendix.

These diffusion coefficients of a wormlike cylinder at infinite dilution are expressed by [19].

$$D_{\|0} = \frac{k_B T}{2\pi \eta^{(S)} L} F_{\|0}^{-1} \tag{B1}$$

$$D_{\perp 0} = \frac{k_B T}{4\pi \eta^{(S)} L} F_{\perp 0}^{-1} \tag{B2}$$

$$D_{r0} = \frac{3k_B T}{\pi \eta^{(S)} L^3} F_{r0}^{-1}$$

$$= \frac{2N_A k_B T}{15\eta^{(S)} M[\eta]} \gamma^{-1} \tag{B3}$$

where $\eta^{(S)}$ is the solvent viscosity, N_A the Avogadro constant, and L and M the contour length and molecular weight of the polymer, respectively. The hydrodynamic factors depending on the thickness and flexibility of the cylinder are calculated by [19]

$$F_{\|0}^{-1} = \frac{2}{1 + (2D_{\perp 0}/D_{\|0})_{[19]}} \left(\frac{3\pi \eta^{(S)} L}{\Xi}\right)_{[168]} \tag{B4}$$

$$F_{\perp 0}^{-1} = \frac{2}{1 + (2D_{\perp 0}/D_{\|0})_{[19]}^{-1}} \left(\frac{3\pi \eta^{(S)} L}{\Xi}\right)_{[168]} \tag{B5}$$

$$F_{r0} = \frac{45M}{2\pi N_A L^3} [\eta]_{[169]} \gamma_{[124]} \tag{B6}$$

where Ξ is the translational frictional coefficient.

The quantities $D_{\perp 0}/D_{\|0}$, $3\pi \eta^{(S)} L/\Xi$, $[\eta]$, and γ on the right hand sides may be calculated using the equations given in the references indicated by the subscript. In [19], we proposed to approximate the ratio $D_{\perp 0}/D_{\|0}$ for the

Table B1. Equations in the quoted references needed for the computation of the correction factors for frictional coefficients at infinite dilution

Quantity	Equation number(s) in the Ref.	Remarks[a]
$\left(\frac{3\pi \eta_0 L}{\Xi}\right)_{[168]}$	(49), (51)	$L \to N$, $d \to d/2q$ $\sigma = 2.278$
$(2D_{\perp 0}/D_{\|0})_{[19]}$	(A8)	$p_e \to L_e/d_e$ $L \to N$, $d \to d/2q$
$[\eta]_{[169]}$	(23), (25)	$p \to L/d$, $\varepsilon = 1$
$\gamma_{[124]}$	(138), (121), (128)	$p \to L/d$, $\varepsilon = 1$
$\chi_{[124]}$	(133)	$p \to L/d$

[a] Notation in the original papers (left) should be replaced according to this column to express the equations in terms of the present notation (right)

wormlike cylinder by that for the ellipsoid with the axial ratio of L_e/d_e. Furthermore, we [19] approximated γ for the wormlike cylinder by that for the spherocylinder with the axial ratio of L/d given in [124]. The same approximation may be used to calculate another hydrodynamic parameter χ appearing in the zero-shear viscosity. Table B1 lists the equations needed for the computation of $3\pi\eta^{(S)}L/\Xi$, $D_{\perp 0}/D_{\parallel 0}$, $[\eta]$, γ, and χ.

Appendix C: Mean-Field Green Function Method

Translational diffusion of a particle can be described by the Green function. For simplicity, we consider here a one-dimensional diffusion process. Let x be the coordinate of the diffusing particle in its path. If no barriers are present in the path, the particle obeys the usual diffusion equation, and the unperturbed Green function associated with this diffusion equation is given by

$$G_0(x, x'; t, t') = \frac{1}{[4\pi D_0(t - t')]^{1/2}} \exp\left[-\frac{(x - x')^2}{4D_0(t - t')}\right] \quad (C1)$$

where D_0 is the unperturbed diffusion coefficient of the particle. The Green function can be regarded as the conditional probability that the particle at the position x' at time t' moves to the position x at time t.

When some point obstacles or barriers appear in the path, additional perturbation terms must be incorporated in the diffusion equation and then the Green function is no longer given by Eq. (C1). Once we know the perturbed Green function G, the *effective diffusion coefficient D* can be calculated from

$$D = \lim_{t-t' \to \infty} -\frac{1}{2(t-t')} \frac{\partial^2 \hat{G}(k; t, t')}{\partial k^2}\bigg|_{k=0} \quad (C2)$$

where $\hat{G}(k; t, t')$ is the Fourier transform of G defined by

$$\hat{G}(k; t, t') = \int_{-\infty}^{\infty} d(x - x') G(x, x'; t, t') \exp[ik(x - x')] \quad (C3)$$

The diffusion coefficient D_0 in free space is obtained from Eqs. (C1)–(C3), so our task turns to finding the perturbed Green function.

Suppose now a single reflecting point barrier is placed at position R in the path at time t_R and removed at time t_Q. Following Edwards and Evans [106], the Green function perturbed by this barrier can be expressed by

$$G(x, x'; t, t') = \int_{-\infty}^{\infty} dx_2 \int_{-\infty}^{\infty} dx_1 G_0(x, x_2; t, t_Q) Q'(x_2, x_1; t_Q, t_R)$$
$$\times G_0(x_1, x'; t_R, t') \quad (C4)$$

where $Q'(x_2, x_1; t_Q, t_R)$ is the transition probability that the particle at the position x_1 at time t_R moves to the position x_2 at time t_Q. The method of mirror images allows Q' to be expressed as

$$Q'(x_2,x_1;t_Q,t_R) = \theta_r(x_2 - R)\,[G_0(x_2, x_1;t_Q,t_R) + G_0(2R - x_2, x_1;t_Q,t_R)]$$
$$\times \theta_r(x_1 - R) + \theta_l(x_2 - R)\,[G_0(x_2,x_1;t_Q,t_R)$$
$$+ G_0(2R - x_2, x_1;t_Q,t_R)]\theta_l(x_1 - R) \quad \text{(C5)}$$

with two step functions

$$\theta_r(x - R) = \begin{cases} 1 & (x > R) \\ 0 & (x < R), \end{cases} \quad \theta_l(x - R) = \begin{cases} 0 & (x > R) \\ 1 & (x < R) \end{cases} \quad \text{(C6)}$$

The first and second parts in Eq. (C5) represent the solutions to the right and left sides of the barrier, and the first and second G_0 in each part correspond to the direct transition and the transition through reflection at the barrier, respectively (cf. Fig. C1).

After averaging the Fourier transform of Eq. (C4) with respect to the random variables R, t_R, and t_Q, and making some mathematical manipulations, we obtain [106]

$$\langle \hat{G}(k;t,t') \rangle = \exp[-D_0(t-t')k^2] + \langle \Delta\hat{G}(k;t,t') \rangle \quad \text{(C7)}$$

with the perturbed part $\langle \Delta\hat{G}(k;t,t') \rangle$ given by

$$\langle \Delta\hat{G}(k;t,t') \rangle = \frac{8}{3\sqrt{\pi}l}(D_0\tau)^{3/2}k^2 + O(k^4) \quad \text{(C8)}$$

for a sufficiently long time interval $t - t'$. Here l is a path length comparable to the maximum distance that the particle can diffuse during $t - t'$ and τ the mean lifetime of the barrier. The brackets $\langle \cdots \rangle$ represent the average with respect to R, t_Q, and t_R. Inserting Eq. (C7) with Eq. (C8) into Eq. (C2), we get the effective diffusion coefficient in the presence of a single reflecting barrier as

$$D = D_0 - \frac{8}{3\sqrt{\pi}l(t-t')}(D_0\tau)^{3/2} \quad \begin{pmatrix} \text{one dimension,} \\ \text{the first order perturbation} \end{pmatrix} \quad \text{(C9)}$$

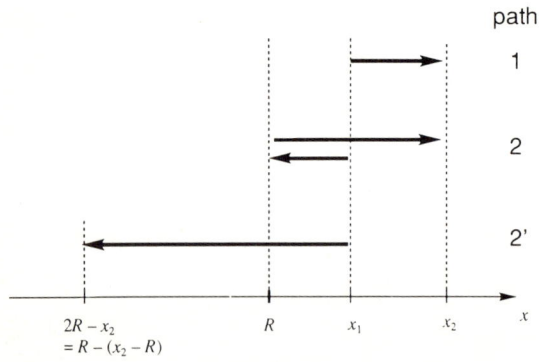

Fig. C1. Schematic explanation of Eq. (C5) in the case that $R < x_1 < x_2$, i.e., $\theta_r(x_i - R) = 1$ and $\theta_l(x_i - R) = 0$ (i = 1, 2); *path 1*, the direct transition; *path 2*, the transition through reflection by the barrier. Note that G_0 for path 2 is equal to that for path 2'

Next we consider the case where more than one barrier (or perturbation element) appears in the diffusion path, and the particle receives multiple perturbations. Teraoka and Hayakawa [107] assumed that the Green function for this case satifies the following Dyson equation:

$$\langle G \rangle = G_0 + G_0 \left\langle \sum_{i=1}^{M} Q_i \right\rangle \langle G \rangle \tag{C10}$$

where Q_i is the i-th perturbation element and M is the number of the elements appearing on the path length l between t and t'. In Eq. (C10) and hereafter we omit the arguments of the Green functions and the integration symbol; the second term on the right hand side of Eq. (C10) involves the same integration as in Eq. (C4). Equation (C10) can be solved recursively for $\langle G \rangle$ to give

$$\langle G \rangle = G_0 + G_0 \left\langle \sum_{i=1}^{M} Q_i \right\rangle G_0 + G_0 \left\langle \sum_{i=1}^{M} Q_i \right\rangle G_0 \left\langle \sum_{j=1}^{M} Q_j \right\rangle G_0 + \cdots \tag{C11}$$

which indicates that Eq. (C10) gives a Green function perturbed by any number of *sequential independent elements*. If the diffusing particle is between two closely located barriers, it can collide with them many times, and the perturbation elements for the multiple collisions may not be independent. Equation (C10) approximates this situtaion by a simple sum of independent perturbations. The Green function $\langle G \rangle$ given by Eq. (C10) is usually referred to as a *mean-field* Green function.

We denote this $\langle G \rangle$ by $\langle G_M \rangle$ in order to indicate explicitly that the Green function is perturbed by M perturbation elements, and then introduce a series of new (hypothetical) Green functions defined by

$$\langle G_m \rangle = G_0 + G_0 \left\langle \sum_{i=1}^{m} Q_i \right\rangle \langle G_m \rangle \quad (m = 1, 2, \ldots, M) \tag{C12}$$

which pick up mean effects of m perturbation elements out of the total M elements. It can be verified that $\langle G_m \rangle$ satisfies the recurrence formula [107]

$$\langle G_m \rangle = \langle G_{m-1} \rangle + \langle G_{m-1} \rangle \langle Q_m \rangle \langle G_m \rangle \quad (m = 1, 2, \ldots, M) \tag{C13}$$

Since we are interested in the diffusion during a long period $t - t'$, M is a very large number, and thus Eq. (C13) may be approximated by

$$\langle G_m \rangle = \langle G_{m-1} \rangle + \langle G_{m-1} \rangle \langle Q_m \rangle \langle G_{m-1} \rangle \quad (m = 1, 2, \ldots, M) \tag{C14}$$

By straightforward extension of the calculation of the first order perturbation, we obtain the following difference equation for the effective diffusion coefficient D_m:

$$\frac{D_m}{D_0} = \frac{D_{m-1}}{D_0} - \frac{8}{3\sqrt{\pi} l(t-t')} \tau^{3/2} D_0^{1/2} \left(\frac{D_{m-1}}{D_0} \right)^{3/2} \tag{C15}$$

Replacing this by a differential equation and solving it under a suitable boundary condition, we obtain the one-dimensional effective diffusion coefficient

$D^{(1)}$ in the form

$$\frac{D^{(1)}}{D_0} \equiv \frac{D_M}{D_0} = \left(1 + \frac{4}{3\sqrt{\pi}} \rho_B^{(1)} \tau^{3/2} D_0^{1/2}\right)^{-2} \quad \begin{pmatrix}\text{one dimension,}\\ \text{multiple perturbation}\end{pmatrix}$$
(C16)

where $\rho_B^{(1)} [= M/l(t - t')]$ is the number of the barriers appearing per unit path length and unit time.

The above formulation can be readily extended to the two-dimensional diffusion of a Brownian particle in the presence of needle-like obstacles with a mean lifetime τ, as has been made by Teraoka and Hayakawa [107], who obtained for the two-dimensional effective diffusion coefficient

$$\frac{D^{(2)}}{D_0} = \left(1 + \frac{2}{3\sqrt{\pi}} \rho_B^{(2)} b \tau^{3/2} D_0^{1/2}\right)^{-2} \quad \begin{pmatrix}\text{two dimension,}\\ \text{multiple perturbation}\end{pmatrix} \quad (C17)$$

where b is the average length of the needle-like barriers and $\rho_B^{(2)}$ is the number of the barriers appearing per unit area and unit time.

It is not difficult to generalize the above formulation of the effective diffusion coefficient to the case in which there appear r different kinds of barriers or perturbation elements in the diffusion space. The result can be used to formulate the effective diffusion coefficient in a multicomponent solution. See [144] for a detailed explanation.

Acknowledgements. The authors are grateful to Professor H. Fujita for his many suggestions which served to improve the manuscript, and to Dr. I. Teraoka for helpful discussion.

12 References

1. Robinson C (1956) Trans Faraday Soc 52: 571
2. Onsager L (1949) Ann NY Acad Sci 51: 627
3. Flory PJ (1956) Proc Roy Soc London A234: 73
4. Doi M, Edwards SF (1986) The Theory of Polymer Dynamics. Clarendon Press, Oxford
5. Flory PJ (1984) Adv Polym Sci 59: 1
6. Odijk T (1986) Macromolecules 19: 2313
7. Semenov AN, Khokhlov AR (1988) Sov Phys Usp 31: 988
8. Ciferri A (1991) Liquid Crystallinity in Polymers. VCH, New York
9. Vroege GJ, Lekkerkerker HNW (1992) Rep Prog Phys 55: 1241
10. Moscicki JK (1985) Adv Chem Phys 58: 631
11. Reiss H, Frisch HL, Lebowitz JL (1959) J Chem Phys 31: 369
12. Cotter MA, Martire DE (1970) J Chem Phys 53: 4500
13. Cotter MA, Martire DE (1970) J Chem Phys 52: 1909
14. Lasher G (1970) J Chem Phys 53: 4141
15. Cotter MA (1974) Phys Rev A 10: 625
16. Cotter MA (1977) J Chem Phys 66: 1098
17. Sato T, Shoda T, Teramoto A (1994) Macromolecules 27: 164
18. Sato T, Teramoto A (1994) In: Teramoto A, Kobayashi M, Norisuye T (ed) Ordering in Macromolecular Systems. Springer-Verlag, Berlin, Heidelberg, p 155

19. Sato T, Takada Y, Teramoto A (1991) Macromolecules 24: 6220
20. Fujita H (1990) Polymer Solutions. Elsevier, Amsterdam
21. Brelsford GL, Krigbaum WR (1991) In: Ciferri A (ed) Liquid Crystallinity in Polymers. VCH, New York, Chapt. 2
22. Norisuye T (1993) Prog Polym Sci 18: 543
23. Fraden S, Maret G, Caspar DLD, Meyer RB (1989) Phys Rev Lett 63: 2068
24. Tang J, Fraden S, Liq Cryst, in press
25. Yanaki T, Norisuye T, Fujita H (1980) Macromolecules 13: 1462
26. Itou S, Nishioka N, Norisuye T, Teramoto A (1981) Macromolecules 14: 904
27. Sato T, Norisuye T, Fujita H (1984) Polym J 16: 341
28. Godfrey JE, Eisenberg H (1976) Biophys Chem 5: 301
29. Record MT, Woodbury CP (1975) Biopolymers 15: 393
30. Arpin M, Strazielle C (1977) Polymer 18: 591
31. Itou T, Chikiri H, Teramoto A, Aharoni SM (1988) Polym J 20: 143
32. Motowoka M, Norisuye T, Teramoto A, Fujita H (1979) Polym J 11: 665
33. Abe A, Tabata S, Kimura N (1991) Macromolecules 24: 6238
34. Conio G, Bianchi E, Ciferri A, Tealdi A, Aden MA (1983) Macromolecules 16: 1264
35. Laivins GV, Gray GD (1985) Macromolecules 18: 1753
36. Itou T, private communication
37. Takahashi Y, Kobatake T, Suzuki H (1984) Rep Prog Polym Phys Jpn 27: 767
38. Bamford CH, Elliott A, Hanby WE (1956) Synthetic Polypeptides. Academic Press, New York
39. Sato T, Teramoto A (1990) Mol Cryst Liq Cryst 178: 143
40. Werbowyj RS, Gray DG (1980) Macromolecules 13: 69
41. Sato T, Teramoto A (1994) Acta Polymerica, 45: 399
42. Landau LD, Lifshitz EM (1980) Statistical Physics. 3rd ed, Pergamon, London
43. Khokhlov AR, Semenov AN (1981) Physica A108: 546
44. Khokhlov AR, Semenov AN (1982) Physica A112: 605
45. Hentschke R (1990) Macromolecules 23: 1192
46. DuPré DB, Yang S (1991) J Chem Phys 94: 7466
47. Chen ZY (1993) Macromolecules 26: 3419
48. Lifshitz IM (1969) Soviet Phys JETP 28: 1280
49. Lifshitz IM, Grosberg AY, Khokhlov AR (1978) Rev Mod Phys 50: 684
50. Parsons JD (1979) Phys Rev A 19: 1225
51. Khokhlov AR, Semenov AN (1985) J Stat Phys 38: 161
52. Khokhlov AR (1991) In: Ciferri A (ed) Liquid Crystallinity in Polymers. VCH, New York, Chapt. 3
53. Lee S-D (1987) J Chem Phys 87: 4972
54. Straley JP (1973) Mol Cryst Liq Cryst 24: 7
55. Kihara T, private communication
56. Kubo K, Ogino K (1979) Mol Cryst Liq Cryst 53: 207
57. Kubo K (1981) Mol Cryst Liq Cryst 74: 71
58. Fujita H (1975) Foundation of Ultracentrifugal Analysis. Wiley-Interscience, New York
59. Kurata M (1982) Thermodynamics of Polymer Solutions. Fujita H (trans), Harwood Academic Publishers, Chur
60. Itou T, Sato T, Teramoto A, Aharoni SM (1988) Polym J 20: 1049
61. Jinbo Y, Sato T, Teramoto A (1994) Macromolecules 27: 6080
62. Scholte TG (1970) J Polym Sci: Part A-2 8: 84
63. Van K, Teramoto A (1985) Polym J 17: 409
64. Itou T, Teramoto A (1988) Macromolecules 21: 2225
65. Itou T, Van K, Teramoto A (1985) J Appl Polym Sci: Appl Polym Symp 41: 35
66. Wee EL, Miller WG (1971) J Phys Chem 75: 1446
67. Miller WG, Wu CC, Wee EL, Santee GLR, Rai JH, Goebel KG (1974) Pure Appl Chem 38: 37
68. Miller WG, Rai JH, Wee EL (1974) Liquid Crystals and Ordered Fluids 2: 243
69. Itou T, Funada S, Shibuya F, Teramoto A (1986) Kobunshi Ronbunshu 43: 191
70. Hermans JJ (1962) J Colloid Sci 17: 638
71. Robinson C, Ward JC, Beevers RB (1958) Discuss Faraday Soc 25: 29
72. Patel DL, Gilbert RD (1983) J Polym Sci: Polym Phys Ed 21: 1079
73. Sato T, Ikeda N, Itou T, Teramoto A (1989) Polymer 30: 311
74. Abe A, Flory PJ (1978) Macromolecules 11: 1122
75. Itou T, Teramoto A (1984) Macromolecules 17: 1419

76. Itou T, Teramoto A (1984) Polym J 16: 779
77. Bawden FC, Pirie NW (1937) Proc Roy Soc London B123: 274
78. Inatomi S, Jinbo Y, Sato T, Teramoto A (1992) Macromolecules 25: 5013
79. Stroobants A, Lekkerkerker HNW, Odijk T (1986) Macromolecules 19: 2232
80. Philip JR, Wooding RA (1970) J Chem Phys 52: 953
81. Odijk T (1990) J Chem Phys 93: 5172
82. Sato T, Teramoto A (1991) Physica A176: 72
83. Sato T, Kakihara T, Teramoto A (1990) Polymer 31: 824
84. Sato T, Norisuye T, Fujita H (1984) Macromolecules 17: 2696
85. Rill R (1986) Proc Natl Acad Sci USA 83: 342
86. Strzelecka TE, Rill RL (1987) J Am Chem Soc 109: 4513
87. Strzelecka TE, Rill RL (1990) Biopolymers 30: 57
88. Strzelecka TE, Rill RL (1991) Macromolecules 24: 5124
89. Kwolek SL, Morgan PW, Schaefgen JR, Gulrich LW (1977) Macromolecules 10: 1390
90. Bair TI, Morgan PW, Killian FL (1977) Macromolecules 10: 1396
91. Aden MA, Bianchi E, Ciferri A, Tealdi A (1984) Macromolecules 17: 2010
92. Yamazaki T, Abe A (1987) Polym. J. 19: 777
93. Abe A, Yamazaki T (1989) Macromolecules 22: 2145
94. Wang H, DuPré DB (1992) J Chem Phys 96: 1523
95. Chapman GE, Campbell ID, McLauchlan KA (1970) Nature 225: 639
96. Yamazaki T, Horiuchi S, Watanabe J, Abe A (1987) Polym Prep Jpn.36: 1766
97. Abe A, Yamazaki T (1989) Macromolecules 22: 2138
98. Berger MN, Tidswell BM (1973) J Polym Sci: Polym Symp 42: 1063
99. Doi M, Edwards SF (1978) J Chem Soc: Faraday Trans 2 74: 560
100. Doi M (1981) J Polym Sci: Polym Phys Ed 19: 229
101. Muthukumar M, Edwards SF (1983) Macromolecules 16: 1475
102. Kratky O, Porod G (1949) Recl Trav Chim Pays-Bas 68: 1106
103. Yamakawa H (1971) Modern Theory of Polymer Solutions. Harper & Row, New York
104. Hoshikawa H, Saito N, Nagayama K (1975) Polym J 7: 79
105. Tagami Y (1969) Macromolecules 2: 8
106. Edwards SF, Evans KE (1982) J Chem Soc: Faraday Trans 2 78: 113
107. Teraoka I, Hayakawa R (1988) J Chem Phys 89: 6989
108. Teraoka I, Hayakawa R (1989) J Chem Phys 91: 2643
109. Sato T, Teramoto A (1991) Macromolecules 24: 193
110. Teraoka I (1988) Ph.D. Thesis, University of Tokyo
111. Teraoka I, Ookubo N, Hayakawa R (1985) Phys Rev Lett 55: 2712
112. Cohen MH, Turnbull D (1959) J Chem Phys 31: 1164
113. Odijk T (1983) Macromolecules 16: 1340
114. Doi M (1983) J Chem Phys 79: 5080
115. Semenov AN (1986) J Chem Soc: Faraday Trans 2 82: 317
116. Kirkwood JG (1949) Recl Trac Chim Pays-Bas 68: 649
117. Kuzuu N, Doi M (1983) J Phys Soc Jpn 52: 3486
118. Ferry JD (1980) Viscoelastic Properties of Polymers. 3rd ed, John Wiley, New York
119. Doi M, Yamamoto I, Kano F (1984) J Phys Soc Jpn 53: 3000
120. Fixman M (1985) Phys Rev Lett 54: 337
121. Fixman M (1985) Phys Rev Lett 55: 2429
122. Bitsanis I, Davis HT, Tirrell M (1988) Macromolecules 21: 2824
123. Bitsanis I, Davis HT, Tirrell M (1990) Macromolecules 23: 1157
124. Yoshizaki T, Yamakawa H (1980) J Chem Phys 72: 57
125. Tirrell M (1984) Rubber Chem Tech 57: 52
126. Auroy P, Hervet H, Leger L (1989) Polym Commun 30: 272
127. Tinland B, Maret G, Rinaudo M (1990) Macromolecules 23: 596
128. Takada Y, Sato T, Teramoto A (1991) Macromolecules 24: 6215
129. Mori Y, Ookubo N, Hayakawa R, Wada Y (1982) J Polym Sci: Polym Phys Ed 20: 2111
130. Zero KM, Pecora R (1982) Macromolecules 15: 87
131. Kubota K, Chu B (1983) Biopolymers 22: 1461
132. Papkov SP, Kulichikhin VG, Kalmykova VD, Malkin AY (1974) J Polym Sci: Polym Phys Ed 12: 1753
133. Baird DG, Ballman RL (1979) J Rheol 23: 505
134. Chu S-G, Venkatraman S, Berry GC, Einaga Y (1981) Macromolecules 14: 939

135. Jain S, Cohen C (1981) Macromolecules 14: 759
136. Enomoto H, Einaga Y, Teramoto A (1984) Macromolecules 17: 1573
137. Enomoto H, Einaga Y, Teramoto A (1985) Macromolecules 18: 2695
138. Milas M, Rinaudo M, Tinland B (1985) Polym Bull 14: 157
139. Mead DW, Larson RG (1990) Macromolecules 23: 2524
140. Takada Y, Sato T, Einaga Y, Teramoto A (1988) Bull Inst Chem Res Kyoto Univ 66: 212
141. Yamaguchi M, Wakutsu M, Takahashi Y, Noda I (1992) Macromolecules 25: 470
142. Sho T, Sato T, Norisuye T (1986) Biophys Chem 25: 307
143. Liu W, Sato T, Norisuye T, Fujita H (1987) Carbohydr Res 160: 267
144. Sato T, Ohshima A, Teramoto A (1994) Macromolecules 27: 1477
145. Fujiyama, T, Sato T, Teramoto A (1995) Acta Polymerica, in press
146. Kiss G, Porter RS (1978) J Polym Sci: Polym Symp 65: 193
147. Kiss G, Porter RS (1980) J Polym Sci: Polym Phys Ed 18: 361
148. Magda JJ, Baek S-G, De Vries KL, Larson RG (1991) Macromolecules 24: 4460
149. Navard P (1986) J Polym Sci: Polym Phys Ed 24: 435
150. Grizzuti N, Cavella S, Cicarelli P (1990) J Rheol 34: 1293
151. Marrucci G, Maffettone PL (1989) Macromolecules 22: 4076
152. Marrucci G, Maffettone PL (1990) J Rheol 34: 1217
153. Marrucci G, Maffettone PL (1990) J Rheol 34: 1231
154. Larson RG (1990) Macromolecules 23: 3983
155. Larson RG, Mead DW (1991) J Polym Sci: Polym Phys Ed 29: 1271
156. Moldenaers P, Fuller GG, Mewis J (1989) Macromolecules 22: 960
157. Burghardt WR, Fuller GG (1991) Macromolecules 24: 2546
158. Picken SJ, Aerts J, Doppert HL, Reuvers AJ, Northolt MG (1991) Macromolecules 24: 1366
159. Larson RG, Doi M (1991) J Rheol 35: 539
160. Ericksen JL (1960) Arch Ration Mech Anal 4: 231
161. Ericksen JL (1961) Trans Soc Rheol 5: 23
162. Leslie FM (1966) Quart J Mech Appl Math 19: 357
163. Leslie FM (1968) Arch Ration Mech Anal 28: 265
164. Semenov AN (1983) Zh Eksp Teor Fiz 85: 549
165. Kuzuu N, Doi M (1984) J Phys Soc Jpn 53: 1031
166. Chen ZY (1994) Macromolecules 27: 2073
167. Sato T, Sato Y, Umemura Y, Teramoto A, Nagamura Y, Wagner J, Weng D, Okamoto Y, Hatada K, Green MM (1993) Macromolecules 26: 4551
168. Yamakawa H, Fujii M (1973) Macromolecules 6: 407
169. Yamakawa H, Yoshizaki T (1980) Macromolecules 13: 633
170. Doi M (1985) J Polym Sci: Polym Symp 73: 93

Editor: H. Fujita
Received January 1995

High Internal Phase Emulsions (HIPEs) – Structure, Properties and Use in Polymer Preparation

N.R. Cameron, D.C. Sherrington
University of Strathclyde, Dept. of Pure and Applied Chemistry,
295 Cathedral Street, Glasgow G1 1XL, UK

High internal phase emulsions (HIPEs) are concentrated systems possessing a large volume of internal, or dispersed phase. The volume fraction is above 0.74, resulting in deformation of the dispersed phase droplets into polyhedra, which are separated by thin films of continuous phase. Their structure, which is analogous to a conventional gas-liquid foam of low liquid content, gives rise to a number of peculiar and fascinating properties including high viscosities and viscoelastic rheological behaviour. Like dilute emulsions, HIPEs are both kinetically and thermodynamically unstable; nevertheless, it is possible to prepare metastable systems which show no change in properties or appearance over long periods of time.

Polymer materials can easily be prepared from HIPEs if one or the other (or both) phases of the emulsion contain monomeric species. This process yields a range of products with widely differing properties. Additionally, as the concentrated emulsion acts as a scaffold or template, the microstructure of the resultant material is determined by the emulsion structure immediately prior to polymerisation.

In this review, the structure, properties, stability and applications of highly concentrated emulsions will be discussed in the first section. Following this, the use of HIPEs to generate novel polymer materials will be the focus of the second part.

1	High Internal Phase Emulsions (HIPEs).	165
	1.1 Introduction.	165
	1.2 HIPE Formation.	165
	1.3 Properties of HIPEs	166
	1.3.1 Geometry and Droplet Packing	166
	1.3.2 Rheology.	173
	1.3.2.1 Theoretical Analyses.	173
	1.3.2.2 Experimental Investigations	179
	1.3.3 Osmotic Pressure.	181
	1.3.4 Continuous Phase Microstructure.	183
	1.4 HIPE Stability	184
	1.5 Non-Aqueous HIPEs	188
	1.6 Applications of HIPEs.	189
2	The Preparation of Polymeric Materials from HIPEs.	189
	2.1 Polymerisation of the Continuous Phase – PolyHIPE	190
	2.1.1 Poly(styrene/divinylbenzene) PolyHIPE Copolymers	190
	2.1.1.1 Preparation.	190

		2.1.1.2 Cellular Structure and Morphology	192
		2.1.1.3 Properties	195
		2.1.1.4 Applications	197
	2.1.2	Other PolyHIPE Systems	201
2.2	Polymerisation of the Dispersed Phase		202
2.3	Polymerisation of Both Phases		207

3 Conclusions ... 209

4 References ... 210

1 High Internal Phase Emulsions (HIPEs)

1.1 Introduction

Lissant [1] classed high internal phase emulsions, or HIPEs, as emulsions containing greater than 70% internal phase volume, although a more precise definition would be those with an internal phase volume above 74.05%. This figure represents the maximum volume occupiable by a number of uniform spheres packed into a given volume in the most efficient manner [2]. Previously it was believed that this figure represented the maximum possible emulsion internal phase volume, without inversion occurring. Quite often this is indeed the case; however, there are now known to be many examples of more concentrated systems.

These concentrated emulsions have been referred to by a number of different names in the literature, including high internal phase ratio emulsions (HIPREs) [1, 3–7], gel-emulsions [8–14] and hydrocarbon gels [15, 16]. In this review, the term HIPE will be used throughout.

The so-called "bi-liquid foams" described by Sebba and Vincent [17–20] will not be discussed since they are not true colloidal emulsions, but resemble conventional foams more closely, possessing polyhedral cells of centimetre dimensions.

In addition, Sebba and others [19, 21–25] have detailed systems known as "polyaphrons", which appear to be o/w HIPEs prepared with very low surfactant concentrations in each phase. Although it is stressed that polyaphrons differ from conventional concentrated emulsions, the present authors and others [7, 16], believe that this is not the case.

1.2 HIPE Formation

The first criterion for the formation of a HIPE is, of course, the presence of two immiscible liquids, one of which is water (or aqueous solution), almost without exception. The nature of the organic, or oil, phase can vary to a considerable extent, although the most stable HIPEs are generally produced with more hydrophobic liquids. However, it is the nature of the surfactant employed to stabilise the HIPE which will ultimately facilitates its formation. Above a certain critical limit of internal phase volume, an emulsion will tend to invert to the opposite type, i.e. an oil-in-water (o/w) emulsion will become the w/o variety, and vice versa. This can be prevented from occurring by careful choice of surfactant, such that it is completely insoluble in the dispersed phase of the emulsion.

The HIPE is formed generally by careful addition of the internal phase to a solution of surfactant in the external phase, under constant agitation. HIPEs may form under other circumstances, however [26]. When a centrifugal field is applied to an emulsion, the droplets can be forced into contact with each other and deformation into polyhedra occurs. The excess continuous phase is forced out of the emulsion, forming a separate phase. This process is known as "creaming". Additionally, emulsions may cream over a period of time under the influence of gravity, forming a layer of HIPE either on top of the dilute emulsion, in the case of oil-in-water emulsions, or below the bulk emulsion for w/o systems.

1.3 Properties of HIPEs

1.3.1 Geometry and Droplet Packing

As was mentioned in the introduction, HIPEs contain an internal phase volume fraction greater than 0.74. Since this is the maximum volume which can be occupied by uniform, undeformed spherical particles, the dispersed phase droplets must either be non-uniform, i.e. polydisperse, or deformed into non-spherical, polyhedral cells.

In a theoretical treatment of the geometry of HIPEs Lissant [3] showed that, for a monodisperse system, the dispersed phase droplets will assume a rhomboidal dodecahedral (RDH) packing from 74% to 94% internal phase volume, with increasing deformation into polyhedra. Above 94%, the packing changes to tetrakaidecahedral (TKDH) (truncated octahedral). It was assumed that monodisperse packing occurred throughout; theoretical calculations showed this system to be more favourable than a polydisperse arrangement [4] (Fig. 1).

The shape of the dispersed phase droplets was investigated experimentally by Lissant and coworker by scanning electron microscopy (SEM) on cured HIPEs of water in a styrene-based resin [5]. At high internal phase volumes, droplets were indeed polyhedral, and appeared to be relatively monodisperse in size.

In a subsequent theoretical analysis, Princen [26] initially used a model of infinitely long cylindrical drops to relate the geometric and thermodynamic properties of monodisperse HIPEs to the volume fraction of the dispersed phase. Thus the analysis could be restricted to a two-dimensional cross-section of the emulsion. Two principle emulsion parameters were considered; the film thickness between adjacent drops (h) and the contact angle (θ) [27–29]. The effects of these variables on the volume fraction, ϕ, both in the presence and absence of a compressive force on the emulsion, were considered. The results indicated that if both h and θ are kept at zero, the maximum volume fraction (ϕ) of the *uncompressed* emulsion is 0.9069, which is equivalent to $\phi = 0.7405$ in real emulsions with spherical droplets (cf. Lissant's work). If θ is zero (or constant) and h is increased, the maximum value of ϕ decreases; on the other hand, increasing θ with zero or constant h causes ϕ to increase above the value 0.9069, again at zero compression. This implies that, in the presence of an appreciable contact angle, without any applied compressive force, values of ϕ in excess of the maximum value for undeformed droplets can occur. Thus, the dispersed phase

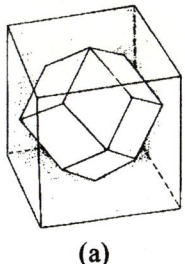
(a)

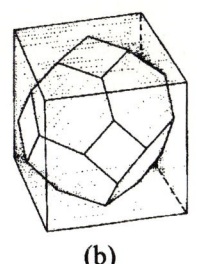
(b)

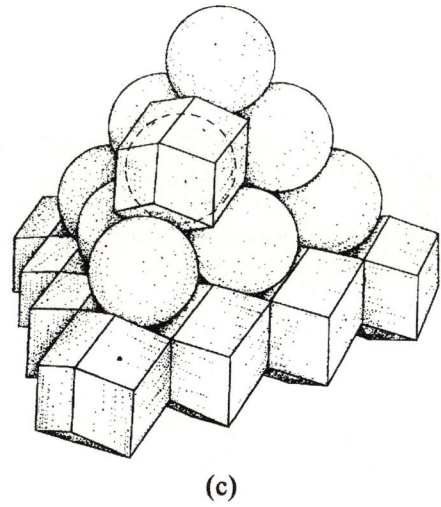
(c)

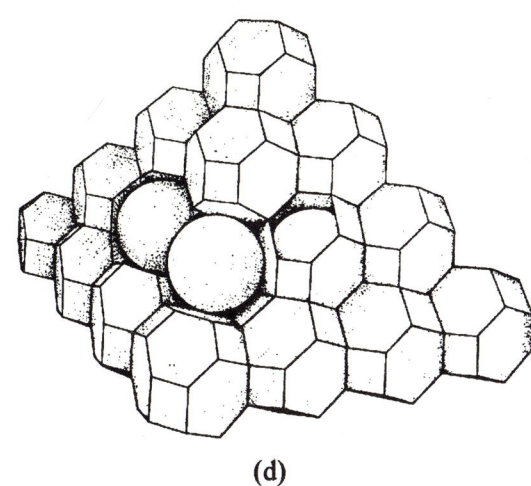
(d)

Fig. 1. Rhomboidal dodecahedron (RDH) (**a**), tetrakaidecahedron (TKDH) (**b**), rhomboidal dodecahedral packing (**c**) and tetrakaidecahedral packing (**d**)

droplets will *spontaneously* deform into polyhedra, with the removal of some of the continuous phase in the process. In this model of cylindrical droplets, each assumes a hexagonal form.

Extending this simple, two-dimensional analysis to include real (monodisperse) emulsions, Princen [30] subsequently demonstrated that the above also held true. Again the results showed that a finite film thickness will tend to reduce the maximum value of ϕ, in uncompressed emulsions, to below 0.7405. For example, emulsion droplets of 1 μm diameter, separated by an equilibrium film thickness of 50 nm, limits ϕ to 0.64. Similarly, a finite contact angle (with zero film thickness) allows values of ϕ in excess of 0.7405, in the absence of a compressive force. As θ increases, the droplets are progressively deformed into truncated spheres, resembling rhomboidal dodecahedra (again cf. Lissant's work); this is the three-dimensional analogue of the hexagonal deformation mentioned above. When $\theta = 30°$, $\phi = 0.964$; at this phase volume, the films separating the droplets reach the sides of the rhombi in which they are inscribed, and the continuous phase exists solely along the edges and in the corners of each rhomboidal cell (Fig. 2).

At this point, the concept of the "linear collapsed Plateau border" is introduced. The Plateau border is the area of bulk continuous phase between three adjacent droplets or cells in an emulsion or foam respectively. The collapsed border is, therefore, an extremely thin version, which can be represented macroscopically, as the line of intersection of three films of zero thickness, at angles of 120°.

As θ is increased above 30°, the droplets occupy more and more of their inscribing rhombi, and the continuous phase is contained in increasingly smaller volumes in the corners of each cell until, at a given value of θ, the droplets reach the eight tetrahedral corners of each cell. Four Plateau borders now converge at a point under the tetrahedral angle of 109.47°, giving rise to the phenomenon of the "collapsed tetrahedral Plateau border". This occurs at $\theta = 35.26°$. The remaining continuous phase is found in the other six corners of the rhombus, and ϕ is estimated to be 0.985. The emulsion now closely resembles a polyhedral foam [31] (Fig. 3).

However, it may be the case that the rhomboidal dodecahedron (RDH) is inadequate for describing the system at high phase volumes. Princen suggests that, up to $\phi = 0.964$, the RDH may be valid; above this value, however, a rearrangement occurs to a regular pentagonal dodecahedron (RPD). The analysis again leads to the same equations for the RPD, with a maximum phase volume ratio of 0.9925, implying that the RPD is a reasonable model for the droplet shape in uncompressed HIPEs with ϕ values very close to unity.

Support for this postulation came from work done on the shape of the ideal foam cell [32–40]. Ross and co-worker [34, 35] proposed three "minimal" geometric structures, i.e. those which will subdivide space with minimum partitional area. These were the pentagonal dodecahedron, the minimal tetrakaidecahedron, originally suggested by Thomson (Lord Kelvin), and the β-tetrakaidecahedron (Fig. 4).

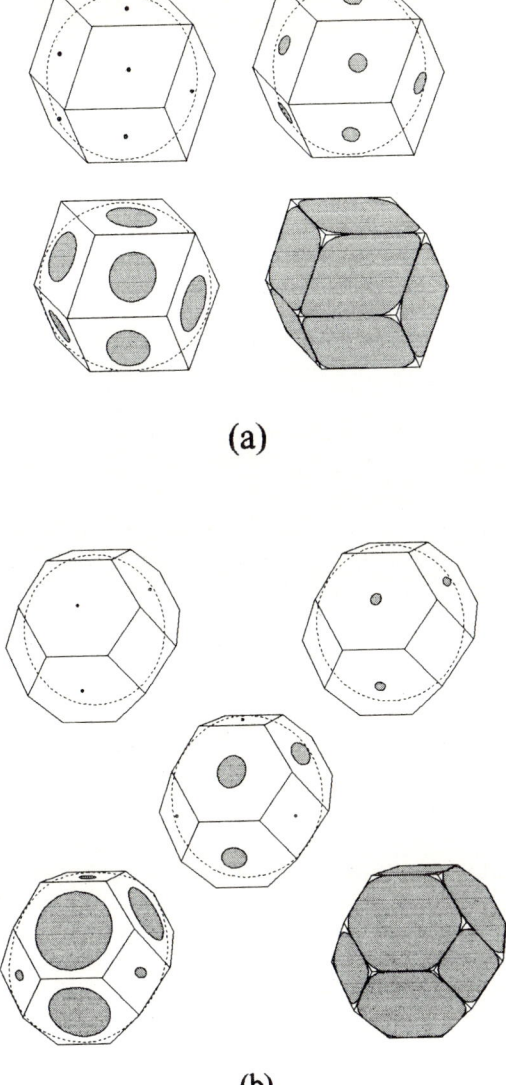

Fig. 2. Dispersed phase droplet inscribed in polyhedra: rhomboidal dodecahedron (**a**) and tetrakaidecahedron (**b**)

The pentagonal dodecahedron, however, is not entirely space-filling, i.e. a close-packed array of such figures has a number of interstitial voids. On the other hand, Kelvin's tetrakaidecahedron and the β-tetrakaidecahedron are. The latter requires 4% more surface area, so a system of such figures would spontaneously rearrange to the more stable array of Kelvin cells. Thus, it would seem that Kelvin's tetrakaidecahedron is the ideal candidate; nevertheless, this is not observed in real systems! Pentagonal faces are shown on foam cells. These

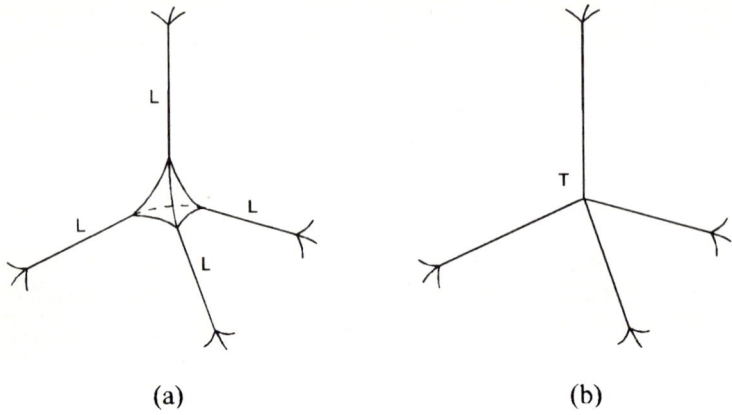

Fig. 3. Collapsed linear Plateau border, L (a) and collapsed tetrahedral Plateau border, T (b)

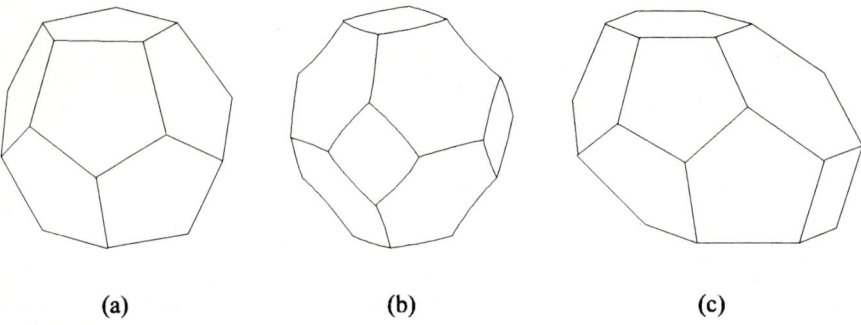

Fig. 4. Regular pentagonal dodecahedron (RPD) (a), Kelvin's minimal tetrakaidecahedron (Kelvin's cell) (b) and β-tetrakaidecahedron (c)

do not change into quadrilateral and hexagonal faces, implying that the close-packing of truncated octahedra does not occur.

This can be explained on the basis of the close-packing of uniform spheres; each sphere has twelve nearest neighbours. The tetrakaidecahedral packing requires fourteen nearest neighbours, and geometrical constraints will not allow this transition to occur. The pentagonal dodecahedral packing, on the other hand, gives each cell twelve nearest neighbours, so, in transforming from a spherical to a polyhedral foam, pentagonal dodecahedra predominate. Some of the cells must adapt to fill the voids left in such an arrangement, resulting, ultimately, in a polydisperse array.

Princen et al. [36] also observed this phenomenon. In addition, they determined experimentally [37] that in a real, polydisperse system, the increase in surface area in changing from spheres to polyhedra is less than in a system involving monodisperse Kelvin tetrakaidecahedra, implying that an array containing a degree of polydispersity is more favourable.

However, Weaire et al. [41] have recently shown that it is possible to produce monodisperse dry foams containing Kelvin polyhedra. Upon addition of liquid, a structural rearrangement occurs at $\phi \sim 0.87$, to a system made up of mainly pentagonal faces. Thus, it would seem that a transition from pentagonal dodecahedral to tetrakaidecahedral packing may take place on reduction of foam liquid content.

The same group has since demonstrated the existence of a periodic space-filling structure which has a lower surface energy than Kelvin's tetrakaidecahedron [42, 43]. The unit cell of this structure consists of eight polyhedra, possessing either twelve or fourteen faces: two pentagonal dodecahedra and six polyhedra with twelve pentagonal plus two hexagonal faces [44]. This structural unit is observed in inorganic clathrate systems. Investigations into the structure of real foams showed the existence of Kelvin cells at the container surfaces; further into the foam bulk, fragments which appeared to resemble the new structure were observed [45].

Regarding compressed emulsions, however, the above analysis cannot be applied since the droplets are distorted by the compressive force into complex, irregular shapes. Variables such as the area of the film between adjacent droplets and the disjoining pressure become extremely difficult to calculate. Nevertheless, calculations can be performed for the particular case of compressed, extremely concentrated systems, where ϕ approaches unity. This is because, as $\phi \to 1$, the droplets more or less resemble the polyhedron of the unit cell.

The results indicated, rather surprisingly, that the relationships between the applied pressure and phase volume are very similar for both RDH and RPD packings, and also for any value of θ and h (i.e. zero or finite). Thus, a general expression was given.

An experimental study was performed to determine the applicability of the theory. Oil-in-water (o/w) emulsions, stabilised with anionic surfactants, were prepared, with known quantities of added electrolyte, and were creamed by either gravitation or centrifugation. The results can be summarised as follows: at low electrolyte concentrations, where h would have a finite value, ϕ was less than 0.74. Over a range of concentrations, where it was assumed that both θ and h were negligible, $\phi = 0.74$ (± 0.02). The emulsions were found to be polydisperse, so this did not appear to affect the volume fraction to a great extent. In addition, ϕ was found to be independent of the method of cream formation.

Above a certain electrolyte concentration, measurable contact angles were observed. At this point, major differences occurred depending on the creaming method employed. Gravitationally-produced creams displayed a decreasing phase volume fraction with increasing contact angle, due to the formation of rigid, open flocs, whereas those produced by centrifugation showed increasing ϕ with increasing θ, as expected. However, values of ϕ were lower than expected due to the presence of pockets of continuous phase trapped in the rigid cream, which were extremely difficult to remove, even by centrifugation. The highest value of ϕ obtainable was 0.91, despite a contact angle of 65°. It should be stressed that all emulsions studied were uncompressed.

Das et al. [46] have studied moderately concentrated emulsions ($0.7 < \phi < 0.9$), from both a theoretical and an experimental standpoint. Both polydisperse and distorted monodisperse systems were considered; for the former, it is possible to achieve a value of 0.89 for ϕ with undistorted spheres in a tridisperse (i.e. three sphere sizes) packing.

For distorted monodisperse systems, a regular pentagonal dodecahedral packing was again assumed. The total surface area was found once more to be less for polydisperse systems than monodisperse, at a given value of ϕ. This is in agreement with calculations and observations made by Princen and Kiss [36]. Experimental evidence (at $\phi = 0.88$) indicated that, initially, a random polydisperse distribution of spheres is produced but with prolonged stirring this alters gradually to a distorted, more monodisperse array. This is due to an increase in efficiency of emulsification and, consequently, an increase in total interfacial area with concomitant reduction in radius of the largest spherical droplets. Thus, distortion of the dispersed phase droplets is increased.

This stirring effect on emulsion geometry and properties has also been noted in our laboratory and those of other workers [47]; it indicates the importance of the length of stirring time on HIPE parameters, a condition which is infrequently monitored. However, work done by Lissant [6] on the structure of HIPEs, in which an initial internal phase of vinyl chloride is polymerised to give solid polymer particles, takes account of the effect of stirring. Those emulsions which were sheared sufficiently seemed to display a relatively monodisperse droplet distribution, whereas those in which stirring was more difficult (due to high viscosities) were more polydisperse.

In a separate study, Mannheimer [48] noted that the flow properties of o/w HIPEs were affected by the length of stirring time; the emulsions became stiffer with prolonged stirring. In addition, it was shown [49] that, in polydisperse systems as ϕ increases, smaller droplets must be distorted to a greater degree than larger droplets; the former are more difficult to deform as they are "harder". Furthermore, highly distorted small droplets would tend to coalesce. Thus the tendency is towards monodispersity with increasing ϕ. Freeze-fracture SEM studies of emulsions appeared to lend weight to these postulations.

As can be seen, many factors affect the geometry of HIPEs. Generally, a degree of polydispersity and some cell distortion is shown in real systems and very rarely, if ever, will a truly monodisperse system be observed[1]. The extent of deviation from monodispersity will depend on the experimental conditions and on the physical properties of the HIPE.

---

[1] Recently, Bibette has described a procedure for the production of truly monodisperse emulsions and HIPEs; this will be described in section 1.3.3.

1.3.2 Rheology

Perhaps the most important and striking features of high internal phase emulsions are their rheological properties. Their viscosities are high, relative to the bulk liquid phases, and they are characterised by a yield stress, which is the shear stress required to induce flow. At stress values below the yield stress, HIPEs behave as viscoelastic solids; above the yield stress, they are shear-thinning liquids, i.e. the viscosity varies inversely with shear rate. In other words, HIPEs (and high gas-fraction foams) behave as non-Newtonian fluids.

In much of the work on rheology, foams and HIPEs have been considered as analogous. The expressions derived are applicable to both systems, only the actual values are different. Consequently, workers in this area choose to study either emulsions or foams (or both) and so, in this section, the rheological properties of HIPEs and high gas-fraction (or "dry") foams will be discussed jointly.

The purpose of this section is to give a brief overview of the rheology of highly concentrated emulsions and foams, in as simple terms as possible. For comprehensive reviews in this field, see Refs. 16 and 50.

1.3.2.1 Theoretical Analyses

Princen [51] and others [52, 53] have again used the 2D-model of monodisperse, cylindrical cells as a starting point for a detailed analysis of the rheological properties of HIPEs and foams. When the system is subjected to a shear strain, the unit cells are stressed and deformation occurs. As the strain increases, the cells become more and more deformed; however, the films must continue to meet at angles of 120° to remain stable. At the yield stress, two intersections merge and four films meet at a single point. This configuration is unstable, so a rearrangement occurs to another hexagonal array, where each cell has "jumped" one position in the shear direction, relative to its lower neighbours. This process is repeated again and again, resulting in flow of the system (Fig. 5).

Stress/strain behaviour in the elastic region, i.e. below the yield stress, as a function of volume fraction, ϕ, contact angle, θ, and film thickness, h, was examined [51]. The yield stress, τ_o, and shear modulus, G, were both found to be directly proportional to the interfacial tension and inversely proportional to the droplet radius. The yield stress was found to increase sharply with increasing ϕ, and usually with increasing θ. A finite film thickness also had the tendency to increase the yield stress. These effects are due to the resulting increase in droplet deformation which induces a higher resistance to flow, as the droplets cannot easily slip past one another.

Khan and Armstrong [52] showed that the critical strain, and therefore the yield stress, was dependent on the initial orientation of the cells, for shearing and elongational deformations; indeed, for one particular orientation under extensional deformation, a yield stress was not observed. Beyond the yield point,

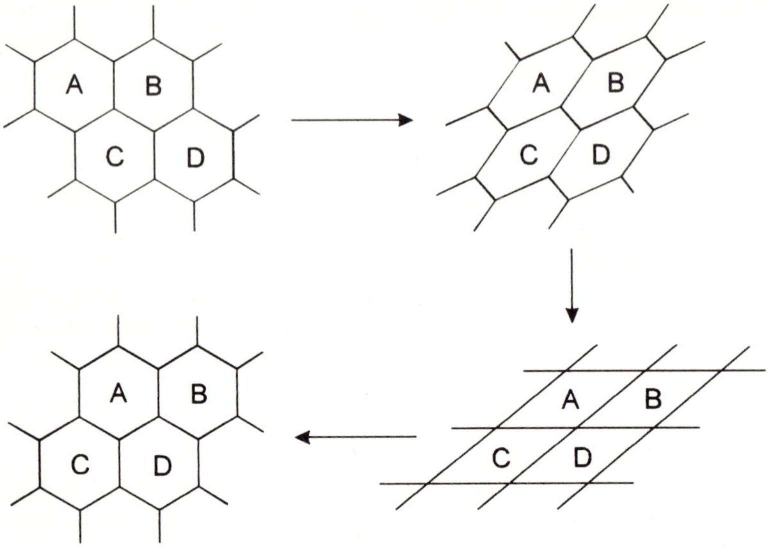

Fig. 5. Distortion of monodisperse hexagonal cells under a shear stress

there is no steady state response of the system to increasing strain; instead, the response is periodic. This periodicity is only shown for certain initial cell orientations, as demonstrated by Kraynik and Hansen [53].

Below the yield point, however, stress/strain behaviour was found to be independent of initial cell orientation, due to the threefold symmetry of the hexagonal cellular array [54]. This allows a correlation between shearing and extensional deformations to be made [55], namely that shear can be considered as elongation followed by rotation. Thus, information on one type of deformation can be obtained by solving expressions for the other.

In a later study [56], the effect of gas volume fraction (ϕ) on foam rheology was investigated. Two models were considered: one in which the liquid was confined to the Plateau borders, with thin films of negligible thickness; and the second, which involves a finite (strain-dependent) film thickness. For small deformations, no differences were observed in the stress/strain results for the two cases. This was attributed to the film thickness being very much smaller than the cell size. Thus, it was possible to neglect the effect of finite film thickness on stress/strain behaviour, for small strains.

Decreasing ϕ, i.e. increasing the liquid content of the foam, had the effect of decreasing the yield stress of the system. The cells become more rounded, and so can move past each other more easily. Also, with decreasing ϕ, initial cell orientation had a considerable effect on the stress/strain relationship, for deformations below the yield point. Indeed, only when the initial orientation, θ, is aligned with the shear direction does the system return to its original configuration, on removal of the deforming strain.

Another important rheological property of dry foams and highly concentrated emulsions is G, the shear modulus. Princen and Kiss [57] demonstrated that this property was dependent on ɸ, the volume fraction of the system. Previously, Stamenovic et al. [58] and, much earlier, Derjaguin and coworker [59], had derived an expression for the shear modulus of foams of volume fraction very close to unity. The value was found to depend on the surface tension of the liquid phase (in foams), for the particular case of $\phi \simeq 1$. However, Princen demonstrated that the values of G obtained were overestimated by a factor of two. This error was attributed to the model used by Stamenovic and coworker, which failed to maintain the equilibrium condition that three films always meet at angles of 120° during deformation.

In a subsequent theoretical study, Stamenovic [60] obtained an expression for the shear modulus independent of foam geometry or deformation model. The value of G was reported to depend only on the capillary pressure, which is the difference between the gas pressure in the foam cells and the external pressure, again for the case of $\phi \simeq 1$. Budiansky et al. [61] employed a foam model consisting of 3D dodecahedral cells, and found that the ratio of shear modulus to capillary pressure was close to that obtained by Princen, but within the experimental limits given by Stamenovic and Wilson.

The viscous properties of HIPEs and high gas fraction foams have also been studied extensively, using a two dimensional, monodisperse, hexagonal cell model. Khan and Armstrong [52] showed that, under steady shear flow (i.e. beyond the yield point of the system), the foam viscosity was inversely proportional to shear rate. At high rates of shear, a constant viscosity value was approached. Gas fraction, ɸ, was assumed to be very close to unity.

In a later investigation, Kraynik and Hansen [62] demonstrated that the shear rate and liquid film viscosity greatly affect the rheological properties of foams. They studied the effect on foam properties and structure with variation of capillary number, Ca, which is the ratio of viscous to surface tension forces in the liquid films, and is given by:-

$$Ca = \frac{\sqrt{3}\phi\mu a\gamma'}{4\sigma}$$

where ɸ is the gas phase volume, μ is the liquid film viscosity, a is the length of a cell side, γ' is the shear rate and σ is the interfacial tension. Thus, Ca varies directly with film viscosity and shear rate.

It was found that increasing Ca caused the yield stress and yield strain to increase, along with cell deformation at the yield point. At sufficiently high values of Ca, cell distortion is so severe that film thinning and rupture can occur, resulting in mechanical failure of the foam (Fig. 6). This implies the presence of a shear strength for foams and HIPEs. The initial orientation of the cells was also found to affect the stress/strain behaviour of the system in the presence of viscous forces [63]. For some particular orientations, periodic flow was not observed for any value of Ca.

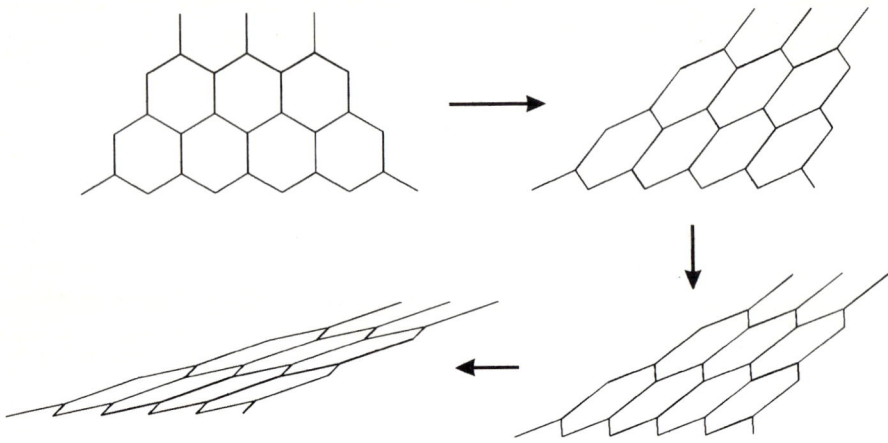

Fig. 6. Distortion of monodisperse hexagonal cells at high values of Ca

This work shows that high shear rates are required before viscous effects make a significant contribution to the shear stress; at low rates of shear the effects are minimal. However, Princen claims that, experimentally, this does not apply. Shear stress was observed to increase at moderate rates of shear [64]. This difference was attributed to the use of the dubious model of all continuous phase liquid being present in the thin films between the cells, with Plateau borders of no, or negligible, liquid content [65]. The opposite is more realistic i.e. most of the liquid continuous phase is confined to the Plateau borders. Princen used this model to determine the viscous contribution to the overall foam or emulsion viscosity, for extensional strain up to the elastic limit. The results indicate that significant contributions to the effective viscosity were observed at moderate strain, and that the foam viscosity could be several orders of magnitude higher than the continuous phase viscosity.

A subsequent analysis [66] also employed this model, with the inclusion of results for the shear strain. The dependence of the viscous effects on initial foam orientation was also noted. Further work [67] on monodisperse "wet" foams, where φ is between 0.9069 and 0.9466, demonstrated that, under shear flow, the foam viscosity increased with increasing φ (decreasing liquid content). In contrast, for small deformations, the viscous contribution to the overall stress was found to be independent of liquid content.

Until now, the theoretical discussion has focused on monodisperse two-dimensional model systems. However, some studies have been performed on polydisperse systems, notably by Weaire et al. [68–72] The evolution of a soap froth of random cell sizes and shapes, known as a Voronoi network, was simulated by computer [68] (Fig. 7). The condition that three films must always meet at angles of 120° was again used. Cells with more than six sides were found

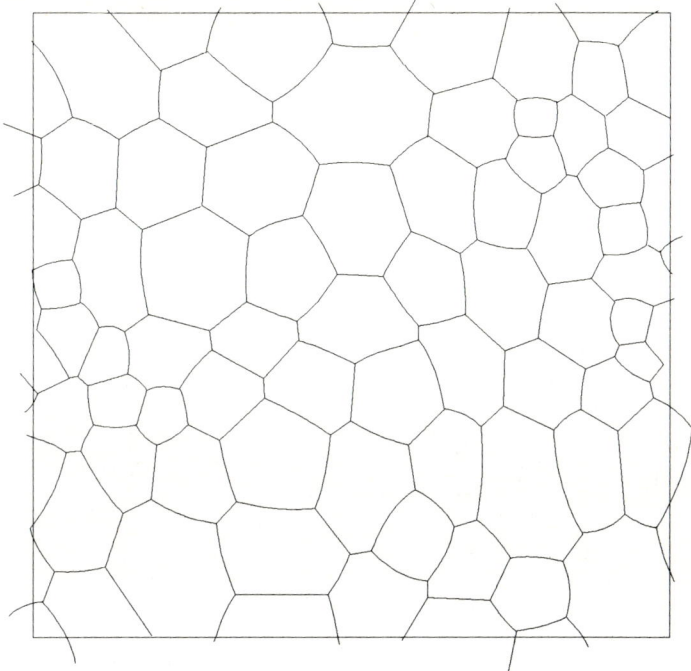

Fig. 7. Voronoi network structure

to grow at the expense of those with fewer than six sides. The shrinking cells lose sides in the process (T_1 process), until they eventually become three-sided. These small cells then disappear (T_2 process), with the gas diffusing into neighbouring larger cells. The overall effect of this is to increase the average cell size of the system (Fig. 8).

The mechanical properties [69] of the system, under an extensional strain, were also investigated. An elastic response was observed for small strains, with the change to plastic behaviour for larger strains. The onset of plastic behaviour is the point where the smaller cells start to lose sides, and corresponds to the yield point in monodisperse hexagonal systems. Rather interestingly, both extensional and shear strain have been found to induce order in random 2D foams [73]. Cells with more than six sides are deformed to a greater extent than those with fewer than six sides, and so tend to lose sides more easily. Since the number of cells is constant, the neighbouring smaller cells gain sides. Thus, the overall tendency is towards a greater number of six-sided cells.

Further work [71, 72] on the variation of the shear modulus of the system with altering structure from perfectly ordered monodisperse to increasingly polydisperse, concluded that there was very little change whatsoever. The hexagonal monodisperse model gave an upper limit to the shear modulus, with a slight decrease being displayed with increasing polydispersity. The yield stress

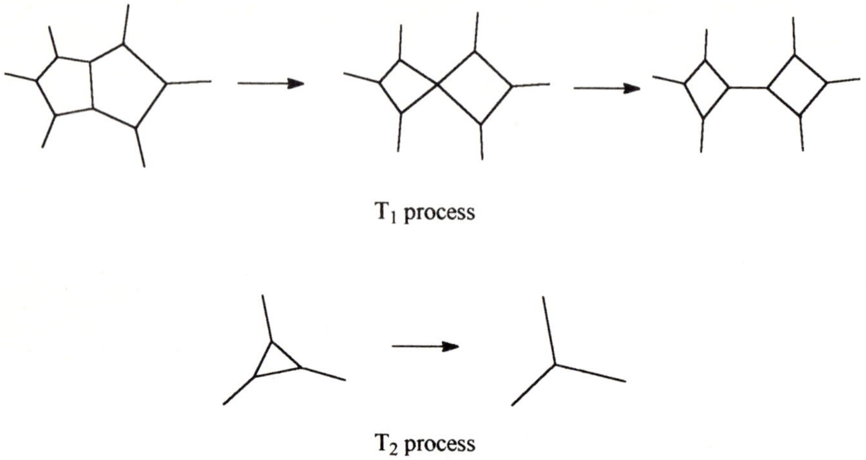

T₁ process

T₂ process

Fig. 8. Topological rearrangements in random 2D foams

[72] also showed little variation with increasing polydispersity, again dropping very slightly. These results suggest that the dependence of the mechanical properties of two-dimensional foams on the structure of the foam is small, and that the monodisperse hexagonal system appears to be a reasonable model.

Kraynik et al. [74] have taken this a stage further by considering polydisperse hexagonal systems. It was found that the elastic response of the system was unaffected by polydispersity, as long as a hexagonal structure was maintained. As with the hexagonal monodisperse system, the elastic limit, or yield stress, depended strongly on orientation. However, those orientations in the uniform array which displayed no yield stress, gave finite values in the polydisperse system, under an extensional deformation. Under shear, the yield stress was a discontinuous function of orientation, as in the monodisperse case. For both forms of strain, the yield stress in polydisperse systems was found to be lower, in keeping with the results of Weaire et al. Finally, the shear modulus upper limit for monodisperse hexagonal systems was expanded to include polydisperse hexagonal systems.

The theoretical analysis for two-dimensional foams and emulsions has recently been expanded to three dimensions [38], with Kelvin's minimal tetrakaidecahedron as the unit cell. The system is subjected to a uniaxial extensional strain. As the elastic limit, or yield point, is approached, the cell shape tends towards a rhombic dodecahedron; however, at the yield point, the shrinking quadrilateral faces of the polyhedron have finite (albeit small) area.

Using planar tetrakaidecahedra as the model, on the other hand, causes the square faces to shrink to zero area at the yield point. The unit cell therefore resembles a true rhombic dodecahedron. The elastic response was found to be anisotropic (i.e. dependent on initial cell orientation) for the planar model, up to the elastic limit. This is in contrast to the monodisperse 2D case, which is isotropic [75]. Beyond the yield point, viscosity cannot be determined for uniaxial extensional deformations since the foam structure is not periodic; the cell does not regain its initial shape. However, under shear flow, the foam is strain periodic, and solutions are obtainable.

1.3.2.2 Experimental Investigations

Experimental studies on the rheological properties of HIPEs and foams have been performed by several groups of workers. Ford and coworker [76] investigated the effects on viscosity of water-in-trimethylbenzene HIPEs, stabilised by pesticidal surfactants, with variation of a number of parameters. Increasing the volume fraction, ϕ, caused the viscosity to increase, as did an increase in the efficiency of the mixing system used. The nature of the emulsifier also affected the viscosity; those which produced rigid interfacial films caused an enhancement. Additionally, increasing surfactant concentration gave increasing emulsion viscosity.

Other researchers have also observed an increase in HIPE viscosity with increasing phase volume ratio [77]; however, the effects of droplet size, polydispersity or continuous phase viscosity were not investigated. Further studies [78] revealed that the viscosity increased for smaller mean droplet radii; this effect was found to be greater at higher internal phase ratios. The total interfacial area increases as droplet size decreases, so viscosity also increases as more energy is required to deform the larger network of thin films [79].

Solans et al. [80] studied the effect of shear rate on shear stress, τ, for w/o HIPEs of $\phi = 0.99$. Above a certain shear rate, the shear stress was seen to rise to a maximum then quickly fall to zero, with time. This was attributed to the breakdown of the emulsion at higher rates of shear.

Princen [64] discovered that the yield stress, τ_o, was strongly dependent on ϕ, increasing sharply with increasing phase volume. τ_o was also found to depend linearly on the surface tension. Variation with the mean droplet radius, however, did not match theoretical predictions; this was reportedly due to the presence of a finite film thickness between adjacent droplets.

The shear modulus, G, was also found to depend strongly on ϕ [57]; this conclusion was verified by experiment. In a previous study, Stamenovic and coworker [58] obtained a value for the shear modulus of about 84% of the predicted value. A further experimental study by Ebert and coworkers [81] gave values of G in good agreement with Princen's theories; however, the yield stress exhibited strong deviations from the theory. This was attributed to the use of a steady-state system by Princen for his experiments, compared to the dynamic system used by the latter researchers. However, Princen subsequently revised his yield stress equation after improving his experimental techniques, giving more accurate results [82]. This highlights a serious problem in determining the rheological properties of HIPEs and foams; namely, different results can be obtained with different experimental methods.

Another major problem which plagues rheology experiments on HIPEs and foams is "wall-slip", which is bulk flow of the system caused by lubrication of the container walls by a thin layer of separated continuous phase. Mannheimer [83], in his experiments on the flow properties through pipes of an o/w HIPE of about 97% phase volume, postulated that this layer formed from the breakdown of the thin films surrounding the emulsion cells, due to increasing strain, and contributed significantly to flow of the system above the yield stress. Indeed, the flow properties depended greatly on the nature of the pipe material itself; in PTFE pipes, no yield value was displayed. This was due to the instantaneous formation of a layer of collapsed HIPE at the wall surface, due to strong hydrophobic interactions between the oil dispersed phase and the pipe material.

Princen [57, 64, 82] and others [84] also noted the presence of wall-slip in rheological experiments on HIPEs and foams. However, instead of attempting to eliminate this phenomenon, Princen [64] employed it to examine the flow properties of the boundary layer between the bulk emulsion and the container walls, and demonstrated the existence of a wall-slip yield stress, below that of the bulk emulsion. This was attributed to roughness of the viscometer walls. Princen and Kiss [57], and others [85], have also showed that wall-slip could be eliminated, up to a certain finite stress value, by roughening the walls of the viscometer. Alternatively [82, 86], it was demonstrated that wall-slip can be corrected for and effectively removed from calculations. Thus, viscometers with smooth walls can be used. This is preferable, as the degree of roughness required to completely eradicate wall-slip is difficult to determine.

Thus, it is highly evident that wall-slip must be taken into account when investigating the rheological properties of HIPEs and foams [87]. Failure to do so will result in false and irreproducible results.

Pons et al. have studied the effects of temperature, volume fraction, oil-to-surfactant ratio and salt concentration of the aqueous phase of w/o HIPEs on a number of rheological properties. The yield stress [10] was found to increase with increasing NaCl concentration, at room temperature. This was attributed to an increase in rigidity of films between adjacent droplets. For salt-free emulsions, the yield stress increases with increasing temperature, due to the increase in interfacial tension. However, for emulsions containing salt, the yield stress more or less reaches a plateau at higher temperatures, after addition of only 1.5% NaCl.

The shear modulus, G_0, was calculated from values of storage (G') and loss (G'') moduli, obtained from dynamic rheological measurements [88]. G_0 displayed a maximum with increasing temperature. This was attributed to the combination of a sharp increase in interfacial tension at lower temperatures, which will increase G_0, with the resulting increase in droplet radius, due to coalescence, at higher temperatures, lowering G_0. Similarly, Otsubo and Prud'homme [89] observed an increase in shear modulus with increasing interfacial tension and decreasing drop size, for o/w HIPEs. Increasing phase volume caused G_0 to increase; a maximum value was expected at high ϕ values, as this results in a large increase in droplet radius. The shear modulus also reached a maximum value with increasing oil-to-surfactant ratio, R (i.e. decreasing surfactant concentration in the continuous phase). This was again due to a combination of two effects; a minimum in droplet radius and an increase in interfacial tension with increasing R. Adding salt to the aqueous phase caused an increase in interfacial tension and no discernible effect on droplet radius, giving a maximum value of G_0 with increasing salt concentration. This effect was shown at all temperatures.

In a separate experimental study, Aronson and Petko [90] also observed an increase in yield value with increasing salt concentration, for w/o HIPEs. However, the interfacial tension of the emulsions was seen to *decrease* with addition of a number of electrolytes to the aqueous phase, in contrast to observations made by Pons et al. The reason for this discrepancy is not clear, but may be due to different interactions between the different surfactants and salts used by each group. It is hinted that the increase in yield value on addition of salt, as observed by Aronson et al., is due to a decrease in average droplet size; however, this was not examined extensively, and determination of droplet diameters was by optical microscopy only.

What is clear, however, is the effect of salt addition on the change in yield stress of the emulsions with time. Those prepared without added electrolyte show a marked decay on storage, over as little as 24 hours, whereas emulsions of salt solutions in oil retain their initial yield values for a number of days. This is attributed to the inhibition of coarsening of the HIPEs, caused by the presence of electrolyte in the aqueous phase; the effect is clearly shown in photomicrographs by Aronson and Petko. The stabilisation of emulsions (HIPEs) by salt addition will be discussed in greater detail in the section on HIPE stability.

Anklam et al. [91] have attempted to measure the extensional rheological properties of w/o emulsions and HIPEs, using a nozzle-type viscometer. However, the results showed a dependence on the nozzle size used, and long relaxation times. Experiments on other non-Newtonian fluids indicated that it was not possible to obtain reliable results with this kind of instrument.

1.3.3 Osmotic Pressure

If a concentrated emulsion is separated from a volume of continuous phase by a moveable, semi-permeable membrane i.e. one which is permeable to all components of the continuous phase but impermeable to the dispersed phase droplets, continuous phase will migrate into the HIPE to relieve the stress of deformation of the dispersed droplets. This can be prevented by applying a pressure to the membrane, squeezing out the excess continuous phase; this is equivalent to the osmotic pressure, π, of the system (Fig. 9).

Princen [92] has demonstrated, theoretically, that π varies directly with interfacial tension and inversely with the mean droplet radius, for polydisperse systems. Variation of the volume fraction, ϕ, from 0.74 to unity causes π to be increased from zero to infinity [93]. Results are obtainable for the limiting case

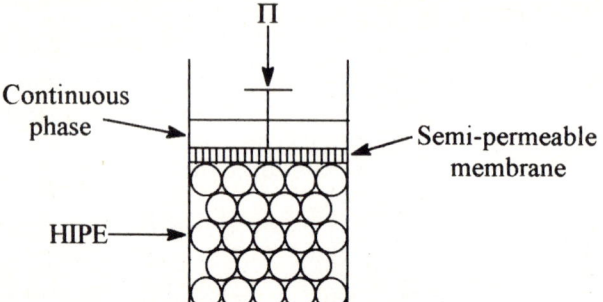

Fig. 9. Osmotic pressure in HIPEs

of very high volume fraction ($\phi \to 1$); however, this is not so for low to intermediate values of ϕ since it is unknown how the droplets will deform under the corresponding pressures.

Experimental data [94] were in good agreement with the theory at high volume fractions ($0.99 < \phi < 1$); at low to intermediate ϕ values, empirical equations were fitted to the results. An interesting point is that extrapolation of the plot of ϕ against osmotic pressure to zero π gives $\phi = 0.712$ i.e. less than the volume fraction for monodisperse, undistorted spheres. This implies that droplet packing is less efficient in polydisperse systems.

As mentioned previously, Bibette [95] has developed a very elegant method for the "purification" of coarse, polydisperse emulsions to produce monodisperse systems. This technique is based on the attractive depletion interaction between dispersed phase droplets, caused by an excess of surfactant micelles in the continuous phase. A phase separation occurs under gravity, between a cream layer and a dilute phase; since the extent of the separation increases with increasing droplet diameter, a separation based on size occurs. By repeating this process, emulsions of very narrow size distribution can be produced.

Bibette has used this method to study the effect of osmotic pressure on the stability of thin films in concentrated o/w emulsions [96], by means of an osmotic stress technique. The emulsion is contained in a dialysis bag, which is immersed in an aqueous solution of surfactant and dextran, a water-soluble polymer. The bag is permeable to water and surfactant, but impermeable to oil and polymer. The presence of the polymer causes water to be drawn out of the emulsion, increasing the phase volume ratio and the deformation of the dispersed droplets (Fig. 10).

The polymer concentration sets the osmotic pressure of the system. This is increased until film rupture occurs, resulting in coalescence, which is easily seen under the microscope since the original emulsions are monodisperse.

Since monodisperse creams of a range of droplet sizes can readily be prepared, it is possible to study the effect of droplet size on the critical osmotic pressure required for film rupture, π^*. This was found to increase with increasing droplet size. The critical osmotic pressure is, in effect, the disjoining pressure; as smaller droplets have higher disjoining pressures (due to a smaller radius of curvature),

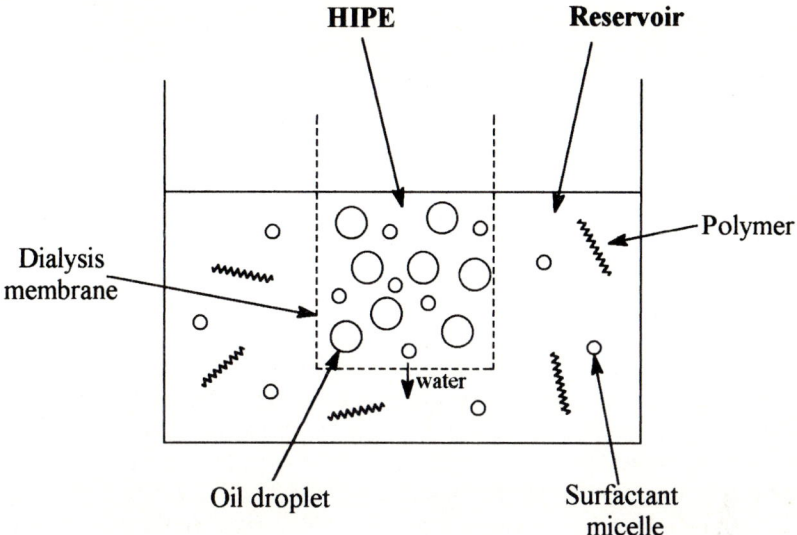

Fig. 10. The osmotic stress technique

smaller applied pressures are required to cause coalescence. This leads to a critical droplet diameter, σ^*, below which two droplets will instantly coalesce on contact, in the absence of any applied pressure.

The effect of surfactant concentration on critical osmotic pressure was also studied [97]. Below a critical surfactant concentration, emulsions are always unstable due to incomplete coverage of the oil-water interfaces. Above this, π^* increases with increasing surfactant concentration until the critical micelle concentration (CMC) is reached, above which it remains more or less constant.

The contribution of double-layer forces to the osmotic pressure of HIPEs was also investigated [98]. These forces arise from the repulsion between adjacent droplets in o/w HIPEs stabilised by ionic surfactants. It was observed that double-layer repulsive forces significantly affected π for systems of small droplet radius, high volume fraction and low ionic strength of the aqueous continuous phase. The discrepancies between osmotic pressure values observed by Bibette [97] and those calculated by Princen [26] were tentatively attributed to this effect.

1.3.4 Continuous Phase Microstructure

The nature of the continuous phase of HIPEs has been the subject of considerable debate recently. Solans et al. [9] investigated the structure of very highly concentrated ($\phi = 0.99$) water-in-oil emulsions and discovered the presence of a liquid crystalline layer, in addition to aqueous and oil phases, on breaking the emulsion by ultracentrifugation. It was proposed that this layer formed the films surrounding the aqueous phase droplets; thus, the emulsion was a w/lc/o system. Additionally, it was suggested that the oil phase consisted of a w/o microemulsion.

Water-in-fluorocarbon emulsions, stabilised with fluorinated nonionic surfactants, were investigated by small angle neutron scattering (SANS) spectroscopy [8, 99]. The results indicated that the continuous oil phase comprised an inverse micellar solution, or water-in-oil microemulsion, with a water content of 5 to 10%. However, there was no evidence of a liquid crystalline layer at the w/o interface. A subsequent study using small angle x-ray scattering (SAXS) spectroscopy gave similar results [100].

Pulsed gradient spin-echo (PGSE) NMR techniques have also been employed to study the structure of the oil phase [12]. This gives an idea of the mobility of each component in the HIPE, and showed that, for stable emulsions and HIPEs, the oil phase was indeed a reverse micellar solution which solubilises water. Further work using PGSE NMR has shown that water can diffuse between aqueous droplets in concentrated emulsions [101]. Presumably this involves solubilisation of the water molecules by the micellar oil phase.

Very recently, ESR techniques have been employed to study the packing of surfactant molecules at the oil/water interface in w/o HIPEs [102, 103]. By including an amphiphilic ESR probe, which is adsorbed at the oil/water interfaces, it is possible to determine the microstructure of the oil phase from the distribution of amphiphiles between the films surrounding the droplets and the reverse micelles. It was found that most of the surfactant is located in the micelles, over a wide range of water fraction values. However, when the water content is very high ($\phi \sim 0.99$), most, if not all, of the emulsifier is present on the surface of the water droplets of the emulsion, to stabilise the large interfacial area created.

1.4 HIPE Stability

A number of factors greatly influence the stability of high internal phase emulsions, including the nature of the surfactant, its concentration, the nature of the continuous phase, the temperature and the presence of salts in the aqueous phase.

Ford and coworker [104] have studied HIPEs of water-in-xylenes, stabilised by a variety of surfactants, and postulated three properties which an emulsifier should possess in order to form stable w/o HIPEs of high volume fraction: a) a lowering of the interfacial tension between water and oil phases, b) the formation of a rigid interfacial film and c) rapid adsorption at the interface.

The most important factor was suggested to be the formation of a rigid film at the interface. This was achieved when an interaction (electrostatic or hydrogen-bonding) could be visualised between adjacent surfactant molecules. For example, the use of a long-chain amine with a surface active acid, e.g. oleic acid, could lead to a mixed film with significant strength due to electrostatic interactions between the polar head-groups. The polarity of the organic phase was also varied; more hydrophilic solvents required a more hydrophobic surfactant to form a stable HIPE, whereas increasing the hydrophobicity of the oil-phase resulted in the need for a hydrophilic surfactant for emulsion stability.

Williams [105] investigated the effect on stability of water-in-styrene/divinylbenzene HIPEs, stabilised with nonionic surfactants, on addition of a range of cosurfactants. Generally, stability was reduced, with higher degrees of coalescence being observed with cosurfactant addition. The stability appeared to be inversely related to the HLB number of the cosurfactant.

A considerable amount of experimental work has been carried out on the so-called "gel emulsions" of water/nonionic surfactant/oil systems [9–14, 80, 106, 107]. These form in either the water-rich or oil-rich regions of the ternary phase diagrams, depending on the surfactant and system temperature. The latter parameter is important as a result of the property of nonionic surfactants known as the HLB temperature, or phase inversion temperature (PIT). Below the PIT, nonionic surfactants are water-soluble (hydrophilic; form o/w emulsions) whereas above the PIT they are oil-soluble (hydrophobic; form w/o emulsions). The systems studied were all of very high phase volume fraction, and were stabilised by nonionic polyether surfactants.

The HLB temperature was found to be the most important factor in the formation of stable emulsions. In each case, w/o HIPEs [9, 11, 80] would only form at temperatures above the HLB temperature of the systems, while o/w HIPEs [14] formed below the PIT. The nature of the oil phase was also found to be of importance to the formation of stable w/o HIPEs [11]; aromatic liquids, for example, did not produce highly concentrated emulsions. With aliphatic oils, the stability was observed to vary with chemical nature. This was due to the different HLB temperatures for each liquid.

A similar effect was noted in separate investigations by another group [108]. Oil-in-water HIPEs, where the oil phase contained aromatic or halogenated liquids, were difficult or impossible to form, with nonionic surfactants. This was postulated to be as a result of interactions between the polar ethylene oxide groups of the surfactant and the aromatic or halogenated solvents, which are more polar than hydrocarbons. Water-in-oil systems also displayed this tendency [21]; however, w/o HIPEs with m-xylene as the oil phase [13] could be produced with monolaurin as nonionic emulsifier, due to stronger intermolecular interactions at the interface.

Apart from anomalous situations where surfactant interacts with the organic phase, the stability of HIPEs is linked to the interfacial tension of the system. Ruckenstein and coworkers [109] showed that the maximum volume of hydrocarbon which could be incorporated in an o/w HIPE increased with increasing surfactant concentration, presumably due to a concomitant decrease in the interfacial tension. Solans et al. [9] claimed that the interfacial tension between the aqueous phase and the liquid-crystalline surfactant layer in their highly concentrated w/o HIPEs was exceedingly low, giving the surfactant phase a very high spreadability around the aqueous phase droplets. Kizling and Kronberg [21] similarly noted that the nonionic polyether surfactants which gave very stable w/o concentrated emulsions also formed lamellar liquid crystalline layers. However, in some cases the lateral interactions between liquid crystalline surfactant molecules may be too strong, preventing HIPE formation.

Another way of looking at this mechanism is to consider the polarity difference between oil and aqueous phases in absence of surfactant. Generally, the more hydrophobic is one phase and the more hydrophilic the other phase, the more stable are the emulsions [110]. Thus, the greater the interfacial tension between oil and water phases, in absence of surfactant, the greater the stability of the HIPE.

The viscosity of the continuous phase was also found to affect the stability of highly concentrated emulsions [108]. It was demonstrated that increasing the viscosity of the continuous phase, either by using a more viscous organic liquid (for w/o systems) or increasing the nonionic surfactant concentration (for o/w systems), gave low maximum volume fractions. This is because the high viscosity prevents efficient mixing of the system and lowers the amount of dispersed phase which can be incorporated in the emulsion.

Increasing temperature has the effect of decreasing emulsion stability; this has been demonstrated by Kunieda et al. [11, 14], among others, and is due to the increase of the rate of coalescence of the dispersed phase droplets with increasing thermal energy. Pons et al. [100] also noted that a temperature increase caused an increase in average droplet size due to increasing interfacial tension.

Another process which leads to HIPE instability is gravitational syneresis, or creaming, where the continuous phase drains from the thin films as a result of density differences between the phases. This produces a separated layer of bulk continuous phase and a more concentrated emulsion phase. The separated liquid can be located either above or below the emulsion, depending on whether the continuous phase is more or less dense, respectively, than the dispersed phase. This process has been studied by Princen [111] who suggests that it can be reduced by a number of parameters, including a high internal phase volume, small droplet sizes, a high interfacial tension and a small density difference between phases.

The addition of salts to the aqueous phase of concentrated emulsions can have profound effects on their stabilities. Water-in-oil HIPEs are generally stabilised by salt addition [10, 12, 13, 21, 80, 90, 112]; however, the nature of the salt used was found to be important [13]. Salts which decrease the cloud point of the corresponding nonionic surfactant aqueous solutions, i.e. which have a salting-out effect, were more active. The interactions of the surfactant molecules at the oil/water interface were increased due to dehydration of the hydrophilic ethylene oxide groups on addition of salt. This was verified experimentally [113] by an ESR method, which demonstrated that the surfactant molecules at the oil/water interface become more ordered if the salt concentration is increased.

Kizling and coworker [21] suggested that salts in the aqueous phase stabilised w/o HIPEs by two means. First, the Ostwald ripening process is inhibited due to the decreased solubility of the aqueous solution in the continuous oil phase. Secondly, the attractive forces between adjacent aqueous droplets are lowered, as a result of the increase in refractive index of the aqueous phase towards that of the oil phase. When the refractive indices of the two phases are matched, the attractive forces are at a minimum and highly stable, transparent emulsions are formed. The attractive force, A, is given by:

$$A = a \left(\frac{\varepsilon_1 - \varepsilon_2}{\varepsilon_1 + \varepsilon_2} \right)^2 + b \frac{(n_1^2 - n_2^2)^2}{(n_1^2 + n_2^2)^{3/2}}$$

where n_i and ε_i are the refractive index and dielectric constant, respectively, of phase i, and a and b are constants.

Further evidence for the theory of stabilisation due to enhanced intermolecular surfactant interactions was presented recently [112]. Two different surfactants were employed to stabilise w/o HIPEs; sorbitan monooleate, a 'monomeric' surfactant, and a polymeric surfactant. Salt addition enhanced the stability of the HIPEs, but more so in the presence of the polymeric surfactant. Again, the interactions between the salt and the surfactant were held responsible. The polymer contained ionic groups, which enabled it to interact strongly with the salt; sorbitan monooleate, however, possesses groups which can only participate in hydrogen-bonding.

A stabilising effect in the presence of salt was also noted by Aronson and Petko [90]. Addition of various electrolytes was shown to lower the interfacial tension of the system. Thus, there was increased adsorption of emulsifier at oil/water interface and an increased resistance to coalescence. Salt addition also increased HIPE stability during freeze-thaw cycles. Film rupture, due to expansion of the water droplets on freezing, did not occur when aqueous solutions of various electrolytes were used. The salt reduced the rate of ice formation and caused a small amount of aqueous solution to remain unfrozen. The dispersed phase droplets could therefore deform gradually, allowing expansion of the oil films to avoid rupture [114].

Finally, some studies have been performed on the addition of salt to the aqueous phase of oil-in-water HIPEs [109]. For systems stabilised by ionic surfactants, increasing salt concentration reduces the double-layer repulsion between droplets; however, stability is more or less maintained, probably due to steric and polarisation repulsions. Above a sufficiently high salt concentration, emulsions become unstable due to salting-out of the surfactant into the oil-phase. For nonionic surfactants, the situation is similar, except that there are no initial double-layer forces. In addition, Babak [115] found that increasing the electrolyte concentration reduced the barrier to coagulation between emulsion droplets, and therefore increased coalescence. Generally, therefore, stability of o/w HIPEs is not enhanced by salt addition.

1.5 Non-Aqueous HIPEs

Non-aqueous (or-oil-in-oil) emulsions, where the phases are two immiscible organic liquids, have received relatively little attention in the literature. Riess et al. [116–119] have studied the stabilisation of waterless systems with block and graft copolymers, where one of the liquids is a good solvent for one of the blocks and a non-solvent for the other, and vice versa. Thus, poly(styrene-b-methylmethacrylate) copolymers could emulsify acetonitrile/cyclohexane mixtures, and poly(styrene-b-isoprene) was effective for DMF/hexane systems [116]. These, however, are not HIPE systems.

Sharma has published a considerable volume of work on non-aqueous emulsions involving ethylene glycol as the polar organic phase and either benzene [120] or chlorobenzene [121] as the non-polar organic phase. The surfactants used were anionic or nonionic, low moleculer weight (i.e. non-polymeric) materials. More recently, attention has focused on the formation of non-aqueous microemulsions [122], usually with formamide as the polar organic phase, and the first example of liposome formation in a non-aqueous solvent was reported [123]. Again, these emulsions were not highly concentrated.

Non-aqueous HIPEs have received even less attention; indeed, to date, there have been only two publications dealing with this subject, to the authors' knowledge [124, 125]. These describe the preparation of highly concentrated emulsions of jet engine fuel in formamide, for use as safety fuels in military applications. The emulsifier system used was a blend of two nonionics, with an optimal HLB value of 12.

Experiments on the stability of the HIPEs indicated that one of the most important factors was the solubility of the emulsifier in the continuous (formamide) phase. Thus, the higher the surfactant solubility, the more stable the emulsion. The emulsifier concentration was also important; stability increased to a maximum, then decreased, with increasing surfactant concentration. Surprisingly, the HLB number did not appear to have much effect on the stability of the emulsions, over the range studied (11 to 14). This was attributed to the high concentration of emulsifier in the continuous phase, although the narrow HLB value range is probably also a factor.

The non-aqueous HIPEs showed similar properties to their water-containing counterparts. Examination by optical microscopy revealed a polyhedral, polydisperse microstructure. Rheological experiments indicated typical shear rate vs. shear stress behaviour for a pseudo-plastic material, with a yield stress in evidence. The yield value was seen to increase sharply with increasing dispersed phase volume fraction, above about 96%. Finally, addition of water to the continuous phase was studied. This caused a decrease in the rate of decay of the emulsion yield stress over a period of time, and an increase in stability. The added water increased the strength of the interfacial film, providing a more efficient barrier to coalescence.

Recently, the authors have developed a number of non-aqueous HIPE systems [126], details of which will be published shortly in the primary literature.

1.6 Applications of HIPEs

There exists, in the literature on high internal phase emulsions, a small number of publications on possible applications of HIPEs, involving a diverse range of topics. The production of petroleum gels as safety fuels is one such example [124, 125]; this was mentioned in the section on non-aqueous HIPEs. The main advantage over conventional fuels is the prevention of spillage, which reduces the risk of fire in an accident. Also, studies on the flash-point of emulsified fuels [127] showed a considerable increase, compared to the liquid state, for commercial multicomponent fuels. In addition, there may be an enhancement of the efficiency of combustion of the fuel on emulsification, as it is known that a small amount of water in fuel can improve its performance [19].

Another potential area of use for HIPEs is in the recovery of oil from tar and oil sands, suggested by Sebba [19, 24]. In this process, an oil-in-water HIPE, e.g. of kerosene, is added to the solid material, and the mixture is agitated. The oil quickly dissolves in the kerosene, and the aqueous surfactant solution wets the surface of the solid particles, preventing readhesion of oil. Addition of a small amount of water causes the oil/kerosene mixture to form a separate layer, which can easily be removed.

Similarly, o/w HIPEs may be used as detergents [19, 24]. Grease can be removed from a surface by dissolution in the internal oil phase; the aqueous surfactant solution then wets the surface, preventing the oil from returning to the surface and allowing it to be rinsed clean with water. It could also be possible to remove soil from fabrics by application of o/w HIPEs, again making use of their detergent properties.

Water-in-oil HIPEs, involving emulsifiers with pesticidal properties, have been examined as potential agricultural sprays [76]. The variation in viscosity with a number of parameters, including internal phase volume and efficiency of mixing, was studied. A more highly viscous fluid was desired, as it was believed that this would lead to the production of larger droplets on spraying, which would drift less and give more accurate crop coverage.

Highly concentrated emulsions are also evident in everyday applications. A classic example is mayonnaise, in which a large volume of vegetable oil is emulsified in a small amount of vinegar, using lecithin from egg-yolk as the emulsifier. In addition, HIPEs are most probably found in many cosmetic products, especially gels and creams. However, little information is available on products of commercial importance, so one can only speculate on their exact nature and composition.

2 The Preparation of Polymeric Materials from HIPEs

Perhaps one of the most important applications of HIPEs is their ability to be used as template systems for the synthesis of a range of polymeric materials. By polymerisation of either, or both, phases of the emulsion, novel materials with a variety of fascinating properties can be produced.

2.1 Polymerisation of Continuous Phase – PolyHIPE

If a high internal phase emulsion is prepared in which the continuous phase contains one or more monomeric species, and polymerisation is initiated, a novel type of highly porous material is produced. Polymers of this type are referred to as PolyHIPE, using the nomenclature devised by Unilever scientists [128].

The range of monomers which can be employed is largely dictated by the physical chemistry of the emulsion system. For instance, monomers must be sufficiently hydrophobic to allow the formation of stable w/o HIPEs. In addition, most systems which have been studied have used polymerisation methods which require either an initiation step, or addition of a catalyst. This is due to the fact that the first step in the preparation of the polymer is the preparation of HIPE; this can only proceed satisfactorily in the absence of any significant degree of polymerisation. Thus, it can be seen that radical addition polymerisation is suitable for the synthesis of PolyHIPE polymers, whereas condensation polymerisation can be more problematical. Also, the latter reactions often generate water as the by-product, hence the aqueous component of the HIPE is inhibiting to the polycondensation.

2.1.1 Poly(styrene/divinylbenzene) PolyHIPE Copolymers

2.1.1.1 Preparation

By far the most studied PolyHIPE system is the styrene/divinylbenzene (DVB) material. This was the main subject of Barby and Haq's patent to Unilever in 1982 [128]. HIPEs of an aqueous phase in a mixture of styrene, DVB and nonionic surfactant were prepared. Both water-soluble (e.g. potassium persulphate) and oil-soluble (2, 2'-azo-bis-isobutyronitrile, AIBN) initiators were employed, and polymerisation was carried out by heating the emulsion in a sealed plastic container, typically for 24 hours at 50°C. This yielded a solid, crosslinked, monolithic polymer material, with the aqueous dispersed phase retained inside the porous microstructure. On exhaustive extraction of the material in a Soxhlet with a lower alcohol, followed by drying in vacuo, a low-density polystyrene foam was produced, with a permanent, macroporous, open-cellular structure of very high porosity (Fig. 11).

The porous structure of the polymer can clearly be seen from its scanning electron micrograph (SEM) (Fig. 12). The cell sizes are usually smaller than in conventional gas-blown polystyrene foams and of higher spherical symmetry (Fig. 13).

High Internal Phase Emulsions (HIPEs)

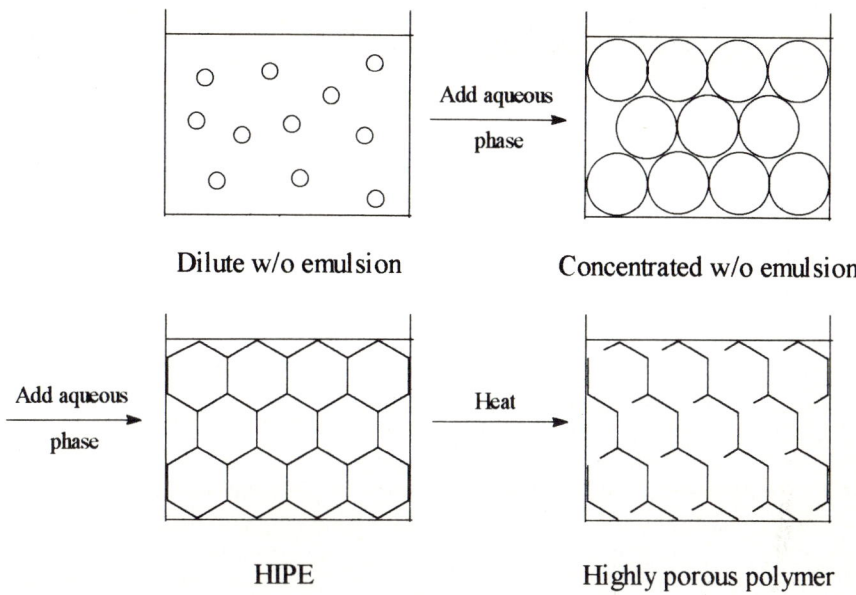

Fig. 11. Preparation of poly(styrene/DVB) PolyHIPE

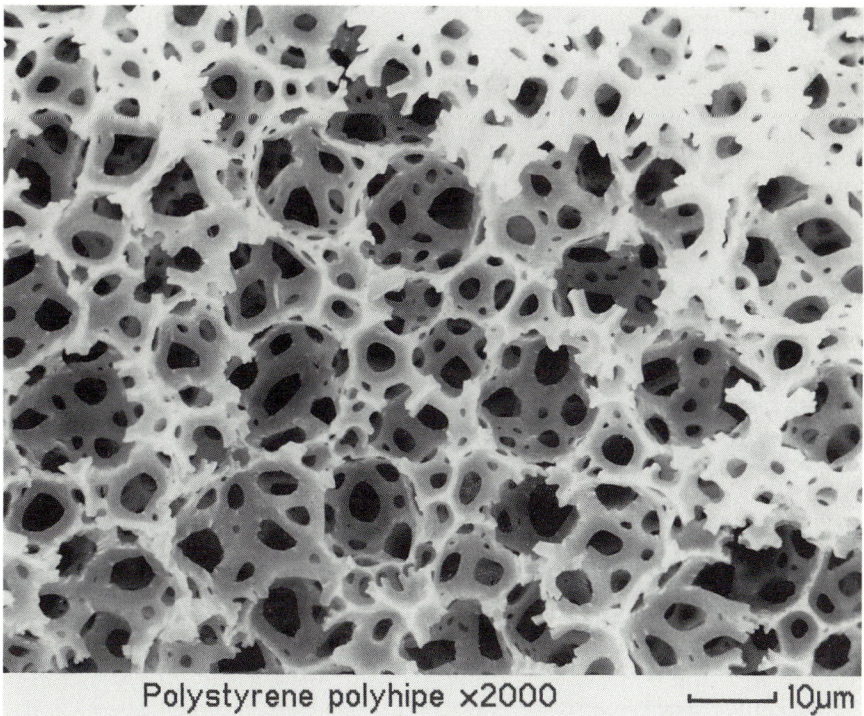

Fig. 12. Scanning electron micrograph (SEM) of poly(styrene/DVB) PolyHIPE

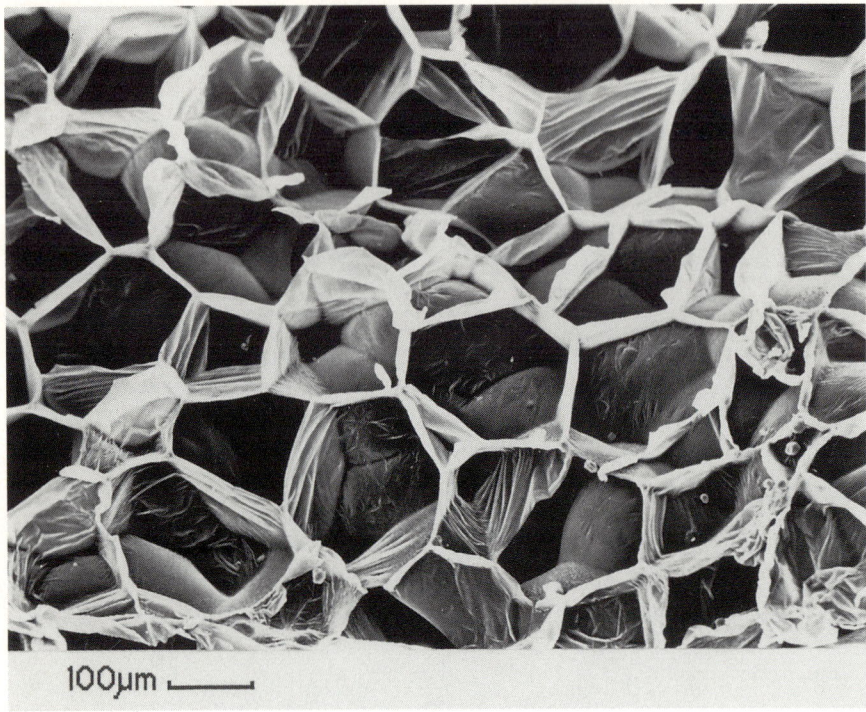

Fig. 13. Scanning electron micrograph (SEM) of gas-blown polystyrene foam

Researchers at Unilever [128] discovered that the surfactant used to form the HIPEs must be of low HLB value (between 2 and 6), as would be expected for w/o emulsions. The optimum surfactant was sorbitan monooleate (Span 80), which has an HLB value of 4.3. However, the HLB number of the surfactant is not the only criterion for the preparation of stable HIPEs; the chemical nature was also found to be of importance [105].

The concentration of the surfactant in the monomer phase was found to be critical to the formation of a stable polymer foam [129, 130]. At least 4% surfactant, relative to the total oil phase, was required for PolyHIPE formation, whereas formulations containing above 80% resulted in the formation of an unconnected or closed-cell material. Surfactant levels between 20 and 50% were deemed to be optimum at all internal phase volumes. Additionally, Litt et al. [131] demonstrated that block copolymer surfactants can be used to prepare water-in-styrene HIPEs. From these, highly porous uncrosslinked polystyrene PolyHIPE materials were synthesised.

2.1.1.2 Cellular Structure and Morphology

Williams [129] discovered that the surfactant level had a profound effect on the cellular structure of PolyHIPE materials. Below about 5% surfactant, the polymers produced had a closed-cell structure, i.e. each cell was discrete from its neighbours. The aqueous phase in this material remained trapped inside the structure, resulting in a high density polymer. Above approximately 7% surfactant, however, an open-cell foam was produced, with an entirely interconnected microstructure. The aqueous phase could easily be removed yielding a dry, low density polymer matrix.

Surprisingly, the surfactant concentration was found to be more important than the phase volume in determining the final cellular structure; a material prepared from a HIPE of as much as 97% internal phase volume and 5% surfactant still gave a closed-cell polymer [129].

It seems that increasing the surfactant concentration causes thinning of the films between adjacent droplets of dispersed phase. Above a certain level, the films become so thin that on polymerisation, holes appear in the material at the points of closest droplet contact. A satisfactory explanation for this phenomenon has not yet been postulated [132]. It is evident, however, that the films must be intact until polymerisation has occurred to such an extent as to lend some structural stability to the monomer phase; if not, large-scale coalescence of emulsion droplets would occur yielding a poor quality foam. In general, vinyl monomers undergo a volume contraction on polymerisation (i.e. the bulk density increases) and in the limits of a thin film, this effect may play a role in hole formation, especially at higher conversions in the polymerisation process.

Williams et al. have also investigated the effect of variation of the DVB content of the monomer phase on the cellular structure of the resulting foam [130]. The phase volume and surfactant and initiator concentrations were kept constant while the DVB content was increased from 0 to 100%; this caused a drop in average cell size from 15 μm to 6 μm. The increased hydrophobicity of DVB compared to styrene probably results in a more stable emulsion, giving a slower rate of droplet coalescence and smaller average cell size.

The salinity of the aqueous phase also had a drastic effect on the properties of the polymer. Changing the initiator from potassium persulphate to AIBN, a non-electrolyte initiator, caused a large increase in average cell diameter. Using AIBN and adding an inert salt, potassium sulphate (K_2SO_4) to the aqueous phase, in increasing concentrations, caused a reduction in cell size of an order of magnitude. This is evidently a direct result of the stabilising effect of salts on w/o HIPEs, the reasons for which were mentioned in the section on HIPE stability.

Because of the macrocellular structure of PolyHIPE polymers, the materials produced as described so far possess only modest surface areas, i.e. less than $5 \, m^2 g^{-1}$, and the cell walls are essentially 'solid' with no discrete internal morphology of their own. This is shown clearly in a low resolution transmission electron micrograph (Fig. 14a) of a 5% crosslinked polystyrene PolyHIPE with a pore volume of 90%. Using the methodology developed to generate porous morphology in conventional styrene-divinylbenzene resin beads, Sherrington et al. [133] have produced PolyHIPE materials with an additional porous structure within the walls of the macrocells. Thus, on increasing the crosslink ratio to

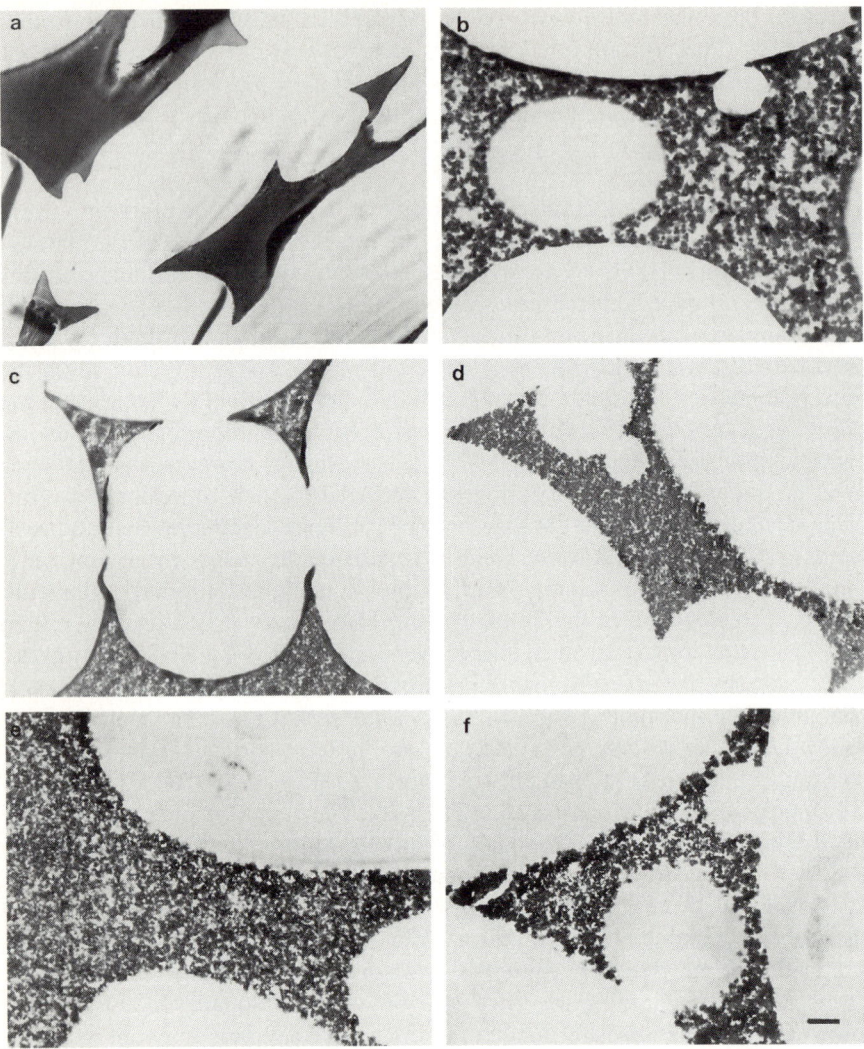

Fig. 14. Transmission electron micrograph (TEM) of poly(styrene/DVB) PolyHIPE prepared with a porogenic solvent in the continuous phase; scale bar 1.5 μm

20% and keeping all other parameters the same, the surfactant used in HIPE formation starts to act as a porogen (pore-forming component) and a discrete morphology begins to appear in the cell walls (Fig. 14c). Deliberate addition of a precipitating porogen, namely petroleum ether, with the crosslink ratio maintained at 20%, further enhances the wall porosity (Fig. 14b). Finally, using very high levels of crosslinker, 50% (Fig. 14d and 14e) and 80% (Fig. 14f), together with relatively large levels of solvating porogen (toluene), conditions known to generate an extensive and fine porous structure in conventional resin beads, likewise develops PolyHIPE materials with an extremely well developed fine porous morphology within the cell walls. This in turn has a profound effect on the overall surface area of the material (see below).

2.1.1.3 Properties

PolyHIPE materials possess many peculiar properties as a result of their unique cellular structure. Referring specifically to open-cell polymers (which have received the most attention in the literature), they are characterised by a very low dry bulk density, typically less than 0.1 g cm^{-3}, due to their highly porous, interconnected structure.

The cell sizes can range from about 5 to 100 µm, depending on the parameters previously mentioned. These are relatively large, compared to pore sizes found in poly(styrene/DVB) resins. Consequently, PolyHIPE materials have lower internal surface areas, usually less than 5 m^2 g^{-1}, as determined by the BET method [133]. However, if a porogenic solvent is included in the monomeric oil phase of the concentrated emulsion, polymers with a much higher surface area can be obtained. Good solvents for linear polystyrene, such as toluene, have the greatest effect on surface area; a 1:1 mixture of toluene to total monomers, with a concomitant high level of crosslinker (80%), produces materials with surface areas up to 350 m^2 g^{-1} (Table 1). Bearing in mind that these materials also have very low bulk density (< 0.1 g cm^{-3}) and very high pore volumes (90%) their porous structure is truly remarkable.

Another important property of open-cell PolyHIPE materials is their ability to absorb large quantities of solvent, by capillary action [128]. Simply immersing a piece of the material in the liquid causes absorption, with displacement of the air from inside the matrix. This occurs until all voids are filled. The nature of the liquid will affect the volume which can be taken up [133]. Methanol, which is a non-swelling solvent for crosslinked polystyrene, is absorbed to a lesser

Table 1. Properties of poly(styrene/DVB) polyHIPE prepared with a porogenic solvent in the continuous phase

PolyHIPE polymer	HIPE composition			Surface area (m^2 g^{-1})[a]
	Styrene and Et-styrene (vol. %)	DVB (vol. %)	Monomer/ porogen	
X5PV90[b]	95	5	–	3.8
X20PV90	80	20	–	22
X20PV90 (0.5 PE)	80	20	1/0.5 PE	33
X55PV90 (0.5 T)	45	55	1/0.5 T	137
X55PV90 (1 T)	45	55	1/1 T	264
X80PV90 (1 T)	20	80	1/1 T	354

[a] From N$_2$ absorption by BET method; [b] X = crosslinker concentration, PV90: organic phase/aqueous phase = 10/90 (90% phase volume); PE = petroleum ether; T = toluene.

extent than toluene, which causes the walls of the foam to swell. The absorption of very polar liquids, such as water, may be quite low, due to incompatibility with the hydrophobic polymer.

Similarly, it is possible to pump liquids through blocks of PolyHIPE, taking advantage of the fully interconnected microstructure. The rather large cell size of the material means that back pressures are relatively low. This is potentially of great importance in liquid chromatography, and will be discussed later.

Poly(styrene/DVB) PolyHIPE polymers have similar overall mechanical properties to conventional gas-blown polystyrene foams, although the (normally) smaller cell size and increased spherical symmetry of the former yields higher compressive strengths. The stress/strain behaviour shows an initial linear elastic region, the slope of which determines Young's modulus. At higher stresses, the foam structure collapses at the crush strength value, and eventually fails [129]. With decreasing foam density, the crush strength decreases, and the material becomes less brittle and more compressible (i.e. Young's modulus decreases) (Fig. 15).

The mechanical properties are also affected by the surfactant concentration in the emulsion precursor. Maxima in both crush strength and Young's modulus were shown at the surfactant concentration for optimum emulsion stability. Foams prepared from 100% styrene were found to have much lower compressive moduli than those containing DVB [130]. This was attributed to plasticisation of the polymer by the surfactant.

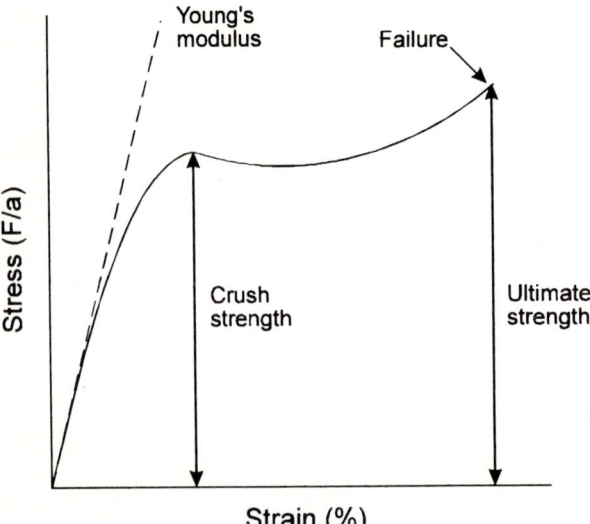

Fig. 15. Stress/strain curve of a typical poly(styrene/DVB) PolyHIPE material

2.1.1.4 Applications

PolyHIPE has found a successful application in the field of solid phase peptide synthesis (SPPS), where the highly porous microstructure acts as a support material for a polyamide gel [134]. The polystyrene matrix is functionalised to give vinyl groups on its internal surfaces, and is then impregnated with a DMF solution of N,N'-dimethylacrylamide, acryloylsarcosine methyl ester, crosslinker and initiator. Polymerisation grafts the soft gel onto the rigid support, giving a novel composite material (Fig. 16).

The combination of the polyamide gel and the porous polystyrene matrix produces a material which has a high loading capacity and favourable mechanical properties, so can be used in a column in an automated process. For example, sequence 65–74 of the acyl carrier protein (ACP) was successfully synthesised in high yield and purity with this support. Additionally, the performance is greatly enhanced, compared to conventional support materials.

Recently, PolyHIPE monolithic materials have been developed as carriers for flavin (10-ethyl-isoalloxazine) in a collaboration between the groups of Sherrington and Challa [135]. Flavin was immobilised by three methods: a) direct covalent bonding to chloromethylated PolyHIPE, b) deposition of a polyelectrolyte complex of a flavin-containing polycation and poly(sodium styrene sulphonate) on the internal surface of poly(styrene/DVB) PolyHIPE and c) complexation of a flavin-containing polycation with sulfonated poly(styrene/DVB) PolyHIPE (Fig. 17). The aerobic oxidation of 1-benzyl-1, 4-dihydronicotinamide (BNAH) was used as the test reaction to investigate the catalytic activities of the flavin-containing supports (Fig. 18). The material prepared by the third method was found to give the best results. The catalyst

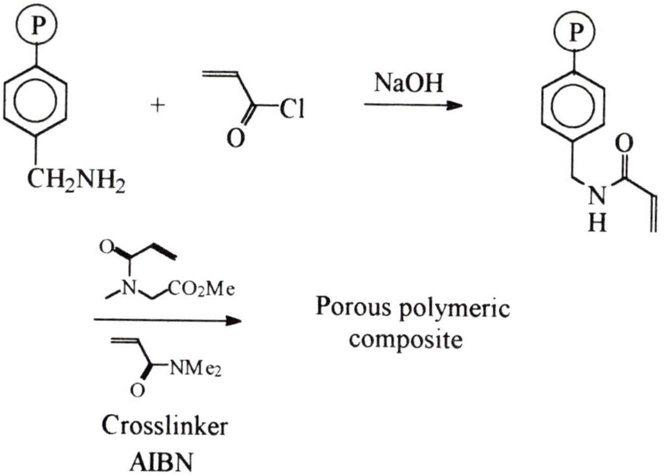

Fig. 16. Preparation of PolyHIPE composite for solid phase peptide synthesis (SPPS)

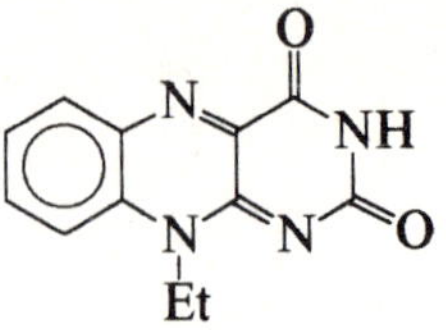

Flavin (10-ethyl-isoalloxazine)

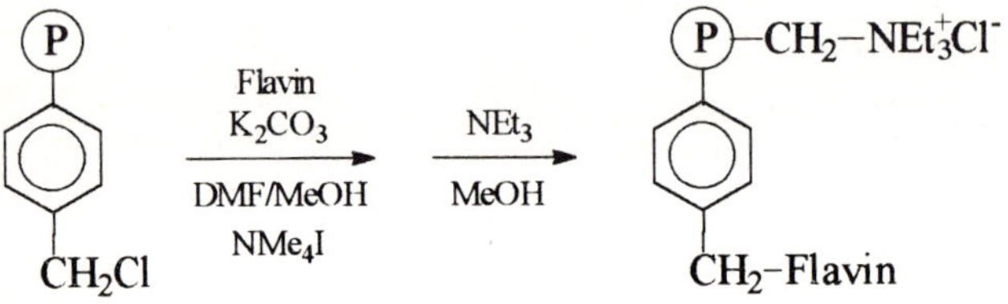

Reaction of flavin with chloromethylated PolyHIPE (Method 1).

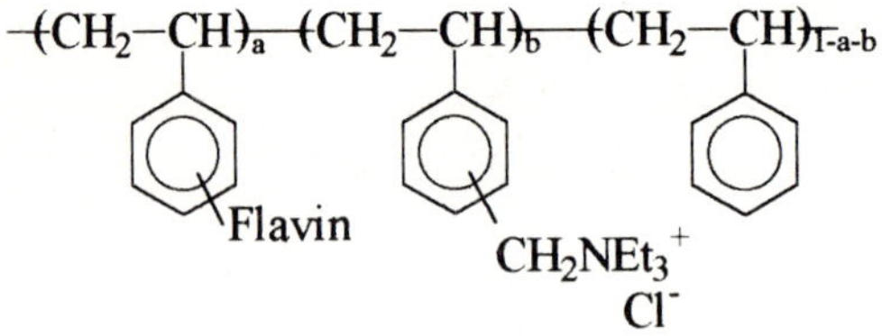

Flavin-containing polycation used in methods 2 and 3.

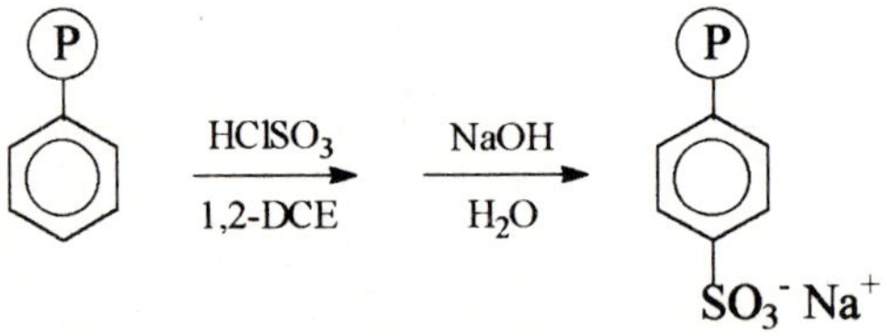

Sulphonation of PolyHIPE for method 3.

Fig. 17. Structure of flavin and preparation of flavin-containing PolyHIPE materials

was stable over prolonged periods, and high flowrates in column reactors were feasible.

PolyHIPE, in granular form, has recently been employed as a support for bicatalyst systems [136]. Styrene/DVB porous polymers were prepared with free vinyl groups (acryloyl, allyl and vinylbenzyl) on the surfaces of the cavities. Impregnation with solutions of quaternary onium monomers, with subsequent polymerisation, resulted in grafting onto the pendant double bonds, to give a surface-quaternised material (Fig. 19).

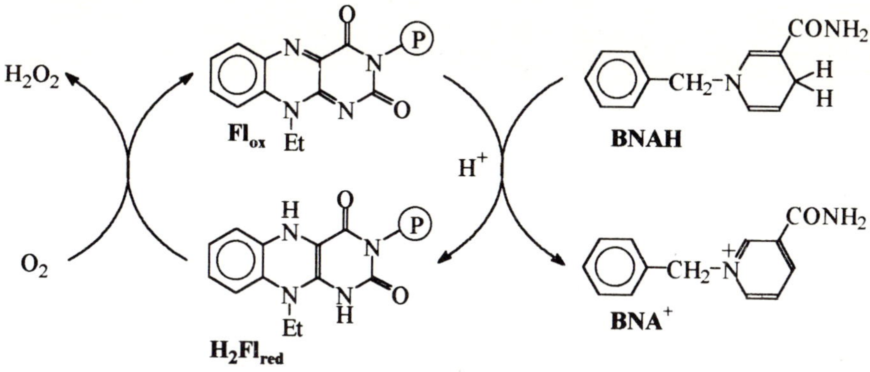

Fig. 18. Aerobic oxidation of BNAH with immobilised flavin

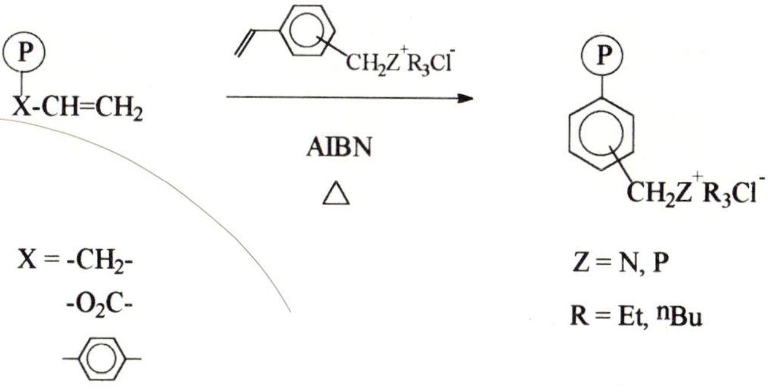

Fig. 19. Preparation of quaternary onium-containing PolyHIPE polymers

The counter-ions of some of the quaternary onium groups were exchanged with an anionic phosphine compound, which was then used to complex palladium. Thus, a polymer material containing phase transfer catalyst and transition-metal catalyst groups was obtained (Fig. 20). The Heck-type vinylation reaction [137] was used to examine the catalytic activity of the heterogeneous system. The polymer-supported catalyst was found to compare favourably with the homogeneous system (Fig. 21).

Poly(styrene/DVB) PolyHIPE materials have also been employed as supports for conducting polymers. A thin coating of polypyrrole was deposited on the internal surface of the porous polymer by initially imbibing a non-aqueous solution of oxidant, partial drying and subsequent addition of pyrrole solution [138, 139]. Higher spreadability of the pyrrole solution was obtained by using 3-alkylpyrrole surfactants to form the concentrated emulsions [140], giving composite materials with higher conductivities. However, increasing the concentration of the pyrrole surfactant caused destabilisation of the emulsion. Porous, conducting polythiophene composite materials have been prepared by a similar

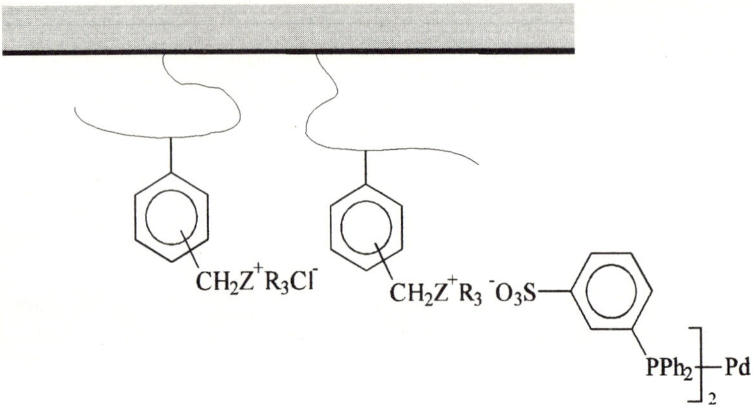

Fig. 20. PolyHIPE bicatalyst support material

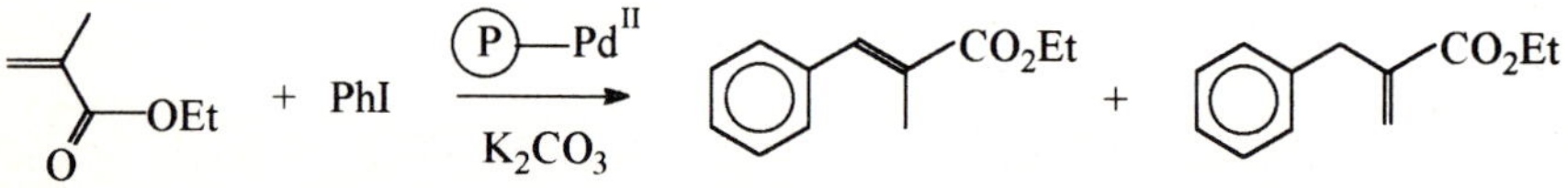

Fig. 21. Heck-type vinylation reaction with PolyHIPE-supported Pd catalyst

method [141]. Earlier, Riess et al. succeeded in forming a silver film on the interior of PolyHIPE polymers, with a view to producing materials with good thermal insulation [142].

PolyHIPE foams have also found uses in high energy physics experiments, such as inertially confined fusion (ICF) [132]. Open microcellular polymeric materials are used in such studies to simulate gases, and are required to meet two criteria: they must possess small cell sizes and densities between air and solid polymer. The low density materials are conveniently represented by PolyHIPE polymers, and a considerable effort has been put into reducing their cavity sizes. The higher density polymers are easily prepared by impregnating open-cell poly(styrene/DVB) PolyHIPE with a heptane solution of styrene, DVB and AIBN, followed by polymerisation [143]. Upon drying, a composite material of higher density is obtained.

The ability of PolyHIPE materials to absorb liquids has been exploited in experiments on their potential use as carrier materials for the safe transport of hazardous or flammable liquids [128]. A poly(styrene/DVB) sample was able to absorb twenty times its own weight of liquid paraffin, simply by immersing the material in the liquid. However, the problem in this application is the subsequent removal of the liquid from the polymer. This can only be achieved by vacuum distillation, which is very difficult with high boiling liquids.

Other applications suggested for PolyHIPE materials include their use as inert matrices for the immobilisation of cells and enzymes [144]. Ruckenstein and coworker [145] have immobilised lipase on crosslinked polystyrene Poly-HIPE, via hydrophobic interactions with the polymer. The activity of the immobilised enzyme in the hydrolysis of triacylglycerides was higher than the free enzyme. Also, the micro-organism *Phanerochaeta chryosporium* has been anchored to PolyHIPE materials [146]. This species produces the enzyme ligninase, which degrades a wide range of compounds including hazardous materials such as polychlorinated biphenols (PCBs). The activity in the degradation of 2-chlorophenol was again found to be higher with the immobilised system.

Another more esoteric use for polystyrene PolyHIPE materials was suggested by Williams et al., and involves the intact capture of microparticles of cosmic dust [147].

2.1.2 Other PolyHIPE Systems

The idea of the preparation of porous polymers from high internal phase emulsions had been reported prior to the publication of the PolyHIPE patent [128]. About twenty years previously, Bartl and von Bonin [148,149] described the polymerisation of water-insoluble vinyl monomers, such as styrene and methyl methacrylate, in w/o HIPEs, stabilised by styrene-ethyleneoxide graft copolymers. In this way, HIPEs of approximately 85% internal phase volume could be prepared. On polymerisation, solid, closed-cell monolithic polymers were obtained. Similarly, Riess and coworkers [150] had described the preparation of closed-cell porous polystyrene from HIPEs of water in styrene, stabilised by poly(styrene-ethyleneoxide) block copolymer surfactants, with internal phase volumes of up to 80%.

Horie et al [151, 152] discovered that the emulsifier system previously employed by Bartl and coworker was ineffective in the preparation of concentrated emulsions of water in styrene/unsaturated polyester mixture. Methyl methacrylate/styrene-grafted poly(ethyleneoxide) and vinyl acetate-grafted PEO were also inadequate; however, addition of a base such as sodium hydroxide or triethanolamine, to the unsaturated polyester, formed salts with free carboxylic acid groups, giving excellent in situ stabilisation of the water-in-oil emulsions. Consequently, it was possible to form a HIPE containing 90% water, which on copolymerisation of the continuous phase, gave closed-cell materials containing entrapped water. These "water-extended" materials were intended as a cheap route to polyester resins with enhanced heat-resistant properties.

PolyHIPE materials have also been prepared by polycondensation in high internal phase emulsions [153]. Thus, a resorcinol-formaldehyde (RF) porous copolymer was synthesised from an o/w HIPE of cyclohexane in an aqueous solution of resorcinol, formaldehyde and surfactant. Addition of an acid catalyst to the emulsion, followed by heating, resulted in copolymerisation. Other systems prepared included urea-formaldehyde, phenol-formaldehyde, melamine-formaldehyde and a polysiloxane-based elastomeric species.

The resorcinol-formaldehyde polymers have been used to prepare highly porous carbon materials, by controlled pyrolysis in an inert atmosphere [144, 154]. The microstructure of the carbon is an exact copy of the porous polymer precursor. Poly(methacrylonitrile) (PMAN) PolyHIPE polymers have also been used for this purpose. These monolithic, highly porous carbons are potentially useful in electrochemical applications, particularly re-chargeable batteries and super-capacitors. The RF materials, with their very high surface areas, are particularly attractive for the latter systems.

Open-cell PolyHIPE materials have also been prepared from hydrophilic methacrylates which, on hydrolysis, yield hydrophilic polymethacrylic acid-based species [155]. Stable HIPEs containing high levels of glycidyl methacrylate can also be formed, from which porous polymers can be made. These have considerable potential for further exploitation due to the reactive epoxide group [156].

2.2 Polymerisation of the Dispersed Phase

The dispersed phase of high internal phase emulsions may also be used to prepare polymeric materials; in this case, conversion of monomer dispersed droplets to polymer results in latexes or particulates.

Lissant [6] prepared polymer particles by polymerisation of HIPEs of vinyl chloride monomer (VCM) in water. Concentrated emulsions with high phase volume ratios, i.e. above 0.9, gave agglomerates of particles with a relatively monodisperse size distribution. The particles themselves appeared to be hollow. The explanation given for this was that poly(vinylchloride) is considerably more dense than VCM; since polymerisation is probably initiated on the surface of the droplets, due to the use of a water-soluble initiator, polymer will grow until the surface is covered, resulting in a rigid sphere encapsulating unreacted monomer. Polymerisation will continue inside the shell, with concomitant shrinkage, until all the monomer is consumed, resulting in hollow spheres. The presence of the rigid shell prevents additional initiator from reaching unreacted monomer, and this therefore results in a high polymer molecular weight.

Polystyrene latexes were similarly prepared by Ruckenstein and Kim [157]. Highly concentrated emulsions of styrene in aqueous solutions of sodium dodecylsulphate, on polymerisation, yielded uncrosslinked polystyrene particles, polyhedral in shape and of relative size monodispersity. Interestingly, Ruckenstein and coworker found that both conversions and molecular weights were higher compared to bulk polymerisation. This was attributed to a gel effect, where the mobility of the growing polymer chains inside the droplets is reduced, due to increased viscosity. Therefore, the termination rate decreases.

High molecular weight monodisperse polystyrene latexes have been prepared by this method [158]. A number of factors were found to influence the size and dispersity of the particles. The size decreased with increasing surfactant concentration and decreasing internal phase volume, and a more monodisperse latex was produced at higher phase volumes. The addition of decane to the oil phase increased its hydrophobicity, resulting in a reduction in particle size and polydispersity.

Copolymer particles can also be prepared from HIPEs [159]. Thus, a HIPE dispersed phase consisting of styrene and methacrylic acid was polymerised to give copolymers. The surface concentration of carboxylic acid groups increased linearly with concentration of methacrylic acid in the feed. The small amount of water present in the concentrated emulsion, relative to conventional emulsion polymerisation, reduces the loss of methacrylic acid, which is highly water-soluble.

The polystyrene latexes produced from concentrated emulsions have been used as carriers for the controlled release of herbicides [160]. The release of 2-(2,4-dichlorophenoxy) propionic acid (2,4-DP) was found to depend on the water concentration, increasing with increasing dilution of the latex. High conversion to polymer was required to prevent a large initial release of herbicide on dilution; however, a significant initial burst was still observed at almost complete conversion. This was reportedly due to dissolution of 2,4-DP at, or near, the surface of the latex particles.

Other latexes which have been produced by this method include poly(butyl methacrylate), poly(butyl acrylate) and poly(styrene/DVB) [161]. Additionally, polymer blends were produced by mixing, under high shear, HIPEs of partially polymerised monomer, followed by completion of polymerisation. The conversion prior to blending had to be less than 5%, to allow efficient mixing of the highly viscous emulsions. The materials thus produced resembled agglomerates of latex particles, due to copolymerisation at the points of contact of partially polymerised droplets.

Dispersed phase polymerisation of HIPEs has also been used to prepare polymer-supported quaternary onium phase transfer catalysts [162]. One strategy involved the polymerisation of a concentrated emulsion of vinyl benzyl chloride (VBC) in water and subsequent quaternisation of the polymer resin with tertiary amines and phosphines (Fig. 22).

In another method, VBC is added to a highly polymerised HIPE of a styrene/DVB mixture in water, followed by complete polymerisation. This gave an agglomerated material with a core-shell morphology. The outer, poly(VBC)

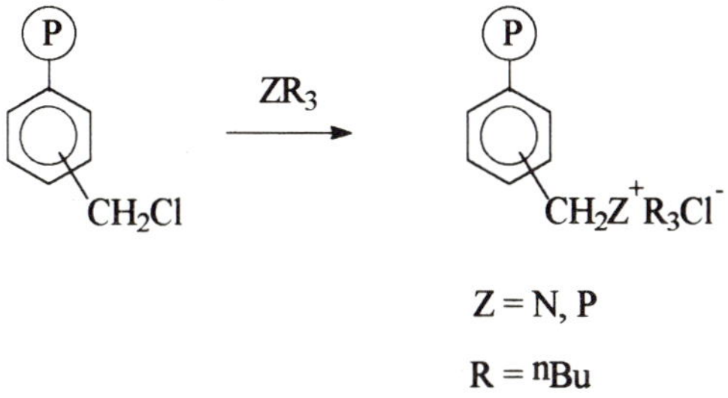

$Z = N, P$

$R = {}^nBu$

Fig. 22. Preparation of quaternary onium catalysts supported on latexes derived from HIPEs

layer was subsequently quaternised. The third route employed concentrated emulsions of styrene in an aqueous solution of a quaternary onium monomer. The ionic monomer should have been located on the surface of the styrene droplets, with the vinylbenzyl groups oriented towards the interior; copolymerisation with styrene should therefore occur. However, HIPEs did not form above certain concentrations of salt comonomer, due to ion-pair interactions with the anionic surfactant. Catalysts prepared by this method, therefore, contained low levels of quaternary onium groups. The catalytic activities of the three polymers were studied in the alkylation reaction of Meldrum's acid (isopropylidene malonate) (Fig. 23).

Moderate yields of product, similar to those found in the homogeneous reaction, were obtained with catalysts prepared by the second method. Those prepared via the third strategy had very low activities due to low catalyst loadings.

Core-shell agglomerate materials, similar to those prepared by the second method above, have also been employed as supports for bicatalyst systems [163]. Porous polymers, having a crosslinked polystyrene core and a polyVBC inhomogeneous shell, were chemically modified in a multi-step procedure to give a range of polymers containing phase-transfer catalytic sites and phosphorus-bound palladium complexes, employing similar chemistry to the example given in the previous section (see Sect. 2.1.1.4). The catalysts were again used in vinylation reactions, and were found to be much more active than the corresponding homogeneous system. Cooperative effects between the two catalytic groups, present in very close proximity on the polymer support, are said to produce this enhanced activity.

Other 'nutshell' materials have been synthesised [164]. Hydrophobic latex particles containing a crosslinked poly(VBC) core and a macroporous poly(styrene/DVB) shell were prepared from concentrated o/w emulsions. Similarly, hydrophilic porous particles of crosslinked acrylamide surrounding a linear poly(ethyleneoxide) core were formed from w/o HIPEs. The poly(VBC) cores of the hydrophobic particles were quaternised and used to bind $[Co(CO)_4]^-$ anions, whereas the hydrophilic latexes were employed in the immobilisation of lipase.

Conducting composite polymer materials have also been prepared from the dispersed phase of concentrated emulsions. Polyurethane/polypyrrole composites [165] were obtained by blending an aqueous suspension of polypyrrole with a HIPE of a chloroform solution of polyurethane in aqueous surfactant

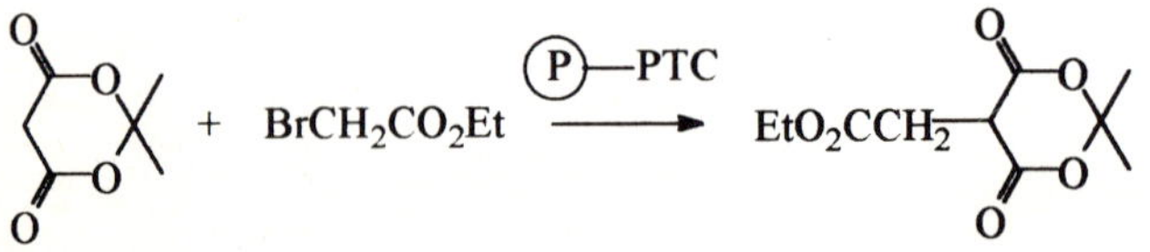

Fig. 23. Alkylation of Meldrum's acid with phase transfer catalysts bound to core-shell particles

(sodium dodecyl sulphate, SDS) solution. The composite particles precipitated from the heterogeneous system, and were pressed to make a porous material. A poly(methyl methacrylate) (PMMA) /polypyrrole composite [166] with a higher conductivity was prepared by a similar route. In this case, aqueous $FeCl_3$ solution was added to an o/w HIPE of PMMA/pyrrole/$CHCl_3$ in aqueous SDS. The oxidant ($FeCl_3$) caused pyrrole polymerisation on the droplet surfaces.

Polypyrrole/poly(ethylene-co-vinyl acetate) conducting composites with improved mechanical properties were prepared by a similar method [167]. In addition, polyaniline/polystyrene [168] and polyaniline/poly(alkyl methacrylate) [169] composites have been synthesised. A solution of persulphate in aqueous HCl was added to an o/w HIPE of polymer and aniline in an organic solvent, dispersed in aqueous SDS solution, causing aniline polymerisation. Films were processed by hot- or cold-pressing.

A rubber-like copolymer/carbon fibre composite material has also been prepared [170]. Carbon fibres were added directly to o/w highly concentrated emulsions of block copolymers, such as styrene/butadiene triblocks (SBS), in toluene, followed by precipitation in methanol, drying and hot-pressing. The surfactant was found to aid adhesion between the polymer and carbon fibres. The materials obtained had fairly even distributions of carbon fibres, good mechanical properties and conductivities which increased with increasing carbon fibre length.

Rubber-toughened polystyrene composites were obtained similarly by polymerising the dispersed phase of a styrene/SBS solution o/w HIPE [171], or a styrene/MMA/(SBS or butyl methacrylate) o/w HIPE [172]. The latter materials were found to be tougher; however, all polymer composites had mechanical properties comparable to bulk materials. Other rubber composite materials have been prepared from PVC and poly(butyl methacrylate) (PBMA) [173], via three routes: a) blending partially polymerised o/w HIPEs of vinylidene chloride (VDC) and BMA, followed by complete polymerisation; b) employing a solution of PBMA in VDC as the dispersed phase, with subsequent polymerisation; and c) blending partially polymerised VDC HIPE with BMA monomer, then polymerisation. All materials obtained possessed mixtures of both homopolymers plus some copolymer, and had better mechanical properties than the linear copolymers. The third method was found to produce the best material.

The range of monomers which are available to prepare latexes from HIPEs is limited to those which are sufficiently hydrophobic to form stable o/w concentrated emulsions. However, Ruckenstein and coworker [174] have shown that this range can be extended by prepolymerising the monomer, in bulk, to low conversion, then using the resulting viscous liquid to form the HIPE. By this method, PMMA latexes could be formed whereas attempts to emulsify MMA monomer in aqueous solution resulted in phase separation.

Another method of lending extra stability to monomeric o/w HIPEs, to enable the preparation of novel latexes, is to include small amounts of fumed silica particles in the oil phase [175]. These particles are relatively hydrophobic, due to the lack of hydroxyl groups on their surface, allowing them to disperse easily in the organic phase. In the emulsion, some are located at the interface, where, possibly, they become hydroxylated on contact with water. Repulsion between silanol groups on the surfaces of adjacent particles may enhance emulsion stability. Crosslinked latexes of PMMA, poly(allylchloride) and poly(3,4-dichloro-1-butene) were synthesised in this manner. The latter two were subsequently quaternised to yield phase transfer catalyst materials, which were used in the oxidation of toluene to benzoic acid, mediated by perruthenate anions [176,177]. The latter are bound to catalytic sites on the polymer (Fig. 24). The perruthenate anions are generated in situ by oxidation of ruthenium chloride with persulphate and 4-N-methylmorpholine-N-oxide (NMO).

Water-in-oil concentrated emulsions have also been utilised in the preparation of polymer latexes, from hydrophilic, water-soluble monomers. Kim and Ruckenstein [178] reported the preparation of polyacrylamide particles from a HIPE of aqueous acrylamide solution in a non-polar organic solvent, such as decane, stabilised by sorbitan monooleate (Span 80). The stability of the emulsion decreased when the weight fraction of acrylamide in the aqueous phase exceeded 0.2, since acrylamide is more hydrophobic than water. Another point of note is that the molecular weights obtained were lower compared to solution polymerisation of acrylamide. This was probably due to a degree of termination by chain transfer from the tertiary hydroxyl groups on the surfactant head group.

Crosslinked polyacrylamide latexes encapsulating microparticles of silica and alumina have also been prepared by this method [179]. Three steps are involved: a) formation of a stable colloidal dispersion of the inorganic particles in an aqueous solution containing acrylamide, crosslinker, dispersant, and initiator; b) HIPE preparation with this aqueous solution as the dispersed phase; and c) polymerisation. The latex particles are polyhedral in shape, shown clearly by excellent scanning electron micrographs, and have sizes of between 1 and 5 μm.

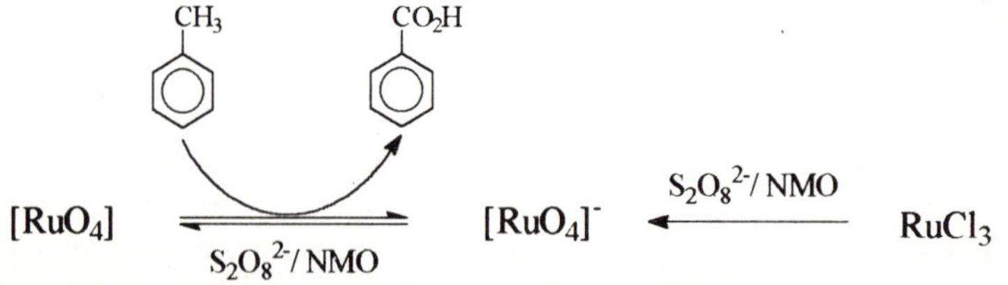

Fig. 24. Oxidation of toluene with perruthenate anions bound to polymer-supported phase transfer catalysts

2.3 Polymerisation of Both Phases

If both continuous and dispersed phases of highly concentrated emulsions contain monomeric species, it is possible to obtain hydrophilic/hydrophobic polymer composite materials. Polyacrylamide/polystyrene composites have been prepared in this manner [180], from both w/o and o/w HIPEs containing aqueous acrylamide and a solution of styrene in an organic solvent.

The composite materials have been used to form selective membranes for the separation of liquid mixtures [181]. The membranes should consist of a polymer which is soluble in the liquid component(s) to be separated, as the dispersed phase-derived polymer, and a continuous phase-derived polymer which is insoluble in all components of the liquid mixture. Thus, membranes consisting of polystyrene in polyacrylamide will separate toluene from cyclohexane, and those comprising polyacrylamide in crosslinked polystyrene can be used for water removal from ethanol. Due to the very thin films of polymer which separate the polyhedral dispersed phase cells, the permeation rates, which are measured by pervaporation, are relatively high.

The rate of permeation was found to depend on a number of parameters. In the separation of toluene from toluene/cyclohexane mixtures, with a polystyrene in polyacrylamide membrane, the rate increased with increasing temperature and toluene concentration, and decreasing polyacrylamide content [182]. Selectivity, on the other hand, increased with decreasing temperature and increasing polyacrylamide content; therefore selectivity is inversely related to permeation rate. Similar results were found in experiments on the separation of water from water/ethanol mixtures by hydrophilic-hydrophobic membranes [183].

Other hydrophilic/hydrophobic membranes have been generated from HIPEs of aqueous sodium acrylate solution in divinylbenzene [184]. The emulsions were more stable than corresponding styrene-containing systems, leading to more favourable membrane preparation. In addition, the use of a polyelectrolyte as the permselective medium increased the affinity for water, giving higher selectivities from water/ethanol mixtures. Selectivity was found to decrease only slightly with increasing temperature.

The mechanical properties of these membranes were improved by including a crosslinker, methylene bisacrylamide, in the aqueous phase, and by using a styrene/butyl acrylate (BA) mixture as the continuous phase [185]. The styrene/BA mixture had to be prepolymerised to low conversion to allow HIPE formation. The permeation rate of the membrane was improved by including a porogen (hexane) in the organic phase, generating a permanent porous structure [186]. The pervaporation rate was indeed increased, however a drop in selectivity for water from water/ethanol mixtures was also observed.

Gelatin-PMMA composite materials can also be prepared via the HIPE pathway [187]. Concentrated emulsions of methyl methacrylate in aqueous gelatin/surfactant solutions, upon polymerisation at 50°C, yielded composite membrane materials. The gelatin lends considerable stability to the emulsions, which will not form in its absence. The membranes swelled in water and certain organic solvents (polar and non-polar). At low gelatin content, they were completely soluble in toluene, but became insoluble at higher gelatin levels. It was suggested that a gelatin network forms at higher concentrations.

Rather peculiarly, it was also found that a slow gelatin crosslinking reaction occurred at room temperature over a number of days [188]. After drying, the resulting membranes were found to swell in water, indicating crosslinking had taken place. This crosslinking reaction appeared to occur only in the presence of monomer and initiator, AIBN; absence of initiator, or substitution of MMA for a non-polymerisable organic liquid, could not produce crosslinked gelatin. A radical mechanism is therefore suggested. A small amount (3–4%) of PMMA was incorporated in the crosslinked gelatin network.

Replacing MMA with the more hydrophobic styrene produced more stable gelatin HIPEs [189], from which other composite materials were produced by

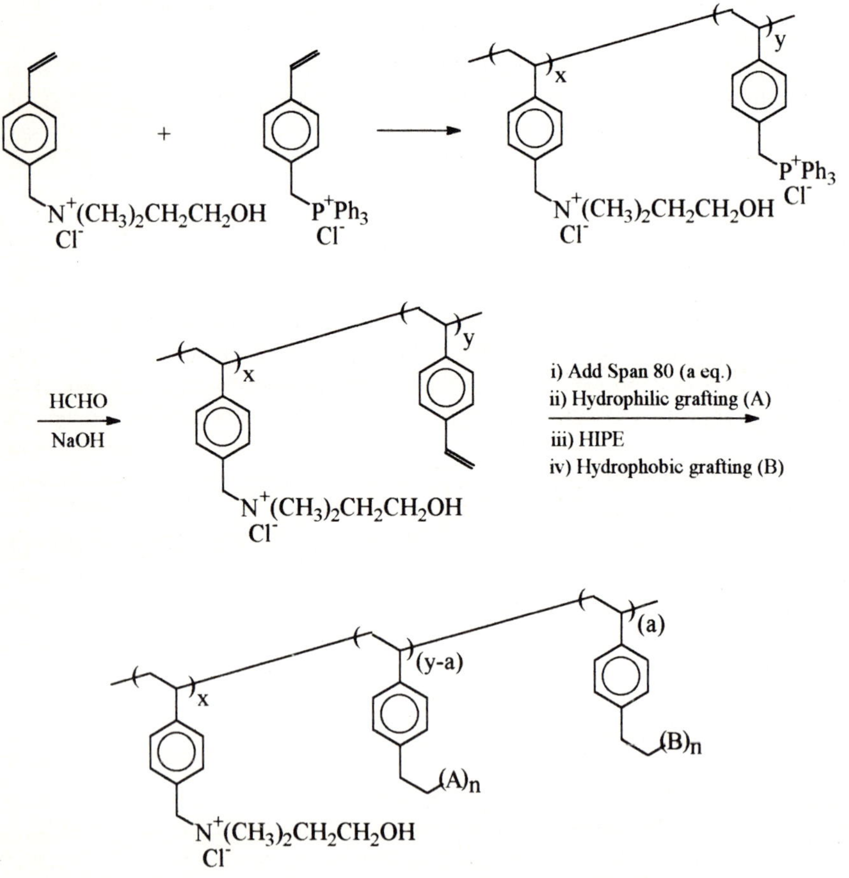

Fig. 25. Preparation of amphiphilic copolymer membrane with hydrophilic and hydrophobic side chains

polymerisation. These membranes also exhibited swelling in water and various organic solvents. The swelling ability in water could be altered by changing the polymerisation time and temperature.

Other amphiphilic membrane materials have been produced by the HIPE polymerisation route. A comb-like polymer, possessing hydrophilic and hydrophobic sidechains anchored to an amphiphilic backbone [190] was thus prepared, by the following strategy. A quaternary ammonium and a quaternary phosphonium monomer were copolymerised to yield a water-soluble copolymer; the phosphonium groups were then converted to vinylstyrene units via the Wittig reaction. On addition of nonionic surfactant (Span 80) to an aqueous solution of the polymer, some surfactant molecules were adsorbed on the available vinylstyrene groups. The remaining free double bonds were copolymerised with a hydrophilic monomer (e.g. acrylamide). A small amount of hydrophobic monomer, such as styrene, was then added with vigorous stirring, generating a HIPE. This was then polymerised; the remaining vinylstyryl groups, protected initially by surfactant adsorption, were copolymerised. The resulting polymer therefore possessed both hydrophilic and hydrophobic sidechains (Fig. 25).

3 Conclusions

High internal phase emulsions (HIPEs) can be generated from two immiscible liquids, one of which is usually, but not necessarily, water or an aqueous solution. Such emulsions have a dispersed phase volume fraction greater than 0.74, which is the value for the close-packing of uniform spheres. Therefore, the dispersed phase droplets are deformed into non-spherical shapes. The production of the concentrated emulsion is highly dependent on the nature and concentration of the surfactant present in the continuous phase. The emulsions have a microstructure resembling a gas-liquid foam of low liquid content, with polyhedral droplets of dispersed phase separated by a network of very thin films of continuous phase. Ideally, at very high internal phase volume fractions, the cells would be size-monodisperse and would resemble the space-filling polyhedral structure of lowest surface area; however, a degree of polydispersity is almost always encountered in real systems, and a variety of polyhedral shapes are observed.

A number of peculiar properties are displayed, including rheology characterised by viscoelasticity. Viscosities are far higher than that of either bulk phase; this is a result of the large amount of energy required to deform the network of thin films of the continuous phase. A yield stress is observed, below which HIPEs behave as elastic solids and will not flow. Resistance to flow occurs from the inability of compressed droplets to easily slip past each other. Above the yield stress, the emulsions are shear-thinning liquids. An osmotic pressure, due to the stress of deformation of the dispersed phase droplets, is also shown.

HIPE stability depends greatly on a number of parameters, including the nature and concentration of the surfactant, the nature and viscosity of each liquid phase, system temperature, mean droplet size, interfacial tension between the phases, strength of the interfacial film and the presence of added electrolyte in the aqueous phase. The formation of a rigid interfacial film is thought to be of paramount importance to the stability of HIPEs.

A wide range of polymeric materials can be prepared from HIPEs. Polymerisation of the continuous phase yields highly porous cellular polymers with a monolithic structure. These are known as PolyHIPE polymers, and possess a number of unique properties including, in most cases, an interconnected cellular structure and a very low dry-bulk density. Their very high porosity favours their use as supports for catalytic species, precursors for porous carbons and inert matrices for the immobilisation of enzymes and micro-organisms.

Other polymer materials which can be prepared include latexes, or particle agglomerates, by dispersed phase polymerisation. These can be either hydrophilic or hydrophobic in nature, or may have core-shell morphologies. They can be employed as support materials for a number of catalyst systems. Polymerisation of both phases of the emulsions produces composite materials, which have found use as selective membranes for the separation of mixtures of liquids with similar physical properties.

4 References

1. Lissant KJ (ed) (1974) Emulsions and Emulsion Technology, Part 1 Marcel Dekker, New York, Chap. 1
2. Ostwald W (1910) Kolloid Z 6: 103; ibid (1910) 7: 64
3. Lissant KJ (1966) J Coll Interf Sci 22: 462
4. Lissant KJ (1970) J Soc Cosmetic Chem 21: 141
5. Lissant KJ, Mayhan KG (1973) J coll Interf Sci 42: 201
6. Lissant KJ, Pearce BW, Wu SH, Mayhan KG (1974) J Coll Interf Sci 47: 416
7. Princen HM (1988) Langmuir 4: 486
8. Ravey J-C, Stébé MJ (1990) Progr Coll Polym Sci 82: 218
9. Solans C, Dominguez JG, Parra JL, Heuser J, Friberg SE (1992) Coll Polym Sci 266: 570
10. Pons R, Solans C, Stébé MJ, Erra P, Ravey J-C (1992) Prog Coll Polym Sci 89: 110
11. Kunieda H, Solans C, Shida N, Parra JL (1987) Coll Surf 24: 225
12. Solans C, Pons R, Zhu S, Davis HT, Evans DF, Nakamura K, Kunieda H (1993) Langmuir 9: 1479
13. Kunieda H, Yano N, Solans C (1989) Coll Surf 36: 313
14. Kunieda H, Evans DF, Solans C, Yoshida M (1990) ibid 47: 35
15. Ebert G, Platz G, Rehage H (1988) Ber Bunsenges Phys Chem 92: 1158
16. Hoffmann H (1990) Adv Coll Interf Sci 32: 123
17. Sebba F (1972) J Coll Interf Sci 40: 468
18. Sebba F (1975) ACS Symp Ser 9: 18
19. Sebba F (1987) Foams and Biliquid Foams – Aphrons, Wiley, New York
20. Vincent B (1975) In: Colloid Science Vol.2 (D.H. Everett, ed.), Chemical Society, London

21. Kizling J, Kronberg B (1990) Coll Surf 50: 131
22. Sebba F (1979) Coll Polym Sci 257: 392
23. Bergeron V, Sebba F (1987) Langmuir 3: 857
24. Sebba F, Chem Ind 1984: 367
25. Paczynska-Lahne B (1990) Prog Coll Polym Sci 83: 196
26. Princen HM (1979) J Coll Interf Sci 71: 55
27. Princen HM (1965) J Coll Interf Sci 20: 156
28. Ivanov IB, Toshev B (1975) Coll Polym Sci 253: 593
29. de Feiter JA, Rijnbout JB, Vrij A (1978) J Coll Interf Sci 64: 258
30. Princen HM, Aronson MP, Moser JC (1980) ibid 75: 246
31. For a detailed description of collapsed linear and tetrahedral Plateau borders, see the following: Princen HM (1984) Coll Surf 9: 47; Princen HM (1984) ibid 9: 67
32. Mannegold E (1953) Schaum, p. 83, Chemie und Technik Verlagsgesellschaft, Heidelberg, cited in Ref. 30
33. Desch CH (1923) Rec Trav Chim 42: 882, cited in: Bikermann JJ (1973) Foams, p. 62, Springer-Verlag, Heidelberg
34. Ross S (1978) Amer J Phys 46: 513
35. Ross S, Prest HF (1986) Coll Surf 21: 179
36. Princen HM, Levinson P (1987) J Coll Interf Sci 120: 172
37. Princen HM, Kiss AD (1987) Langmuir 3: 36
38. Reinelt DA, Kraynik AM (1993) ibid 159: 460
39. Kann KB (1984) Coll J USSR 46: 397
40. Kann KB (1985) ibid 47: 744
41. Weaire D, Pittet N, Hutzler S, Pardal D (1993) Phys Rev Lett 71: 2670
42. Weaire D (1994) Phil Mag Lett 69: 99
43. Weaire D, Phelan R (1994) ibid 69: 107
44. Weaire D (1994) New Scientist 142 (1926) : 34
45. Weaire D, Phelan R (1994) Phil Mag Lett 70: 345
46. Das AK, Ghosh PK (1990) Langmuir 6: 1668
47. Gregory DP, Unilever Research Laboratory, Port Sunlight, private communication.
48. Mannheimer RJ (1972) J Coll Interf Sci 40: 370
49. Mukesh D, Das AK, Ghosh PK (1992) Langmuir 8: 807
50. Aubert JH, Kraynik AM, Rand PB (1986) Sci Am 254: 58; Heller JP, Kuntamukkula MS (1987) Ind Eng Chem Res 26: 318; Kraynik AM (1988) Ann Rev Fluid Mech 20: 325
51. Princen HM (1983) J Coll Interf Sci 91: 160
52. Khan SA, Armstrong RC (1986) J Non-Newtonian Fluid Mech 22: 1
53. Kraynik AM, Hansen MG (1986) J Rheol 30: 409
54. Weaire D (1989) Phil Mag Lett 60: 27
55. Khan SA (1987) Rheol Acta 26: 78
56. Khan SA, Armstrong RC (1989) J Rheol 33: 881
57. Princen HM, Kiss AD (1986) J Coll Interf Sci 112: 427
58. Stamenovic D, Wilson TA (1984) Trans ASME J Appl Mech 51: 229
59. Derjaguin B (1933) Kolloid Z 64: 1; Derjaguin B, Obuchov E (1934) ibid 68: 243
60. Stamenovic D (1991) Trans ASME J Appl Mech 58: 288
61. Budiansky B, Kimmel E (1991) ibid 58: 289
62. Kraynik AM, Hansen MG (1987) J Rheol 31: 175
63. Khan SA, Armstrong RC (1987) J Non-Newtonian Fluid Mech 25: 61
64. Princen HM (1985) J Coll Interf Sci 105: 150
65. Schwartz LW, Princen HM (1987) ibid 118: 201
66. Reinelt DA, Kraynik AM (1989) ibid 132: 491
67. Reinelt DA, Kraynik AM (1990) J Fluid Mech 215: 431
68. Weaire D, Kermode JP (1983) Philos Mag B 48: 245
69. Weaire D, Kermode JP (1984) ibid 50: 379
70. Weaire D, Rivier N (1984) Contemp Phys 25: 59
71. Weaire D, Fu TL, Kermode JP (1986) Phil Mag Lett 54: L39
72. Weaire D, Fu TL (1988) J Rheol 32: 271
73. Weaire D, Bolton F, Herdtle T, Aref H (1992) Phil Mag Lett 66: 293
74. Kraynik AM, Reinelt DA, Princen HM (1991) J Rheol 35: 1235
75. Reinelt DA (1992) J Rheol 37: 1117
76. Ford RE, Furmidge CGL (1967) J Sci Food Agric 18: 419

77. Pal R, Rhodes E (1985) J Coll Interf Sci 107: 301
78. Das AK, Mukesh D, Swayambunathan V, Kotkar DD, Ghosh PK (1992) Langmuir 8: 2427
79. Otsubo Y, Prud'homme RK (1994) Rheol Acta 33: 101
80. Solans C, Azemar N, Parra JL (1988) Prog Coll Polym Sci 76: 224
81. Ebert G, Platz G, Rehage H (1988) Ber Bunsenges Phys Chem 92: 1158
82. Princen HM, Kiss AD (1989) J Coll Interf Sci 128: 176
83. Mannheimer RJ (1972) ibid 40: 370
84. Calvert JR, Nezhati K (1986) Int J Heat Fluid Flow 7: 164
85. Khan SA, Schnepper CA, Armstrong RC (1988) J Rheol 32: 69
86. Yoshimura A, Prud'homme RK (1988) ibid 32: 53
87. Yoshimura A, Prud'homme RK (1988) ibid 32: 575
88. Pons R, Erra P, Solans C, Ravey J-C, Stébé MJ (1993) J Phys Chem 97: 12320
89. Otsubo Y, Prud'homme RK (1994) Rheol Acta 33: 29
90. Aronson MP, Petko MF (1993) J Coll Interf Sci 159: 134
91. Anklam MR, Ware GG, Prud'homme RK (1994) J Rheol 38: 797
92. Princen HM (1986) Langmuir 2: 519
93. Kruglyakov PM, Exerowa DR, Kristov KI (1991) ibid 7: 1846
94. Princen HM (1987) ibid 3: 36
95. Bibette J (1991) J Coll Interf Sci 147: 474
96. Bibette J (1992) Langmuir 8: 3178
97. Bibette J, Morse DC, Witten TA, Weitz DA (1992) Phys Rev Lett 69: 2439
98. Buzza DMA, Cates ME (1993) Langmuir 9: 2264
99. Ravey J-C, Stébé MJ (1989) Physica B 156 & 157: 394
100. Pons R, Ravey J-C, Sauvage S, Stébé MJ, Erra P, Solans C (1993) Coll Surf 76: 171
101. Balinov B, Söderman O, Ravey J-C (1994) J Phys Chem 98: 393
102. Rajagopalan V, Solans C, Kunieda H (1994) Coll Polym Sci 272: 1166
103. Kunieda H, Rajagopalan V, Kimura E, Solans C (1994) Langmuir 10: 2570
104. Ford RE, Furmidge CGL (1966) J Coll Interf Sci 22: 331
105. Williams JM (1991) Langmuir 7: 1370
106. Platz G, Ebert G Viscoelasticity and Anisotropy in Microemulsions In: Polymer Reaction Engineering, Reicher KH, Geiseler W (eds) Hüthig, Heidelberg, 1986, pp.95
107. Ravey J-C, Stébé MJ, Sauvage S (1994) J Chim Phys 91: 259
108. Chen HH, Ruckenstein E (1991) J Coll Interf Sci 145: 260
109. Ruckenstein E, Ebert G, Platz G (1989) ibid 133: 432
110. Chen HH, Ruckenstein E (1990) ibid 138: 473
111. Princen HM (1990) ibid 134: 188
112. Ganguly S, Krishna Mohan V, Jyothi Bhasu VC, Mathews E, Adiseshaiah KS, Kumar AS (1992) Coll Surf 65: 243
113. Rajagopalan V, Solans C, Kunieda H (1994) Coll Polym Sci 272: 1166
114. Aronson MP, Ananthapadmanabhan K, Petko MF, Palatini DJ (1994) Coll Surf 85: 199
115. Babak VG (1994) ibid 85: 279
116. Periard J, Banderet A, Riess G (1970) Polym Lett 8: 109
117. Periard J, Riess G, Neyer-Gomez MJ (1973) Eur Polym J 9: 687
118. Riess G, Periard J, Banderet A Emulsifying Effect of Block & Graft Copolymers – Oil in Oil Emulsions In: G. E. Molau (ed) Colloidal and Morphological Behaviour of Block and Graft Copolymers (Proceedings of an American Chemical Society Symposium held at Chicago, Illinois, Sep. 13, 1970)
119. Riess G (1985) Makromol Chem Suppl 13: 157
120. Sharma MK (1975) J Coll Interf Sci 53: 340; Curr Sci (1975) 44: 770; ibid (1977) 46: 131; Acta Scien Ind (1977) 3: 139; Sci Cult (1977) 43: 456; Indian J Chem Sect A (1977) 15A: 644; Prog Coll Polym Sci (1978) 63: 75; ibid (1978) 63: 90
121. Sharma MK (1978) Prog Coll Polym Sci 63: 87; Curr Sci (1977) 46: 601; Sci Cult (1978) 44: 120; Indian J Chem Sect A (1977) 15A: 684; ibid (1978) 16A: 71
122. Gautier M, Rico I, Ahmad-Zadeh Samii A, de Savignac A, Latter A (1986) J Coll Interf Sci 112: 484; Bergenstähl B, Jönsson A, Sjöblum J, Stenius P, Wärnheim T (1987) Prog Coll Polym Sci 74: 108; Auvray X, Petipas C, Anthore R, Rico I, Lattes A, Ahmad-Zadeh Samii A, de Savignac A (1987) Coll Polym Sci 265: 925; Martino A, Kaler EW (1990) J Phys Chem 94: 1627; Friberg SE, Yang C-C, Gourbran R, Partch RE (1991) Langmuir 7: 1103; Dörfler H-D, Swaboda C (1993) Coll Polym Sci 271: 586; Schubert KV, Busse G, Strey R, Kahlweit M (1993) J Phys Chem 97: 248

123. Meliani A, Perez E, Rico I, Lattes A, Moisand A (1991) New J Chem 15: 871
124. Beerbower A, Nixon J, Wallace TJ (1968) J Aircraft 5: 367
125. Nixon J, Beerbower A (1969) Am Chem Soc Div Petrol Chem Prepr 14: 49
126. Cameron NR (1995) Ph.D. Thesis, University of Strathclyde
127. Ishida H, Iwama A (1984) Combust Sci Tech 36: 51
128. Barby D, Haq Z (1982) Eur Pat 0,060,138 (to Unilever)
129. Williams JM, Wrobleski DA (1988) Langmuir 4: 656
130. Williams JM, Gray AJ, Wilkerson MH (1990) ibid 6: 437
131. Litt MH, Hsieh BR, Krieger IM, Chen TT, Lu HL (1987) J Coll Interf Sci 115: 312
132. Williams JM (1988) Langmuir 4: 44
133. Hainey P, Huxham IM, Rowatt B, Sherrington DC, Tetley L (1991) Macromol 24: 117
134. Small PW, Sherrington DC J. Chem Soc, Chem Commun 1989: 1589
135. Schoo HFM, Challa G, Rowatt B, Sherrington DC (1992) React. Pols. 16: 125
136. Ruckenstein E, Hong L (1992) Chem Mater 4: 122
137. Patel BA, Ziegler CB, Cortese NA, Plevyak JE, Zebovitz TC, Terpko M, Heck RF (1977) J. Org. Chem. 42: 3903; Jeffrey T, J Chem Soc, Chem Commun 1984: 1287
138. Ruckenstein E, Park JS (1991) J Appl Polym Sci 42: 925
139. Ruckenstein E, Park JS (1991) Polym Comp 12: 289
140. Ruckenstein E, Chen J.-H (1991) ibid 43: 1209
141. Ruckenstein E, Park JS (1991) Synth Metals 44: 293
142. Riess G Université de Mulhouse, private communication
143. Williams JM, Wilkerson MH (1990) Polymer 31: 2162
144. Even Jr. WR, Gregory DP (1994) MRS Bull 19: 29
145. Ruckenstein E, Wang X (1993) Biotech Bioeng 42: 821
146. Ruckenstein E, Wang X (1994) ibid 44: 79
147. Williams JM, Wrobleski DA (1989) J Matt Sci Lett 24: 4062
148. Bartl H, von Bonin W (1962) Makromol Chem 57: 74
149. Bartl H, von Bonin W (1963) ibid 66: 151
150. Rogez D, Marti S, Nervo J, Riess G (1975) Makromol Chem 176: 1393
151. Horie K, Mita I, Kambe H (1967) J Appl Polym Sci 11: 57
152. Horie K, Mita I, Kambe H (1968) ibid 12: 13
153. Elmes AR, Hammond K, Sherrington DC (1988) Eur Pat Appl 0,289,238
154. Edwards CJC, Hitchen DA, Sharples M (1988) U.S. Pat. no. 4,755,655
155. Gregory DP Unilever Research Laboratory, Port Sunlight, private communication
156. Elmes AR, Sherrington DC, unpublished results
157. Ruckenstein E, Kim K-J (1988) J Appl Polym Sci 36: 907
158. Sun F, Ruckenstein E (1993) ibid 48: 1279
159. Ruckenstein E, Kim K-J (1989) J Polym Sci Pt A: Polym Chem 27: 4375
160. Kim K-J, Ruckenstein E (1989) J Appl Polym Sci 38: 441
161. Ruckenstein E, Park JS (1990) Polymer 31: 2397
162. Hong L, Ruckenstein E (1992) ibid 33: 1968
163. Hong L, Ruckenstein E (1991/92) React Pols 16: 181
164. Ruckenstein E, Hong L (1992) Chem Mater 4: 1032
165. Ruckenstein E, Chen J-H (1991) Polymer 32: 1230
166. Ruckenstein E, Yang S (1993) ibid 34: 4655
167. Yang S, Ruckenstein E (1993) Synth Metals 60: 249
168. Ruckenstein E, Yang S (1993) ibid 53: 283
169. Yang S, Ruckenstein E (1993) ibid 59: 1
170. Ruckenstein E, Hong L (1994) J Appl Polym Sci 53: 923
171. Ruckenstein E, Li H (1994) ibid 52: 1949
172. Ruckenstein E, Li H (1994) ibid 54: 561
173. Ruckenstein E, Li H (1994) Polymer 35: 4343
174. Ruckenstein E, Sun F (1992) J Appl Polym Sci 46: 1271
175. Hong L, Ruckenstein E (1993) ibid 48: 1773
176. Griffith WP, Ley SV, Whitcombe GP, White AD J Chem Soc, Chem Commun 1987: 1625
177. Sasson Y, Zappi GD, Neumann R (1986) J Org Chem 51: 2880
178. Kim K-J, Ruckenstein E (1988) Makromol Chem, Rapid Commun 9: 285
179. Park JS, Ruckenstein E (1990) Polymer 31: 175
180. Ruckenstein E, Park JS (1988) J Polym Sci Pt.C: Polym Lett 26: 529
181. Ruckenstein E (1989) Coll Polym Sci 267: 792

182. Park JS, Ruckenstein E (1989) J Appl Polym Sci 38: 453
183. Ruckenstein E, Park JS (1990) ibid 40: 213
184. Ruckenstein E, Chen HH (1991) ibid 42: 2429
185. Ruckenstein E, Sun F (1993) J Membrane Sci 81: 191
186. Sun F, Ruckenstein E (1993) ibid 85: 59
187. Xu G, Ruckenstein E (1992) J Appl Polym Sci 46: 683
188. Xu G, Ruckenstein E (1993) ibid 47: 1343
189. Ruckenstein E, Xu G (1993) ibid 47: 1925
190. Ruckenstein E, Hong L (1993) Macromol 26: 1363

Editor: Prof. A. Ledwith
Received: Mai 1995

Author Index Volumes 101-126

Author Index Vols. 1-100 see Vol. 100

Adolf, D. B. see Ediger, M. D..: Vol. 116, pp. 73-110.
Aharoni, S. M. and *Edwards, S. F.*: Rigid Polymer Networks. Vol. 118, pp. 1-231.
Améduri, B. and *Boutevin, B.*: Synthesis and Properties of Fluorinated Telechelic Monodispersed.Compounds. Vol. 102, pp. 133-170.
Amselem, S. see Domb, A. J.: Vol. 107, pp. 93-142.
Andreis, M. and *Koenig, J. L.*: Application of Nitrogen-15 NMR to Polymers. Vol. 124, pp. 191-238.
Angiolini, L. see Carlini, C.: Vol. 123, pp. 127-214.
Anseth, K. S., Newman, S. M. and *Bowman, C. N.*: Polymeric Dental Composites: Properties and Reaction Behavior of Multimethacrylate Dental Restorations. Vol. 122, pp. 177-218.
Armitage, B. A. see O'Brien, D. F.: Vol. 126, pp. 53-84.
Arnold Jr., F. E. and *Arnold, F. E.*: Rigid-Rod Polymers and Molecular Composites. Vol. 117, pp. 257-296.
Arshady, R.: Polymer Synthesis via Activated Esters: A New Dimension of Creativity in Macromolecular Chemistry. Vol. 111, pp. 1-42.

Bahar, I., Erman, B. and *Monnerie, L.*: Effect of Molecular Structure on Local Chain Dynamics: Analytical Approaches and Computational Methods. Vol. 116, pp. 145-206.
Baltá-Calleja, F. J., González Arche, A., Ezquerra, T. A., Santa Cruz, C., Batallón, F., Frick, B. and *López Cabarcos, E.*: Structure and Properties of Ferroelectric Copolymers of Poly(vinylidene) Fluoride. Vol. 108, pp. 1-48.
Barshtein, G. R. and *Sabsai, O. Y.*: Compositions with Mineralorganic Fillers. Vol. 101, pp.1-28.
Batallán, F. see Baltá-Calleja, F. J.: Vol. 108, pp. 1-48.
Barton, J. see Hunkeler, D.: Vol. 112, pp. 115-134.
Bell, C. L. and *Peppas, N. A.*: Biomedical Membranes from Hydrogels and Interpolymer Complexes. Vol. 122, pp. 125-176.
Bennett, D. E. see O'Brien, D. F.: Vol. 126, pp. 53-84.
Berry, G.C.: Static and Dynamic Light Scattering on Moderately Concentraded Solutions: Isotropic Solutions of Flexible and Rodlike Chains and Nematic Solutions of Rodlike Chains. Vol. 114, pp. 233-290.
Bershtein, V. A. and *Ryzhov, V. A.*: Far Infrared Spectroscopy of Polymers. Vol. 114, pp. 43-122.
Bird, R. B. see Curtiss, C. F.: Vol. 125, pp. 1-102.
Bigg, D. M.: Thermal Conductivity of Heterophase Polymer Compositions. Vol. 119, pp. 1-30.
Binder, K.: Phase Transitions in Polymer Blends and Block Copolymer Melts: Some Recent Developments. Vol. 112, pp. 115-134.
Biswas, M. and *Mukherjee, A.*: Synthesis and Evaluation of Metal-Containing Polymers. Vol. 115, pp. 89-124.

Boutevin, B. and *Robin, J. J.*: Synthesis and Properties of Fluorinated Diols. Vol. 102. pp. 105-132.
Boutevin, B. see Amédouri, B.: Vol. 102, pp. 133-170.
Bowman, C. N. see Anseth, K. S.: Vol. 122, pp. 177-218.
Boyd, R. H.: Prediction of Polymer Crystal Structures and Properties. Vol. 116, pp. 1-26.
Bronnikov, S. V., Vettegren, V. I. and *Frenkel, S. Y.*: Kinetics of Deformation and Relaxation in Highly Oriented Polymers. Vol. 125, pp. 103-146.
Bruza, K. J. see Kirchhoff, R. A.: Vol. 117, pp. 1-66.
Burban, J. H. see Cussler, E. L.: Vol. 110, pp. 67-80.

Cameron, N. R. and *Sherrington, D. C.*: High Internal Phase Emulsions (HIPEs) - Structure, Properties and Use in Polymer Preparation. Vol. 126, pp. 163-214.
Candau, F. see Hunkeler, D.: Vol. 112, pp. 115-134.
Capek, I.: Kinetics of the Free-Radical Emulsion Polymerization of Vinyl Chloride. Vol. 120, pp. 135-206.
Carlini, C. and *Angiolini, L.*: Polymers as Free Radical Photoinitiators. Vol. 123, pp. 127-214.
Casas-Vazquez, J. see Jou, D.: Vol. 120, pp. 207-266.
Chen, P. see Jaffe, M.: Vol. 117, pp. 297-328.
Choe, E.-W. see Jaffe, M.: Vol. 117, pp. 297-328.
Chow, T. S.: Glassy State Relaxation and Deformation in Polymers. Vol. 103, pp. 149-190.
Chung, T.-S. see Jaffe, M.: Vol. 117, pp. 297-328.
Connell, J. W. see Hergenrother, P. M.: Vol. 117, pp. 67-110.
Criado-Sancho, M. see Jou, D.: Vol. 120, pp. 207-266.
Curro, J.G. see Schweizer, K.S.: Vol. 116, pp. 319-378.
Curtiss, C. F. and *Bird, R. B.*: Statistical Mechanics of Transport Phenomena: Polymeric Liquid Mixtures. Vol. 125, pp. 1-102.
Cussler, E. L., Wang, K. L. and *Burban, J. H.*: Hydrogels as Separation Agents. Vol. 110, pp. 67-80.

Dimonie, M. V. see Hunkeler, D.: Vol. 112, pp. 115-134.
Dodd, L. R. and *Theodorou, D. N.*: Atomistic Monte Carlo Simulation and Continuum Mean Field Theory of the Structure and Equation of State Properties of Alkane and Polymer Melts. Vol. 116, pp. 249-282.
Doelker, E.: Cellulose Derivatives. Vol. 107, pp. 199-266.
Domb, A. J., Amselem, S., Shah, J. and *Maniar, M.*: Polyanhydrides: Synthesis and Characterization. Vol.107, pp. 93-142.
Dubrovskii, S. A. see Kazanskii, K. S.: Vol. 104, pp. 97-134.
Dunkin, I. R. see Steinke, J.: Vol. 123, pp. 81-126.

Economy, J. and *Goranov, K.*: Thermotropic Liquid Crystalline Polymers for High Performance Applications. Vol. 117, pp. 221-256.
Ediger M. D. and *Adolf, D. B.*: Brownian Dynamics Simulations of Local Polymer Dynamics. Vol. 116, pp. 73-110.
Edwards, S. F. see Aharoni, S. M.: Vol. 118, pp. 1-231.
Erman, B. see Bahar, I.: Vol. 116, pp. 145-206.
Ezquerra, T. A. see Baltá-Calleja, F. J.: Vol. 108, pp. 1-48.

Fendler, J.H.: Membrane-Mimetic Approach to Advanced Materials. Vol. 113, pp. 1-209.

Fetters, L. J. see *Xu, Z.:* Vol. 120, pp. 1-50.
Förster, S. and *Schmidt, M.:* Polyelectrolytes in Solution. Vol. 120, pp. 51-134.
Frenkel, S. Y. see *Bronnikov, S. V.:* Vol. 125, pp. 103-146.
Frick, B. see *Baltá-Calleja, F. J.:* Vol. 108, pp. 1-48.
Fridman, M. L.: see *Terent'eva, J. P.:* Vol. 101, pp. 29-64.

Ganesh, K. see *Kishore, K.:* Vol. 121, pp. 81-122.
Geckeler, K. E. see *Rivas, B.:* Vol. 102, pp. 171-188.
Geckeler, K. E.: Soluble Polymer Supports for Liquid-Phase Synthesis. Vol. 121, pp. 31-80.
Gehrke, S. H.: Synthesis, Equilibrium Swelling, Kinetics Permeability and Applications of Environmentally Responsive Gels. Vol. 110, pp. 81-144.
Godovsky, D. Y.: Electron Behavior and Magnetic Properties Polymer-Nanocomposites. Vol. 119, pp. 79-122.
González Arche, A. see *Baltá-Calleja, F. J.:* Vol. 108, pp. 1-48.
Goranov, K. see *Economy, J.:* Vol. 117, pp. 221-256.
Grosberg, A. and *Nechaev, S.:* Polymer Topology. Vol. 106, pp. 1-30.
Grubbs, R., Risse, W. and *Novac, B.:* The Development of Well-defined Catalysts for Ring-Opening Olefin Metathesis. Vol. 102, pp. 47-72.
van Gunsteren, W. F. see *Gusev, A. A.:* Vol. 116, pp. 207-248.
Gusev, A. A., Müller-Plathe, F., van Gunsteren, W. F. and *Suter, U. W.:* Dynamics of Small Molecules in Bulk Polymers. Vol. 116, pp. 207-248.
Guillot, J. see *Hunkeler, D.:* Vol. 112, pp. 115-134.
Guyot, A. and *Tauer, K.:* Reactive Surfactants in Emulsion Polymerization. Vol. 111, pp. 43-66.

Hadjichristidis, N. see *Xu, Z.:* Vol. 120, pp. 1-50.
Hall, H. K. see *Penelle, J.:* Vol. 102, pp. 73-104.
Hammouda, B.: SANS from Homogeneous Polymer Mixtures: A Unified Overview. Vol. 106, pp. 87-134.
Hedrick, J. L. see *Hergenrother, P. M.:* Vol. 117, pp. 67-110.
Heller, J.: Poly (Ortho Esters). Vol. 107, pp. 41-92.
Hemielec, A. A. see *Hunkeler, D.:* Vol. 112, pp. 115-134.
Hergenrother, P. M., Connell, J. W., Labadie, J. W. and *Hedrick, J. L.:* Poly(arylene ether)s Containing Heterocyclic Units. Vol. 117, pp. 67-110.
Hiramatsu, N. see *Matsushige, K.:* Vol. 125, pp. 147-186.
Hirasa, O. see *Suzuki, M.:* Vol. 110, pp. 241-262.
Hirotsu, S.: Coexistence of Phases and the Nature of First-Order Transition in Poly-N-isopropylacrylamide Gels. Vol. 110, pp. 1-26.
Hunkeler, D., Candau, F., Pichot, C., Hemielec, A. E., Xie, T. Y., Barton, J., Vaskova, V., Guillot, J., Dimonie, M. V., Reichert, K. H.: Heterophase Polymerization: A Physical and Kinetic Comparision and Categorization. Vol. 112, pp. 115-134.

Ichikawa, T. see *Yoshida, H.:* Vol. 105, pp. 3-36.
Ilavsky, M.: Effect on Phase Transition on Swelling and Mechanical Behavior of Synthetic Hydrogels. Vol. 109, pp. 173-206.
Inomata, H. see *Saito, S.:* Vol. 106, pp. 207-232.
Irie, M.: Stimuli-Responsive Poly(N-isopropylacrylamide), Photo- and Chemical-Induced Phase Transitions. Vol. 110, pp. 49-66.
Ise, N. see *Matsuoka, H.:* Vol. 114, pp. 187-232.

Ivanov, A. E. see Zubov, V. P.: Vol. 104, pp. 135-176.

Jaffe, M., Chen, P., Choe, E.-W., Chung, T.-S. and *Makhija, S.*: High Performance Polymer Blends. Vol. 117, pp. 297-328.
Jou, D., Casas-Vazquez, J. and *Criado-Sancho, M.*: Thermodynamics of Polymer Solutions under Flow: Phase Separation and Polymer Degradation. Vol. 120, pp. 207-266.

Kaetsu, I.: Radiation Synthesis of Polymeric Materials for Biomedical and Biochemical Applications. Vol. 105, pp. 81-98.
Kammer, H. W., Kressler, H. and *Kummerloewe, C.*: Phase Behavior of Polymer Blends - Effects of Thermodynamics and Rheology. Vol. 106, pp. 31-86.
Kandyrin, L. B. and *Kuleznev, V. N.*: The Dependence of Viscosity on the Composition of Concentrated Dispersions and the Free Volume Concept of Disperse Systems. Vol. 103, pp. 103-148.
Kaneko, M. see Ramaraj, R.: Vol. 123, pp. 215-242.
Kang, E. T., Neoh, K. G. and *Tan, K. L.*: X-Ray Photoelectron Spectroscopic Studies of Electroactive Polymers. Vol. 106, pp. 135-190.
Kazanskii, K. S. and *Dubrovskii, S. A.*: Chemistry and Physics of „Agricultural" Hydrogels. Vol. 104, pp. 97-134.
Kennedy, J. P. see Majoros, I.: Vol. 112, pp. 1-113.
Khokhlov, A., Starodybtzev, S. and *Vasilevskaya, V.*: Conformational Transitions of Polymer Gels: Theory and Experiment. Vol. 109, pp. 121-172.
Kilian, H. G. and *Pieper, T.*: Packing of Chain Segments. A Method for Describing X-Ray Patterns of Crystalline, Liquid Crystalline and Non-Crystalline Polymers. Vol. 108, pp. 49-90.
Kishore, K. and *Ganesh, K.*: Polymers Containing Disulfide, Tetrasulfide, Diselenide and Ditelluride Linkages in the Main Chain. Vol. 121, pp. 81-122.
Klier, J. see Scranton, A. B.: Vol. 122, pp. 1-54.
Kobayashi, S., Shoda, S. and *Uyama, H.*: Enzymatic Polymerization and Oligomerization. Vol. 121, pp. 1-30.
Koenig, J. L. see Andreis, M.: Vol. 124, pp. 191-238.
Kokufuta, E.: Novel Applications for Stimulus-Sensitive Polymer Gels in the Preparation of Functional Immobilized Biocatalysts. Vol. 110, pp. 157-178.
Konno, M. see Saito, S.: Vol. 109, pp. 207-232.
Kopecek, J. see Putnam, D.: Vol. 122, pp. 55-124.
Kressler, J. see Kammer, H. W.: Vol. 106, pp. 31-86.
Kirchhoff, R. A. and *Bruza, K. J.*: Polymers from Benzocyclobutenes. Vol. 117, pp. 1-66.
Kuleznev, V. N. see Kandyrin, L. B.: Vol. 103, pp. 103-148.
Kulichkhin, S. G. see Malkin, A. Y.: Vol. 101, pp. 217-258.
Kuchanov, S. I.: Modern Aspects of Quantitative Theory of Free-Radical Copolymerization. Vol. 103, pp. 1-102.
Kummerloewe, C. see Kammer, H. W.: Vol. 106, pp. 31-86.
Kuznetsova, N. P. see Samsonov, G. V.: Vol. 104, pp. 1-50.

Labadie, J. W. see Hergenrother, P. M.: Vol. 117, pp. 67-110.
Lamparski, H. G. see O'Brien, D. F.: Vol. 126, pp. 53-84.
Laschewsky, A.: Molecular Concepts, Self-Organisation and Properties of Polysoaps. Vol. 124, pp. 1-86.

Laso, M. see Leontidis, E.: Vol. 116, pp. 283-318.
Lazár, M. and *Rychlý, R.*: Oxidation of Hydrocarbon Polymers. Vol. 102, pp. 189-222.
Lenz, R. W.: Biodegradable Polymers. Vol. 107, pp. 1-40.
Leontidis, E., de Pablo, J. J., Laso, M. and *Suter, U. W.*: A Critical Evaluation of Novel Algorithms for the Off-Lattice Monte Carlo Simulation of Condensed Polymer Phases. Vol. 116, pp. 283-318.
Lesec, J. see Viovy, J.-L.: Vol. 114, pp. 1-42.
Liang, G. L. see Sumpter, B. G.: Vol. 116, pp. 27-72.
Lin, J. and *Sherrington, D. C.*: Recent Developments in the Synthesis, Thermostability and Liquid Crystal Properties of Aromatic Polyamides. Vol. 111, pp. 177-220.
López Cabarcos, E. see Baltá-Calleja, F. J.: Vol. 108, pp. 1-48.

Majoros, I., Nagy, A. and *Kennedy, J. P.*: Conventional and Living Carbocationic Polymerizations United. I. A Comprehensive Model and New Diagnostic Method to Probe the Mechanism of Homopolymerizations. Vol. 112, pp. 1-113.
Makhija, S. see Jaffe, M.: Vol. 117, pp. 297-328.
Malkin, A. Y. and *Kulichkhin, S. G.*: Rheokinetics of Curing. Vol. 101, pp. 217-258.
Maniar, M. see Domb, A. J.: Vol. 107, pp. 93-142.
Matsumoto, A.: Free-Radical Crosslinking Polymerization and Copolymerization of Multivinyl Compounds. Vol. 123, pp. 41-80.
Matsuoka, H. and *Ise, N.*: Small-Angle and Ultra-Small Angle Scattering Study of the Ordered Structure in Polyelectrolyte Solutions and Colloidal Dispersions. Vol. 114, pp. 187-232.
Matsushige, K., Hiramatsu, N. and *Okabe, H.*: Ultrasonic Spectroscopy for Polymeric Materials. Vol. 125, pp. 147-186.
Mays, W. see Xu, Z.: Vol. 120, pp. 1-50.
Mikos, A. G. see Thomson, R. C.: Vol. 122, pp. 245-274.
Miyasaka, K.: PVA-Iodine Complexes: Formation, Structure and Properties. Vol. 108. pp. 91-130.
Monnerie, L. see Bahar, I.: Vol. 116, pp. 145-206.
Morishima, Y.: Photoinduced Electron Transfer in Amphiphilic Polyelectrolyte Systems. Vol. 104, pp. 51-96.
Müllen, K. see Scherf, U.: Vol. 123, pp. 1-40.
Müller-Plathe, F. see Gusev, A. A.: Vol. 116, pp. 207-248.
Mukerherjee, A. see Biswas, M.: Vol. 115, pp. 89-124.
Mylnikov, V.: Photoconducting Polymers. Vol. 115, pp. 1-88.

Nagy, A. see Majoros, I.: Vol. 112, pp. 1-113.
Nechaev, S. see Grosberg, A.: Vol. 106, pp. 1-30.
Neoh, K. G. see Kang, E. T.: Vol. 106, pp. 135-190.
Newman, S. M. see Anseth, K. S.: Vol. 122, pp. 177-218.
Noid, D. W. see Sumpter, B. G.: Vol. 116, pp. 27-72.
Novac, B. see Grubbs, R.: Vol. 102, pp. 47-72.
Novikov, V. V. see Privalko, V. P.: Vol. 119, pp. 31-78.

O'Brien, D. F., Armitage, B. A., Bennett, D. E. and *Lamparski, H. G.*: Polymerization and Domain Formation in Lipid Assemblies. Vol. 126, pp. 53-84.
Ogasawara, M.: Application of Pulse Radiolysis to the Study of Polymers and Polymerizations. Vol. 105, pp. 37-80.
Okabe, H. see Matsushige, K.: Vol. 125, pp. 147-186.

Okada, M.: Ring-Opening Polymerization of Bicyclic and Spiro Compounds. Reactivities and Polymerization Mechanisms. Vol. 102, pp. 1-46.
Okano, T.: Molecular Design of Temperature-Responsive Polymers as Intelligent Materials. Vol. 110, pp. 179-198.
Onuki, A.: Theory of Phase Transition in Polymer Gels. Vol. 109, pp. 63-120.
Osad'ko, I.S.: Selective Spectroscopy of Chromophore Doped Polymers and Glasses. Vol. 114, pp. 123-186.

de Pablo, J. J. see Leontidis, E.: Vol. 116, pp. 283-318.
Padias, A. B. see Penelle, J.: Vol. 102, pp. 73-104.
Penelle, J., Hall, H. K., Padias, A. B. and *Tanaka, H.:* Captodative Olefins in Polymer Chemistry. Vol. 102, pp. 73-104.
Peppas, N. A. see Bell, C. L.: Vol. 122, pp. 125-176.
Pichot, C. see Hunkeler, D.: Vol. 112, pp. 115-134.
Pieper, T. see Kilian, H. G.: Vol. 108, pp. 49-90.
Pospíšil, J.: Functionalized Oligomers and Polymers as Stabilizers for Conventional Polymers. Vol. 101, pp. 65-168.
Pospíšil, J.: Aromatic and Heterocyclic Amines in Polymer Stabilization. Vol. 124, pp. 87-190.
Priddy, D. B.: Recent Advances in Styrene Polymerization. Vol. 111, pp. 67-114.
Priddy, D. B.: Thermal Discoloration Chemistry of Styrene-co-Acrylonitrile. Vol. 121, pp. 123-154.
Privalko, V. P. and *Novikov, V. V.:* Model Treatments of the Heat Conductivity of Heterogeneous Polymers. Vol. 119, pp 31-78.
Putnam, D. and *Kopecek, J.:* Polymer Conjugates with Anticancer Acitivity. Vol. 122, pp. 55-124.

Ramaraj, R. and *Kaneko, M.:* Metal Complex in Polymer Membrane as a Model for Photosynthetic Oxygen Evolving Center. Vol. 123, pp. 215-242.
Rangarajan, B. see Scranton, A. B.: Vol. 122, pp. 1-54.
Reichert, K. H. see Hunkeler, D.: Vol. 112, pp. 115-134.
Risse, W. see Grubbs, R.: Vol. 102, pp. 47-72.
Rivas, B. L. and *Geckeler, K. E.:* Synthesis and Metal Complexation of Poly(ethyleneimine) and Derivatives. Vol. 102, pp. 171-188.
Robin, J. J. see Boutevin, B.: Vol. 102, pp. 105-132.
Roe, R.-J.: MD Simulation Study of Glass Transition and Short Time Dynamics in Polymer Liquids. Vol. 116, pp. 111-114.
Rusanov, A. L.: Novel Bis (Naphtalic Anhydrides) and Their Polyheteroarylenes with Improved Processability. Vol. 111, pp. 115-176.
Rychlý, J. see Lazár, M.: Vol. 102, pp. 189-222.
Ryzhov, V. A. see Bershtein, V. A.: Vol. 114, pp. 43-122.

Sabsai, O. Y. see Barshtein, G. R.: Vol. 101, pp. 1-28.
Saburov, V. V. see Zubov, V. P.: Vol. 104, pp. 135-176.
Saito, S., Konno, M. and *Inomata, H.:* Volume Phase Transition of N-Alkylacrylamide Gels. Vol. 109, pp. 207-232.
Samsonov, G. V. and *Kuznetsova, N. P.:* Crosslinked Polyelectrolytes in Biology. Vol. 104, pp. 1-50.

Santa Cruz, C. see Baltá-Calleja, F. J.: Vol. 108, pp. 1-48.
Sato, T. and *Teramoto, A.*: Concentrated Solutions of Liquid-Crystalline Polymers. Vol. 126, pp. 85-162.
Scherf, U. and *Müllen, K.*: The Synthesis of Ladder Polymers. Vol. 123, pp. 1-40.
Schmidt, M. see Förster, S.: Vol. 120, pp. 51-134.
Schweizer, K. S.: Prism Theory of the Structure, Thermodynamics, and Phase Transitions of Polymer Liquids and Alloys. Vol. 116, pp. 319-378.
Scranton, A. B., Rangarajan, B. and *Klier, J.*: Biomedical Applications of Polyelectrolytes. Vol. 122, pp. 1-54.
Sefton, M. V. and *Stevenson, W. T. K.*: Microencapsulation of Live Animal Cells Using Polycrylates. Vol.107, pp. 143-198.
Shamanin, V. V.: Bases of the Axiomatic Theory of Addition Polymerization. Vol. 112, pp. 135-180.
Sherrington, D. C. see Lin, J.: Vol. 111, pp. 177-220.
Sherrington, D. C. see Steinke, J.: Vol. 123, pp. 81-126.
Sherrington, D. C. see Cameron, N. R.: Vol. 126, pp. 163-214.
Shibayama, M. see Tanaka, T.: Vol. 109, pp. 1-62.
Shoda, S. see Kobayashi, S.: Vol. 121, pp. 1-30.
Siegel, R. A.: Hydrophobic Weak Polyelectrolyte Gels: Studies of Swelling Equilibria and Kinetics. Vol. 109, pp. 233-268.
Singh, R. P. see Sivaram, S.: Vol. 101, pp. 169-216.
Sivaram, S. and *Singh, R. P.*: Degradation and Stabilization of Ethylene-Propylene Copolymers and Their Blends: A Critical Review. Vol. 101, pp. 169-216.
Starodybtzev, S. see Khokhlov, A.: Vol. 109, pp. 121-172.
Steinke, J., Sherrington, D. C. and *Dunkin, I. R.*: Imprinting of Synthetic Polymers Using Molecular Templates. Vol. 123, pp. 81-126.
Stenzenberger, H. D.: Addition Polyimides. Vol. 117, pp. 165-220.
Stevenson, W. T. K. see Sefton, M. V.: Vol. 107, pp. 143-198.
Sumpter, B. G., Noid, D. W., Liang, G. L. and *Wunderlich, B.*: Atomistic Dynamics of Macromolecular Crystals. Vol. 116, pp. 27-72.
Suter, U. W. see Gusev, A. A.: Vol. 116, pp. 207-248.
Suter, U. W. see Leontidis, E.: Vol. 116, pp. 283-318.
Suzuki, A.: Phase Transition in Gels of Sub-Millimeter Size Induced by Interaction with Stimuli. Vol. 110, pp. 199-240.
Suzuki, A. and *Hirasa, O.*: An Approach to Artifical Muscle by Polymer Gels due to Micro-Phase Separation. Vol. 110, pp. 241-262.

Tagawa, S.: Radiation Effects on Ion Beams on Polymers. Vol. 105, pp. 99-116.
Tan, K. L. see Kang, E. T.: Vol. 106, pp. 135-190.
Tanaka, T. see Penelle, J.: Vol. 102, pp. 73-104.
Tanaka, H. and *Shibayama, M.*: Phase Transition and Related Phenomena of Polymer Gels. Vol. 109, pp. 1-62.
Tauer, K. see Guyot, A.: Vol. 111, pp. 43-66.
Terent'eva, J. P. and *Fridman, M. L.*: Compositions Based on Aminoresins. Vol. 101, pp. 29-64.
Theodorou, D. N. see Dodd, L. R.: Vol. 116, pp. 249-282.
Thomson, R. C., Wake, M. C., Yaszemski, M. J. and *Mikos, A. G.*: Biodegradable Polymer Scaffolds to Regenerate Organs. Vol. 122, pp. 245-274.
Tokita, M.: Friction Between Polymer Networks of Gels and Solvent. Vol. 110, pp. 27-48.

Tsuruta, T.: Contemporary Topics in Polymeric Materials for Biomedical Applications. Vol. 126, pp. 1-52.

Uyama, H. see Kobayashi, S. : Vol. 121, pp. 1-30.

Vasilevskaya, V. see Khokhlov, A., Vol. 109, pp. 121-172.
Vaskova, V. see Hunkeler, D.: Vol. 112, pp. 115-134.
Verdugo, P.: Polymer Gel Phase Transition in Condensation-Decondensation of Secretory Products. Vol. 110, pp. 145-156.
Vettegren, V. I. see Bronnikov, S. V.: Vol. 125, pp. 147-186.
Viovy, J.-L. and *Lesec, J.*: Separation of Macromolecules in Gels: Permeation Chromatography and Electrophoresis. Vol. 114, pp. 1-42.
Volksen, W.: Condensation Polyimides: Synthesis, Solution Behavior, and Imidization Characteristics. Vol. 117, pp. 111-164.

Wake, M. C. see Thomson, R. C.: Vol. 122, pp. 245-274.
Wang, K. L. see Cussler, E. L.: Vol. 110, pp. 67-80.
Wunderlich, B. see Sumpter, B. G.: Vol. 116, pp. 27-72.

Xie, T. Y. see Hunkeler, D.: Vol. 112, pp. 115-134.
Xu, Z., Hadjichristidis, N., Fetters, L. J. and *Mays, J. W.*: Structure/Chain-Flexibility Relationships of Polymers. Vol. 120, pp. 1-50.

Yannas, I. V.: Tissue Regeneration Templates Based on Collagen-Glycosaminoglycan Copolymers. Vol. 122, pp. 219-244.
Yamaoka, H.: Polymer Materials for Fusion Reactors. Vol. 105, pp. 117-144.
Yaszemski, M. J. see Thomson, R. C.: Vol. 122, pp. 245-274.
Yoshida, H. and *Ichikawa, T.*: Electron Spin Studies of Free Radicals in Irradiated Polymers. Vol. 105, pp. 3-36.

Zubov, V. P., Ivanov, A. E. and *Saburov, V. V.*: Polymer-Coated Adsorbents for the Separation of Biopolymers and Particles. Vol. 104, pp. 135-176.

Subject Index

Acrylamide 203, 205-208
Acryloyl lipids 56, 57, 59, 71, 79-82
- sarcosine, methyl ester 196
Acyl carrier protein 196
Alumina 205
Attractive forces 186
2,2'-Azo-bis-isobutyronitrile 189, 192, 207

1-Benzyl-1,4-dihydronicotinamide 196
Bicatalyst systems 197-198, 203
Bilayers, permeability 60
Biohybridized materials 35
Biomaterials 2
Biomedical polymers 2
Biomimicking materials 35
Biphenols, polychlorinated 200
Block copolymers 187, 200, 204

Calcium, cytoplasmic 27
Capillary action 194
- pressure 174
Carbon fibers 204
-, porous 201
Catalysts, transition-metal, polymer-supported 198, 203
Cell orientation, initial 172-177
- packing 165, 169-171
Cell-adhesive peptide 35
Chemical potential, polydisperse systems 99
- -, scaled particle theory 97
Clathrates 170
Cloud point, nonionic surfactants 185
Cobalt complexation 203
Comb copolymers 208
Computer simulation 131
Conducting polymers 198-199, 203-204
Contact angle 165, 167, 170-172
Copoly(IPAAm) 18

Core-shell particles 202-203
Cosmic dust 200
Cosurfactants 184
Cream formation 164-165, 170, 185
Creams 181
Critical strain 172
Crush strength, polyHIPE 195
Cystine lipids 66-67

Deformations, extensional 172, 173, 176, 177, 178, 180
-, shearing 172, 173, 176, 177
Dextran 181
Diacetylenic lipids 61-64, 82
Dialysis 181
Dielectric constant 186
Dienoyl lipids 59, 60, 64, 66-68, 70-74, 82
Diffusion, lateral, monomers 56, 59
- coefficient, longitudinal 127-128, 131-133
- -, rotational 125-126, 133-136, 145-147
- -, transverse 123-125, 131-133
N,N'-Dimethylacrylamide 196
Distribution functions, Gaussian trial function 97, 153
- -, Onsager trial function 96-97, 153
Disulfide lipids 66-67
Divinylbenzene 184, 189-200, 202, 206
Dodecahedron 167-170, 177
Domain formation, polymerization induced 69-81
Double-layer forces 182, 186
Droplet diameter, critical 182
- radius 178-181
- - of curvature 181-182

Electric birefringence, rotational diffusion coefficient 135-136, 145-146
Emulsions, fluorocarbon 183

Emulsions, polydisperse 170-171, 181, 187
Energy transfer, polymerization enhancement 80-81
Entropy loss, orientational 96-97
Equation of state 100
ESR spectroscopy 183, 185

Film area 170
- thickness 165, 167, 170, 172, 173, 178
Flavin 196
Flocs 170
Foam cell, ideal 167-170
Foams, biliquid 164
-, gas-blown 189, 195
-, high gas fraction 172
-, mechanical failure 174
Fuels, emulsified 188
Fusion, inertially confined 199
Fuzzy cylinder 121-123

G-protein 74
Gas pressure 174
Gel-effect 201
Gel-emulsion 164, 184
Gelatin 206-208
- crosslinking 207
- /poly(methyl methacrylate) 206-207
- /polystyrene 207-208
Graft copolymers 187, 200
Green function method 123-127, 155-158
- - -, transverse diffusion coefficient 123-125

Heck vinylation reaction 198, 203
Helmholtz free energy, polydisperse systems 99
- - -, scaled particle theory 97
HEMA-styrene triblock copolymers 25
Herbicides 202
HIPE, cell size 173
-, freeze-thaw stability 186
HLB 184, 187, 191
Hole theory 127-128
Hydrocarbon gels 164
Hydrogels 17
-, thermoresponsive 18

Hydrophilic polymers 15
Hydrophilicity 6
Hydrophobicity 6

Interaction potential, electrostatic 113
- -, hard-core 93
Interlamellar attachment (ILA) 76-78
Inverted cubic structures 77
Isopropylidene malonate (Meldrum's acid) 203

Jet engine fuel 187

Kinetic equation, general 119-121
- -, nematic state 149

Lactose-substituted polystyrene (PVLA) 43
Lecithin 188
Light scattering, dynamic 145-147
Ligninase 200
Lipase 200, 203
Lipid conformation 58
- domains 54-56, 61-69, 81
Lipids, anionic 79-81
-, cationic 61-63, 67-69
-, fluorinated 64-66, 82
Liposomes 44
-, non-aqueous 187
Liquid absorption, polyHIPE 194-195, 199
Liquid chromatography 195
Liquid crystals 182, 184, 185

Mayonnaise 188
Meldrum's acid 203
Membrane proteins 73-75
Membranes 180, 206-208
Methacrylic acid 202
Micelle concentration, critical 182
Micelles 181, 183
Microdomain-structured materials 21
Microemulsions 183, 187
Molecules, water, random network 33
Monolaurin 184
Monomers, lateral diffusion 56, 59

Subject Index

Non-Newtonian fluids 172, 180

Oil recovery, HIPE 188
Oleic acid 184
Osmotic compressibility 103-105
- pressure 97-103
- -, critical 181-182
Ostwald ripening 186

Palladium complexation 198, 203
Particle theory, scaled 94-95
Particles, hollow 201
Permeability, polymerized bilayers 60
Persistence length 91-92
Pervaporation 206
Petroleum gels 188
Phase boundary concentration 103-116
Phase diagram, isotropic-nematic 106
- -, ternary systems 110-112
Phase-transfer catalysts, polymer-supported 198, 202-203, 205
Phosphatidylcholine (PC) 55-65, 68-81
Phosphatidylethanolamine (PE) 55, 57, 61, 62, 68, 74, 76-79
Phospholipase 72, 73
Phospholipid copolymers 42
Photopolymerization, vesicle fusion 77-79
Plateau borders 167, 173, 175
Poly(allyl chloride) 205
Poly(γ-benzyl L-glutamate), normal stress difference 147-148, 150-152
-, order parameter 117-118
-, osmotic pressure 102-103
-, zero-shear viscosity 143-144
Poly(butyl acrylate) 202
Poly(butyl methacrylate) 202, 204
Poly(3,4-dichloro-1-butene) 205
Poly(ethylene oxide) 203
- conjugates 16
- /poly(aspartic acid) block copolymer 44
Poly(n-hexyl isocyanate), order parameter 117-119
-, osmotic compressibility 103-104
Poly(N-isopropylacrylamide) 18
Poly(methacrylonitrile) 201
Poly(methyl methacrylate) 204-207
- /polypyrolle 204

Poly(sodium styrene sulfonate) 196
Poly(vinyl benzyl chloride) 202-203
Poly(vinyl chloride)/poly(methyl methacrylate) 204
Polyacrylamide 205, 206
- /polystyrene 206
Polyamide, polyether-segmented 25
Polyamine-graft copolymers 28
Polyaniline/poly(alkyl methacrylate) 204
- /polystyrene 204
Polyaphrons 164
Polyelectrolyte solutions, phase boundary concentration 113-116
Polyether-segmented polyamide 25
PolyHIPE, cell size 192, 194, 195, 199
Polymer networks, crosslinked 58-60, 75-79
Polymerization, degree of 56-57, 69, 75
Polypyrrole 198
- /poly(ethylene-vinyl acetate) 204
Polystyrene, rubber-toughened 204
Polyurethane/polypyrrole 204
Polyurethanes, segmented 21
Porogens 193, 194, 206
Potassium persulfate 189, 192
ζ-Potential 31
Pressure, disjoining 176
-, external 174
Protein adsorption 6, 26, 35
Proteins, peripheral 74
PTFE 179
Pulsed gradient spin-echo NMR 183

Random network concept, water 33
Reptation theory 128-129
Resins, styrene-based 165
Resorcinol/formaldehyde 200-201
Rhodopsin 73-75
Rodlike polymers, second virial coefficient 100

Salt concentration, effects 170-182, 192
Salting-out, nonionic surfactants 185, 186
Scaled particle theory 94-95
Scanning electron microscopy 165, 171, 189, 205
Schizophyllan, osmotic compressibility 105

Schizophyllan, zero-shear viscosity 142-143
Second virial coefficient 93, 100
Sedimentation equilibrium 103-104
Self-consistent mean field 120
Self-diffusion coefficient 131-135
Shear modulus 172, 174-180
- rate 172, 174-178, 187
- strain 172, 175
- strength 174
- stress 175, 178, 187
Shear-thinning 172
Silica 205
Small angle neutron scattering 183
Small angle X-ray scattering 183
Sodium dodecyl sulfate 201, 204
Solid phase peptide synthesis 196
Sorbitan monooleate 186, 191, 205, 208
Sorbyl lipids 56, 57, 59, 71, 74-79, 82
Stiff polymers, second virial coefficient 93
Strain response 173, 178
Stress expression 129-131
Stress/strain behavior 172, 176, 177, 187, 195
Styrene 184, 189-208
- /butadiene/styrene 204
- /divinylbenzene 192
Surface tension 172, 174, 178-186
Surfactants, alkyl pyrrole 198
-, anionic 170, 187, 203
-, block copolymer 191, 200
-, fluorinated nonionic 183
-, ionic/nonionic 184-189, 208
-, pesticidal 178, 188
-, polyether 184, 185
Syneresis see cream formation

TEM 192-193
Ternary systems, phase diagram 110-112
β-Tetrakaidecahedron 167-170, 177
Third virial coefficient 100
Transition-metal catalysts, polymer-supported 198, 203

Uncorked vesicles 67
Ultracentrifugation 183

Vesicle destabilization 76-79
- fusion 77-79
- -, photopolymerization-induced 77-79
Vesicles, pore size 72
-, skeletonized 71
-, uncorked 67
Vinyl chloride 171, 201
Viscoelasticity 172
Viscometers, nozzle-type 180
-, wall roughness 179
Viscosity, continuous phase 175, 178, 185
-, effective 175
-, emulsion 171, 172, 175, 188
- equation, zero-shear viscosity 139-142
-, foam 174-175, 178
-, liquid films 174
Volume contraction, vinyl monomers 192, 201
Voronoi networks 175-176

Wall-slip yield stress 179
Water molecules, random network 33
Water-soluble polymers
Wittig reaction 208
Wormlike chain parameters 91

Xanthan, self-diffusion coefficient 134-135
- solutions, zero-shear viscosity 137-139, 142-143
Xylene 183-184

Yield stress 172-180, 187
Young's modulus 195

Zero-shear viscosity 137-143

Springer-Verlag and the Environment

We at Springer-Verlag firmly believe that an international science publisher has a special obligation to the environment, and our corporate policies consistently reflect this conviction.

We also expect our business partners – paper mills, printers, packaging manufacturers, etc. – to commit themselves to using environmentally friendly materials and production processes.

The paper in this book is made from low- or no-chlorine pulp and is acid free, in conformance with international standards for paper permanency.